Reconciling agricultural production with biodiversity conservation

It is widely recognised that agriculture is a significant contributor to global warming and climate change. Agriculture needs to reduce its environmental impact and adapt to current climate change whilst still feeding a growing population, i.e. become more 'climate-smart'. Burleigh Dodds Science Publishing is playing its part in achieving this by bringing together key research on making the production of the world's most important crops and livestock products more sustainable.

Based on extensive research, our publications specifically target the challenge of climate-smart agriculture. In this way we are using 'smart publishing' to help achieve climate-smart agriculture.

Burleigh Dodds Science Publishing is an independent and innovative publisher delivering high quality customer-focused agricultural science content in both print and online formats for the academic and research communities. Our aim is to build a foundation of knowledge on which researchers can build to meet the challenge of climate-smart agriculture.

For more information about Burleigh Dodds Science Publishing simply call us on +44 (0) 1223 839365, email info@bdspublishing.com or alternatively please visit our website at www.bdspublishing.com.

Related titles:

Advances in Conservation Agriculture Volume 1: Systems and Science
Print (ISBN 978-1-78676-264-1); Online (ISBN 978-1-78676-266-5, 978-1-78676-267-2)

Advances in Conservation Agriculture Volume 2: Practice and Benefits
Print (ISBN 978-1-78676-268-9); Online (ISBN 978-1-78676-270-2, 978-1-78676-271-9)

Climate change and agriculture
Print (ISBN 978-1-78676-320-4); Online (ISBN 978-1-78676-322-8, 978-1-78676-323-5)

Agroforestry for sustainable agriculture
Print (ISBN 978-1-78676-220-7); Online (ISBN 978-1-78676-222-1, 978-1-78676-223-8)

Chapters are available individually from our online bookshop:
https://shop.bdspublishing.com

climate•SMART•publishing
IN AGRICULTURAL SCIENCE

BURLEIGH DODDS SERIES IN AGRICULTURAL SCIENCE
NUMBER 87

Reconciling agricultural production with biodiversity conservation

Edited by Professor Paolo Bàrberi and Dr Anna-Camilla Moonen, Institute of Life Sciences – Scuola Superiore Sant'Anna, Italy

burleigh dodds
SCIENCE PUBLISHING

Published by Burleigh Dodds Science Publishing Limited
82 High Street, Sawston, Cambridge CB22 3HJ, UK
www.bdspublishing.com

Burleigh Dodds Science Publishing, 1518 Walnut Street, Suite 900, Philadelphia, PA 19102-3406, USA

First published 2020 by Burleigh Dodds Science Publishing Limited

Library of Congress Control Number: 2020942298

British Library Cataloguing in Publication Data
A catalogue record for this book is available from the British Library

ISBN 978-1-78676-348-8 (Print)
ISBN 978-1-78676-351-8 (PDF)
ISBN 978-1-78676-350-1 (ePub)
ISSN 2059-6936 (print)
ISSN 2059-6944 (online)

DOI 10.19103/AS.2020.0071

Typeset by Deanta Global Publishing Services, Dublin, Ireland

Contents

Series list

Title	Series number
Achieving sustainable cultivation of maize - Vol 1 From improved varieties to local applications Edited by: Dr Dave Watson, CGIAR Maize Research Program Manager, CIMMYT, Mexico	001
Achieving sustainable cultivation of maize - Vol 2 Cultivation techniques, pest and disease control Edited by: Dr Dave Watson, CGIAR Maize Research Program Manager, CIMMYT, Mexico	002
Achieving sustainable cultivation of rice - Vol 1 Breeding for higher yield and quality Edited by: Prof. Takuji Sasaki, Tokyo University of Agriculture, Japan	003
Achieving sustainable cultivation of rice - Vol 2 Cultivation, pest and disease management Edited by: Prof. Takuji Sasaki, Tokyo University of Agriculture, Japan	004
Achieving sustainable cultivation of wheat - Vol 1 Breeding, quality traits, pests and diseases Edited by: Prof. Peter Langridge, The University of Adelaide, Australia	005
Achieving sustainable cultivation of wheat - Vol 2 Cultivation techniques Edited by: Prof. Peter Langridge, The University of Adelaide, Australia	006
Achieving sustainable cultivation of tomatoes Edited by: Dr Autar Mattoo, USDA-ARS, USA & Prof. Avtar Handa, Purdue University, USA	007
Achieving sustainable production of milk - Vol 1 Milk composition, genetics and breeding Edited by: Dr Nico van Belzen, International Dairy Federation (IDF), Belgium	008
Achieving sustainable production of milk - Vol 2 Safety, quality and sustainability Edited by: Dr Nico van Belzen, International Dairy Federation (IDF), Belgium	009
Achieving sustainable production of milk - Vol 3 Dairy herd management and welfare Edited by: Prof. John Webster, University of Bristol, UK	010
Ensuring safety and quality in the production of beef - Vol 1 Safety Edited by: Prof. Gary Acuff, Texas A&M University, USA & Prof. James Dickson, Iowa State University, USA	011
Ensuring safety and quality in the production of beef - Vol 2 Quality Edited by: Prof. Michael Dikeman, Kansas State University, USA	012
Achieving sustainable production of poultry meat - Vol 1 Safety, quality and sustainability Edited by: Prof. Steven C. Ricke, University of Arkansas, USA	013
Achieving sustainable production of poultry meat - Vol 2 Breeding and nutrition Edited by: Prof. Todd Applegate, University of Georgia, USA	014

Acknowledgements

We wish to acknowledge the following for their help in reviewing particular chapters:

- Chapter 5: Dr Jon Marshall, Independent Researcher, UK; and Dr Duncan Westbury, University of Worcester, UK
- Chapter 6: Professor John Dover, Staffordshire University, UK

Introduction

More intensive, monocultural agriculture has been associated with a decline in diversity of habitat and plant species which leads to corresponding declines in diversity of insect, bird and mammal species. There is mounting evidence that a more biodiverse landscape improves ecosystem services which benefits farmers. This collection summarises the wealth of research on ways of improving biodiversity in agricultural landscapes, with a focus on temperate agriculture.

The first part of this volume reviews landscape approaches to biodiversity conservation, mapping and modelling biodiversity as well as assessing the economic value of biodiversity conservation practices. Part 2 reviews management practices promoting biodiversity such as field margins and hedgerows, ways of reconciling agricultural production and biodiversity in grassland management as well as the role of agroforestry in promoting biodiversity.

Part 1 Methods to study biodiversity in agroecosystems

The book begins with a focus on the challenge of mapping biodiversity in agricultural landscapes at the EU level. Chapter 1 reviews the current surveys of biodiversity in agricultural areas in the EU. It begins by examining established surveys, ongoing pilots and plans for new surveys at the EU level, focusing specifically on the monitoring of farmland birds and butterflies, grasslands, pollinators and soil biodiversity. The chapter also assesses where we stand in the short-medium term in terms of our knowledge of agrobiodiversity in the EU and which gaps still need to be filled to sufficiently describe biodiversity dynamics.

The next chapter looks at modelling biodiversity in agriculture. Models have been widely used in agricultural science to understand complexity, predict the consequence of change and extend knowledge to new scales. 'Crop modelling' began in the 1960s and is now well advanced and applied globally but is restricted to a few economic species. More recently, modelling has been expanded to cover biota as mediators of ecological process or as endpoints for assessment of environmental status. Chapter 2 examines the range of modelling approaches in biodiversity studies, including individual-based approaches that combine within- and between-species diversity, process-based models operating at plot, field and increasingly at landscape scales, integrated system models that seek optimal trade-offs between biodiversity and economic outputs, and global frameworks that combine the biophysical, economic, social and political forces acting on biodiversity. The chapter concludes with a more practical approach known as multi-attribute decision modelling that can be developed and applied by biodiversity managers and planners.

The subject of Chapter 3 is assessing the economic value of agricultural biodiversity. The chapter reviews developments in methods to assess the economic value of agricultural biodiversity, outlines their limitations and proposes a possible, novel way forward. It discusses the different definitions of agrobiodiversity. It then highlights ways of evaluating agrobiodiversity. This is followed by two sections, the first introducing and discussing the ecosystem services framework (ESF) and its limitations, the second outlining the integration of ecosystem interactions in the ESF. The chapter then explores two fundamental problems affecting the evaluation of agricultural biodiversity: the ecosystem services whose value cannot be derived from the market and uncertainty. On the basis of these considerations, the authors propose a novel way forward, the investor perspective, where by giving the natural environment the rights of a legal persona enshrined in a Bill of Rights, the authors address many of the drawbacks of current evaluation methods. The chapter concludes with recommendations for evaluators and decision and policy makers, and with an extended bibliography.

The final chapter of Part 1 examines functional biodiversity for the provision of agroecosystem services. In the context of sustainable agricultural development, the provision of other services beyond production is becoming a priority. Chapter 4 highlights that a functional approach to agrobiodiversity is the best approach to meeting this goal in both research and agricultural practice. The chapter includes an analysis of the development of studies on agriculture-biodiversity relationships and refers to two milestones that had a tremendous impact on research: the United Nation's Convention on Biological Diversity (1992) and the Millennium Ecosystem Assessment (2005). Interest in the functional aspects of agrobiodiversity, i.e. the potential or actual provision of ecosystem services, has boomed in the latest 15 years, yet use of the term 'functional biodiversity' is still scarce. The chapter also provides a trait-based definition of functional biodiversity and a four-step methodology that should shed light on the potential of elements at each agrobiodiversity level - from gene to species and habitat - to provide single or multiple agroecosystem services in any context.

Part 2 Management practices to support agroecosystem services

Part 2 begins by discussing the role of field margins in biodiversity conservation in agroecosystems. Chapter 5 reviews research on ways of modifying the agricultural landscape to reverse the decline in a range of fauna and flora. The chapter starts by looking at the range of options for promoting biodiversity in agricultural landscapes. It then focuses on field margins, their characteristics and types as well as the role of field margins in agroecosystems. The chapter then reviews research on managing field margins to promote insect biodiversity. It also looks at managing field margins to promote rare arable plants (RAP), reptiles, amphibians, birds and mammals.

Chapter 6 considers the role of hedgerows in supporting biodiversity and other ecosystem services in intensively-managed agricultural landscapes. Over the past half century, agricultural intensification has substantially changed agricultural landscapes and farming systems. These changes have been beneficial to provisioning services, i.e. agricultural yields, but detrimental for biodiversity and associated ecosystem services. As a contribution to halting and reversing the decline, it has been suggested that conserving and/or restoring hedgerows would be beneficial. Hedgerows are key features of agricultural landscapes. The chapter begins by highlighting what hedgerows are and their role as a habitat in promoting biodiversity. It then looks at hedgerows and hedgerow networks and goes on to examine how hedgerows contribute to the provision of ecosystem services. The chapter also provides a case study on hedgerow plantation and bocage restoration and how it enhances biodiversity and other ecosystem services. It concludes by providing potential areas for future research as well as resources for further information.

The subject of Chapter 7 is reconciling production and biodiversity in management of pastures and grasslands. Grasslands are crucial for the conservation of biodiversity across the world. Current agricultural practices have increasingly replaced grasslands by crops or managed them more intensively, resulting in a dramatic reduction of biodiversity. Is it possible to reconcile profitable grassland production with conservation of biodiversity? The chapter reviews research on the relationship between biodiversity and the quantity and quality of biomass produced by grasslands. It also shows it is possible to manage grasslands to reconcile production and biodiversity of grasslands. A case study on flowering meadows is also provided, followed by an analysis of potential future trends in research.

The final chapter of the book focuses on the importance of agroforestry systems in supporting biodiversity conservation and agricultural production from a European perspective. Chapter 8 begins by reviewing the contribution of agroforestry to global biodiversity goals. It then discusses agroforestry and the protection of species and habitats, followed by a discussion of agroforestry and the maintenance and restoration of ecosystems. The chapter also analyses the importance of achieving sustainable agriculture and forestry practices as well as preserving global biodiversity.

Part 1

Methods to study biodiversity in agroecosystems

Chapter 1

The challenge of monitoring biodiversity in agricultural landscapes at the EU level

M. L. Paracchini, European Commission, Joint Research Centre (JRC), Italy; S. Condé, European Topic Centre on Biological Diversity – Muséum National d'Histoire Naturelle, France; R. D'Andrimont, European Commission, Joint Research Centre (JRC), Italy; B. Eiselt, European Commission, Eurostat, Luxembourg; O. Fernandez Ugalde, E. Gervasini and A. Jones, European Commission, Joint Research Centre (JRC), Italy; V. Kovacevic, European Commission, DG Environment, Belgium; R. Oppermann, Institut für Agrarökologie und Biodiversität (IFAB), Germany; A. Orgiazzi, M. Van der Velde, C. Polce and C. Rega, European Commission, Joint Research Centre (JRC), Italy; C. Van Swaay, De Vlinderstichting, The Netherlands; and P. Voříšek, Czech Society for Ornithology, Czech Republic

1 Introduction

Following the acknowledgement that biodiversity in agricultural lands globally and in the European Union (EU) has been strongly impacted by the intensification of agricultural practices (Dudley and Alexander, 2017; IPBES, 2019), many efforts have been carried out to revert the trend, starting with

http://dx.doi.org/10.19103/AS.2020.0071.04
Published by Burleigh Dodds Science Publishing Limited, 2020.

agri-environmental schemes becoming compulsory for EU Member States in 1992 (EU Regulation 2078/92) (Batary et al., 2015) aimed at reducing pressures from agriculture in order to meet environmental objectives such as the protection or enhancement of biodiversity, the improvement of soil, water, landscape and air quality, climate change mitigation and adaptation.

Through the Common Agricultural Policy (CAP) cycles that followed, the concern for biodiversity has been embedded into the legislation as a target in general (e.g. protection and enhancement of biodiversity in axis 2 of rural development policy) and in specific terms (e.g. high nature value farming and forestry) (2006/144/EC). Persisting concerns about the fate of biodiversity, which emerged most evidently in the public consultation on modernizing and simplifying the common agricultural policy launched by the European Commission (EC) in 2017[1], have been embedded in the legislative proposal for the CAP post-2020 (COM (2018) 392 final), which identifies as one of its nine priorities to 'contribute to the protection of biodiversity, enhance ecosystem services and preserve habitats and landscapes'.

In parallel, environmental legislation through the decades has targeted rare and threatened species, and rare natural habitats (EEC, 1979; EEC, 1992; European Parliament and the Council of the European Union, 2009): as a result, the European Union currently hosts the largest coordinated network of protected areas in the world, the Natura 2000 Network[2]. By adding the concept of restoration to the protection concept, the legislation of the past two decades has widened the scope, addressing all habitats and not only those more endangered and ecologically valuable. This started with the Commission's proposal to the Gothenburg European Council (EC, 2001), which calls for protecting and restoring habitats and natural systems and halting the loss of biodiversity by 2010, a concept which was reinforced in the EU Biodiversity Strategy to 2020 (EC, 2011). The latter, introducing the concept of ecosystem services, makes all habitats possible targets for restoration. In particular, target 3, which relates to agriculture specifically, defines the goal of maximizing '*areas under agriculture across grasslands, arable land and permanent crops that are covered by biodiversity-related measures under the CAP so as to ensure the conservation of biodiversity and to bring about a measurable improvement in the conservation status of species and habitats that depend on or are affected by agriculture and in the provision of ecosystem services as compared to the EU2010 Baseline, thus contributing to enhance sustainable management*'. The EU Biodiversity Strategy to 2030 reinforces this line of action by dedicating a whole section to bringing nature back to agricultural land (EC, 2020 - Section 2.2.2)[3].

[1] https://ec.europa.eu/agriculture/sites/agriculture/files/consultations/cap-modernising/highlights-public-consul_en.pdf
[2] (https://ec.europa.eu/environment/nature/natura2000/index_en.htm)
[3] https://eur-lex.europa.eu/legal-content/EN/TXT/?qid=1590574123338&uri=CELEX:52020DC0380

Published by Burleigh Dodds Science Publishing Limited, 2020.

Lastly, legislative requirements that contribute to biodiversity preservation also include the so-called 'environmental safeguards' directives, requiring formal environmental assessments to be carried out for projects (under Directive 85/337/EEC[4] and following amendments), and plans/programs (under Directive 2001/42/EC[5]) with potential detrimental effects, including on biodiversity and habitats.

To assess the effectiveness of such efforts different techniques are applied, spanning from the analysis of case study areas (Kettunen and Ten Brink, 2006; Kleijn et al., 2006) to the use of proxies (Alliance Environnement, 2017) or models (Kok et al., 2018). In this frame, there is, overall, a lack of data recorded through monitoring efforts, EU wide assessments are in fact presently relying on a limited set of surveyed data: farmland birds (Gregory et al., 2005)[6], grassland butterflies (EEA, 2013a) and the reporting under the Birds and Habitats Directive (EC, 2015).

Biodiversity decline, and in particular the loss of genetic diversity, is within the nine global-scale processes that are essential to maintain the earth system in a resilient and accommodating state defined by Steffen et al. (2015), one of the two processes laying outside the safe operating space. Despite the urgency to revert the trend and the efforts from the policy side to incorporate the concern, signals are not encouraging (EC, 2020). Better targeting and improved assessments need filling knowledge gaps and using updated and detailed data, covering different taxa. Moreover, in the frame of planning, implementing, monitoring and assessing EU policy, sources of information should cover the entire European Union, and should be based on a harmonized approach for data collection. Establishing surveys is an important way to guarantee that such information becomes available.

The EU Biodiversity Strategy to 2020 defines biodiversity as 'the unique variety of life on our planet', the UN Convention on Biological Diversity as 'the variability among living organisms from all sources including, inter alia, terrestrial, marine and other aquatic ecosystems and the ecological complexes of which they are part; this includes diversity within species, between species and of ecosystems' (UN, 1993). Such definitions suggest a complexity that is probably the reason why it has been so difficult to put in place a large-scale monitoring system that includes surveys of the main components of biodiversity. Surveys can be burdensome and therefore the costs can exceed current financial and administrative capacity, especially when an entire continent should be covered. Nevertheless, initiatives and pilots are ongoing, to enlarge the available data pool.

This chapter reviews where we stand in surveying biodiversity in agricultural areas at the EU level as well as plans to increase monitoring efforts.

[4] https://eur-lex.europa.eu/legal-content/EN/TXT/?uri=CELEX:01985L0337-20090625
[5] https://eur-lex.europa.eu/legal-content/EN/ALL/?uri=CELEX%3A32001L0042
[6] https://pecbms.info/trends-and-indicators/indicators)

The sections that follow describe established surveys, ongoing pilots and plans for new surveys at the EU scale. A multiplicity of information is available at local/ regional/national scale, but this chapter focusses on the long and winding road to wall-to-wall coverage of the European Union. At the end, the point can be made on where we will stand in the short-medium term with our knowledge of agro-biodiversity in the European Union, and which gaps still need to be filled to appropriately and sufficiently describe biodiversity dynamics.

2 Farmland birds and butterflies

2.1 The Pan-European common bird monitoring scheme

Many countries of the European Union are characterized by a long-lasting tradition of bird-watching, on which scientifically grounded countrywide surveys were nested. Countries such as Finland, Sweden and Denmark organized a countrywide monitoring scheme in 1975, the United Kingdom in the 1960s. Others started later on, but nowadays all EU countries except Malta have a monitoring scheme in place.

In 2002, the Pan-European Common Bird Monitoring Scheme (PECBMS) was started as a joint initiative of the European Bird Census Council (EBCC) and BirdLife International, with the aims of: collecting data on European common bird species from national monitoring schemes and calculate European common bird indices and indicators, to raise awareness and to feed the policy process; support the national coordinators in setting up the schemes, and guarantee a harmonized approach in the calculation of the indices; explore the relation between population trends and main driving forces (EBCC, 2019).

Most surveys are carried out through point or line transect counts, where the selection of the plots to be surveyed (each plot containing one or more point or transect) is made either following rigorous statistical procedures (e.g. systematic selection, stratified random selection) or a free choice approach (Table 1). The surveyor visits the assigned location one or more times during the year, in predetermined time windows (e.g. 10 May–20 June as in the Italian survey), and records the individuals seen or heard. A thorough statistical analysis is followed to identify errors and outliers.

The extraordinary component of PECBMS are the thousands volunteers who count the birds in the field, each year, according to a methodology standardized at national level. The data are sent to the national offices, where, using TRIM software made available by PECBMS (van Strien et al., 2001; Statistics Netherlands, 2017), calculate the national species indices and trends. PECBMS combines national species indices with supra-national indices for individual species for the European Union and its main regions (new and old European Union, and West, South, North, Central and East Europe), plus Europe

Table 1 Principal characteristics of national breeding birds monitoring schemes

Country	Generic breeding bird monitoring scheme	Contributes to PECBMS	Start year	Field survey methods	Selection of plots	Number of species
Austria	Yes	Yes	1998	Point counts	Free choice	85
Belgium	Yes	Yes	1990	Point counts	Stratified random, other	134
Bulgaria	Yes	Yes	2004	Line transects	Stratified random	63
Croatia	Yes	Not yet	2015	Point counts	Stratified semi-random	30
Cyprus	Yes	Yes	2006	Line transects	Other	45
Czech Republic	Yes	Yes	1982	Point counts, line transects	Free choice, stratified random	218
Denmark	Yes	Yes	1975	Point counts	Free choice	143
Estonia	Yes	Yes	1983	Point counts	Free choice	90
Finland	Yes	Yes	1975	Point counts, line transects, other	Systematic, other	140
France	Yes	Yes	1989	Point counts	Other	150
Germany	Yes	Yes	1989	Line transects, point counts, territory mapping	Free choice, stratified random	100
Greece	Yes	Yes	2007	Point counts	Stratified random	233
Hungary	Yes	Yes	1999	Point counts	Stratified random	420
Ireland	Yes	Yes	1998	Line transects	Stratified random	55
Italy	Yes	Yes	2000	Point counts	Random	103
Latvia	Yes	Yes	1995	Point counts, territory mapping	Random, systematic, other	90
Lithuania	Yes	Yes	1994	Point counts	Stratified semi-random	70

(Continued)

Table 1 (Continued)

Country	Generic breeding bird monitoring scheme	Contributes to PECBMS	Start year	Field survey methods	Selection of plots	Number of species
Luxembourg	Yes	Yes	2002	Point counts, territory mapping, line transect	Random, stratified random	120
Malta	No					
Netherlands	Yes	Yes	1984	Territory mapping, point counts, line transects	Free choice, random, stratified random	100
Poland	Yes	Yes	2000	Line transects	Stratified random	110
Portugal	Yes	Yes	2004	Point counts	Stratified random	64
Romania	Yes	Yes	2006	Point counts	Stratified semi-random	70
Slovak Republic	Yes	Yes	1994	Point counts	Free choice	100
Slovenia	Yes	Yes	2007	Line transects	Stratified non-random	29
Spain	Yes	Yes	1996	Point counts, line transects	Stratified random, other	200
Sweden	Yes	Yes	1975	Point counts, line transects	Free choice, systematic	180
United Kingdom	Yes	Yes	1966	Territory mapping, line transects	Free choice, stratified random	111

Published by Burleigh Dodds Science Publishing Limited, 2020.

Table 2 Species composing the EU farmland bird index

Alauda arvensis	*Emberiza melanocephala*	*Passer montanus*
Alectoris rufa	*Falco tinnunculus*	*Perdix perdix*
Anthus campestris	*Galerida cristata*	*Petronia petronia*
Anthus pratensis	*Galerida theklae*	*Saxicola rubetra*
Bubulcus ibis	*Hirundo rustica*	*Saxicola torquatus*
Burhinus oedicnemus	*Lanius collurio*	*Serinus serinus*
Calandrella brachydactyla	*Lanius minor*	*Streptopelia turtur*
Carduelis cannabina	*Lanius senator*	*Sturnus unicolor*
Ciconia ciconia	*Limosa limosa*	*Sturnus vulgaris*
Corvus frugilegus	*Melanocorypha calandra*	*Sylvia communis*
Emberiza cirlus	*Miliaria calandra*	*Tetrax tetrax*
Emberiza citrinella	*Motacilla flava*	*Upupa epops*
Emberiza hortulana	*Oenanthe hispanica*	*Vanellus vanellus*

as a whole. All indices are annually updated, and EU indices are regularly sent to and published by EUROSTAT[7]. Thirty-nine species compose the Farmland Bird Index (FBI, Table 2). These are common species, which are dependent from agroecosystems for feeding and nesting, and as such are considered to be a descriptor of the state of agroecosystems. The index is a composite, multispecies index calculated using Monte Carlo simulations as described in Soldaat et al. (2017).

Of the monitoring initiatives presented in this paper, this is the only one that has a wall-to-wall coverage of the European Union, with a sampling density sufficient to derive statistically meaningful information at different scales (EU, national, regional), on different ecosystem types (farmland, forest; montane birds and mire birds for North Europe indicators are under development) or climate change (Gregory et al., 2009).

The information is valuable and widely used, since it is considered that, being at the top of the food chain, birds are indicators of the environment's health. The farmland bird indicator is to date the most widespread biodiversity indicator used in EU policies (impact indicator of the CAP[8]; condition indicator of the Monitoring and Assessment of Ecosystems and their Services - MAES[9]), indicator frameworks (Streamlining European Biodiversity Indicators - SEBI[10]; agri-environmental indicators - AEIs[11]; OECD agri-environmental indicators[12]) and global assessments (e.g. FAO, 2019).

[7] https://ec.europa.eu/eurostat/web/products-datasets/-/t2020_rn130
[8] https://ec.europa.eu/info/sites/info/files/food-farming-fisheries/key_policies/documents/impact-indicator-fiches_en.pdf
[9] https://ec.europa.eu/environment/nature/knowledge/ecosystem_assessment/pdf/5th%20MAES%20re port.pdf
[10] https://www.eea.europa.eu/data-and-maps/indicators/abundance-and-distribution-of-selected-species-8/assessment-1
[11] https://ec.europa.eu/eurostat/web/agriculture/agri-environmental-indicators
[12] https://stats.oecd.org/Index.aspx?QueryId=77269&lang=en

2.2 European breeding bird atlas

The EBCC has organized a second European Breeding Bird Atlas (EBBA2). This project, complementary to the PECBMS, brings information on spatial distribution and abundance of all bird species in Europe including European Russia and Turkey (www.ebba2.info). The outputs of the project which has collected field data from 2013 to 2017 are the maps of distribution and abundance of each species in a grid of 50 km×50 km, and maps of modelled probability of occurrence of selected species in a grid of 10 km×10 km. The atlas also provides maps of change in distribution in comparison with the 1980s, when the first atlas was produced. Information about the spatial distribution of species has been shown to have an extremely high value for policy, nature conservation and research (Herrando et al., 2019). The EBBA2 outputs will be published in 2020.

2.3 Butterflies

The experience of bird monitoring has paved the way to other initiatives aimed at planning and putting in place large-scale surveys. One of these is butterfly monitoring. Butterfly populations are in fact highly sensitive to environmental change, providing an early warning of impacts on ecosystems. Moreover, being insects, they are part of the food chain and they are also important pollinators. The process of establishing monitoring schemes in EU countries and improving the Grassland Butterfly Index is very similar to the PECBMS experience. Some countries had developed their own Butterfly Monitoring Scheme already in the nineties (e.g. Belgium – Flanders, the Netherlands, Finland, Spain – Catalonia), with the United Kingdom at the forefront, having started its monitoring activities in 1976 (Brereton et al., 2009).

Butterflies are counted along line transects, by experts or skilled volunteers. The counts take place several times a year (in the majority of cases 10–20 times), under well-defined weather conditions and times of the day. Transects are distributed either following a statistical procedure for random selection, but in the majority of cases are free choice of volunteers (Brereton et al., 2009). A transect is on average 1 km long, and it may intersect one habitat type or more. The surveyor should walk the transect every week during the butterfly season, counting the individuals seen within an imaginary box of 5 m wide, 5 m high and 5 m ahead of the observer (Sevilleja et al., 2019).

The early 2000s were a topical moment for butterfly monitoring: van Swaay et al. (2006) associated butterfly species to biotopes, highlighting the importance of grasslands, in particular dry grasslands, as the most species-rich biotopes, but at the same time as the biotopes hosting the largest numbers of threatened species. Van Swaay and van Strien (2005) proposed the development of a European grassland butterfly indicator, which followed the protocol of the European common bird indicator (Gregory et al., 2005) and made use of the same TRIM programme to calculate national indices. Finally, Butterfly Conservation Europe was founded, to coordinate monitoring efforts and support national offices in setting up their monitoring schemes (van Swaay et al., 2006).

Through the years that followed, more countries initiated a monitoring scheme, and the European Grassland Butterfly Indicator was updated every two years from 2008 onwards (EEA, 2013a; Van Swaay et al., 2019).

More recently, the European Commission has launched a pilot project (Assessing ButterFlies in Europe – ABLE) to increase the number of monitored countries, expanding the surveys in particular in South and East of the European Union, to have a more representative index. Moreover, ABLE is meant to develop indices for other habitat types besides grasslands, and to support the monitoring network through online resources. The initial period of funding is two years (until November 2020), and the partnership is composed of Butterfly Conservation Europe, the Centre for Ecology and Hydrology (UK), the Helmholtz Centre for Environmental Research (Germany), Dutch Butterfly Conservation (the Netherlands) and Butterfly Conservation (UK) (more information available at www.butterfly-monitoring.net).

3 Grasslands

In 2018 the first pilot of a EU-wide grassland survey was carried out in the frame of the Land Use/Cover Area-Frame Survey (LUCAS). LUCAS is a harmonized land cover and land-use data collection exercise that extends over the whole of the EU's territory. Data are gathered through direct observations made by surveyors on the ground or through photo-interpretation, aimed primarily at collecting information on land use and land cover. The whole LUCAS 2018 survey was based on 337 854 points/observations, of which 238 077 were in-field and 99 777 photo-interpreted (EUROSTAT, 2019).

The survey started in 2009 and through the years different modules were plugged-in to exploit the LUCAS infrastructure and collect data on relevant environmental variables. Two of the modules cover grasslands and soils (for the latter see Section 5).

During spring/summer 2018, 2173 grassland points were surveyed in 26 countries for a total of 134 surveyors involved. In parallel, 36 expert botanists visited a subsample of 730 points, and their data served as control points to evaluate the reliability of the survey (Sutcliffe et al., 2019) (Fig. 1). The points were selected among those that in the previous survey LUCAS 2015 were identified as grasslands, and their distribution is statistically significant at the level of environmental zones.

Figure 1 Expert botanists (a, c) and corresponding surveyors (b, d) transects. © European Union, LUCAS 2018.

The reason for having a grassland survey derives from the awareness that most EU grasslands are not described in any database to a degree that may be used in the policy process. For example, there is no information about their ecological quality. Protected habitats are monitored in the frame of the Habitats Directive (see Section 6) and the list of grassland habitats, though exhaustive, covers about 30% of EU grassland extent. The remaining 70% is described mostly in terms of geographical distribution and biomass production. Information on species is completely lacking.

Points were surveyed on a transect of 20 m × 2.5 m (Fig. 2), if they had at least 50% grassland coverage. Walking alongside the transect line, the surveyor should survey the vegetation of 1.25 m to the left and 1.25 m to the right. The starting point and the transect should be at least 5 m inside the grassland parcel, for the surveyed vegetation to be representative of the parcel and not of the edge. The transect should always originate from a LUCAS point, but the technical specifications contained information on how to reposition the transect if, given the starting point and the required orientation, it was not completely contained in the parcel.

The timing of the survey was carefully planned, according to grassland phenology in the European Union. To allow the identification of the plants, the surveyor had to visit the site before the first cut or before the start of grazing. The

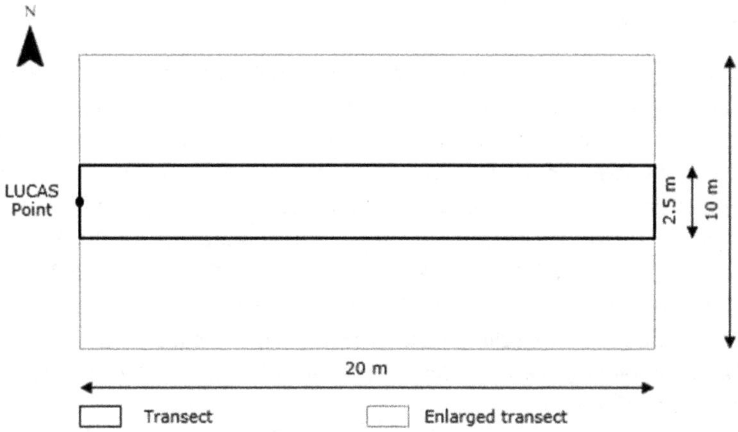

Figure 2 Diagram of the theoretical location of the transect and the enlarged transect with respect to the LUCAS point (Eurostat 2018).

surveys were therefore carried out following a latitudinal time gradient, from mid-April in Mediterranean South until mid-July in Boreal Scandinavia North.

The parameters list was the product of a consultation of EC services in need of information, to support environmental, agricultural and energy EU policies. The results of the consultation were translated into parameters to be surveyed.

These included information on:

- site characteristics (altitude, exposition, inclination, type of surface, moisture),
- a general description of vegetation (vigour, layers, presence of graminoids, forbs, woody plants layer, bare ground, age - older or younger than 5 years etc.),
- a more specific description of the flowering forbs (number of flowering forb species, flower density),
- identification of EUNIS habitat type (https://eunis.eea.europa.eu/about), and
- key species (plant indicator species) and structural plant species.

Concerning the last bullet point, 41 key species or key species groups were identified, and grouped in 10 lists with 20 species groups each for the different biogeographic zones characterizing the European Union. The revised list in the forthcoming surveys is reduced to 12 key species groups, identical for all the biogeographic zones, as this number was considered to be representative and suitable for non-botanist surveyors receiving a specific training to be able to recognize individual species. Such key species lists are assumed to be a direct indicator of the grasslands' ecological quality. A grassland module is currently planned as part of LUCAS 2022, on 20000 points.

4 Pollinators

The raising awareness of the decline of insects (Hallmann et al., 2017; Sanchez-Bayo and Wyckhuys, 2019) and pollinators in particular (Goulson et al., 2015; Potts et al., 2010), the quantification of the losses and of the impacts on ecosystem services, the increased understanding of the causes and the awareness of knowledge gaps (IPBES, 2016) are factors that have strongly influenced the availability of funds for research as well as the adoption by the European Commission, on 1 June 2018, of the EU initiative on pollinators (EC, 2018).

The main aims of the initiative are to improve the scientific knowledge about insect pollinator decline, tackle its main known causes and strengthen collaboration between all the actors concerned. The initiative is organized around three priorities:

i improving knowledge of pollinator decline, its causes and consequences,
ii tackling the causes of pollinator decline, and
iii raising awareness, engaging society-at-large and promoting collaboration.

One of the main knowledge gaps that is identified concerns species distribution and abundance of pollinators, it is estimated in fact that half of the bee species in Europe is data deficient (IEEP & IUCN, 2018). Scattered monitoring programmes for bees and butterflies are in place, but not much is known about all other pollinator species. Under the first priority it is therefore planned that a common EU monitoring scheme is developed. Once implemented, this scheme should generate data that will enable the full assessment of the problem and the effectiveness of mitigation actions. Action 1A of the pollinators Initiative explicitly states 'The Commission will devise and test an EU-wide pollinator monitoring scheme to ensure the provision of good quality data for assessing the status and trends of pollinator species in the EU and developing a pollinator indicator. A technical expert group will be set up to support this work'.

Discussions have started (UN WCMC, 2017; IEEP and IUCN, 2018) and an expert group has been set up and work on the monitoring scheme[13] and the

[13] Draft report title "Proposal for an EU Pollinator Monitoring Scheme", authors: Potts S. (University of Reading, UK), Dauber J. (Thünen Institute, Germany), Hochkirch A. (Trier University, Germany), Oteman B. (Dutch Butterfly Conservation, Netherlands), Roy D. (UK Centre for Ecology & Hydrology, UK), Ahrne K. (Swedish University of Agricultural Sciences, Sweden), Biesmeijer K. (Naturalis Biodiversity Center, Netherlands), Breeze T. (University of Reading, UK), Carvell C. (UK Centre for Ecology & Hydrology, UK), Ferreira C. (IUCN, Belgium), Fitzpatrick U. (National Biodiversity Data Centre, Ireland), Isaac N. (UK Centre for Ecology & Hydrology, UK), Kuussaari M. (Finnish Environment Institute, Finland), Ljubomirov T. (Bulgarian Academy of Sciences, Bulgaria), Maes J. (European Commission Joint Research Centre, Italy), Ngo H. (IPBES, Germany), Pardo A. (Centre for Ecology, Evolution and Environmental Changes, Portugal), Polce C. (European Commission Joint Research Centre, Italy), Quaranta M. (Council for Agricultural Research and Agricultural Economy Analysis, Italy), Settele J. (Helmholtz-Centre for Environmental Research - UFZ, Germany), Sorg M. (Entomological Society Krefeld, Germany), Stefanescu C. (Granollers Natural Sciences Museum, Spain), and Vujic A. (University of Novi Sad, Serbia)

proposal for the indicator. Common monitoring and standardization of data and approaches to monitoring are seen as essential requirements under this premise; suggested methods for monitoring are a combination of passive methods (that do not rely on attracting insects), such as standardized transect walks, and active methods (that rely on attracting insects) such as pan-trapping.

An example of standardized transect is the bumblebee transect count method, which is similar to the butterfly method, and it is structured as follows: volunteers walk fixed transects of 1–2 km (divided into 4–10 sections with different habitat types) each month from March to October (on sunny days between 11:00 and 17:00) recording all bumblebee species in an imaginary 4 m × 4 m box on either side and in front, if necessary by catching bees in net or pot (Bumblebee Conservation Trust, 2017; Comont and Dickinson, 2018).

Interesting features are that data from new recorders are not used in the first two years to ensure data quality by allowing the recorder to develop sufficient skills in identification, and to put in place a mentoring system of experienced recorders tutoring new recorders.

The bee pan-trapping method has been standardized at the EU level by the FAO (2016): it uses bowl or pan traps, which are small plastic bowls or cups, coloured white, fluorescent blue or fluorescent yellow, filled with water mixed with a small amount of detergent, which acts as a surfactant. Twenty-four bowls should be placed on a line or transect 5 m apart, alternating the three colours and left for 24 hours or a fixed shorter period (to be noted), avoiding heavy shade. Catches are sieved and placed in 70% alcohol in container or sample bag (large-bodied non-Hymenoptera insects need to be removed first) which can be sent to the laboratory for identification. Samples should be stored in a freezer if they are not processed within one or two days. The processing of the samples includes washing, drying, pinning, labelling and maintaining the specimens. Lastly, the data from the specimens should be entered into a database system, together with validation and double-checking procedures.

The species to be monitored should include both common (with regional-specific lists) and key indicator species, and cover different pollinator groups, including wild bees, hoverflies and butterflies.

In parallel, a monitoring scheme for the products of the beehive (pollen, nectar, honeybee bread and honey) could be rolled out to monitor pesticide residues and other environmental pollutants.

5 Soil biodiversity

Soil represents a complex habitat sustaining a huge diversity of organisms that are structured by and embedded within the physical matrix (Geisen et al., 2019). Despite its importance for a range of ecosystems and ecosystem services, from nutrient cycle regulation to soil erosion control (Barrios, 2007), it is recognized that our knowledge of the soil habitat is limited (Jeffery et al., 2010). Currently, there are no specific policy measures or designated protection areas in the European Union targeting soil biodiversity. In a perspective of filling this gap, a pilot campaign was launched within the soil module of the 2018 'Land Use/Cover Area frame statistical Survey' (LUCAS Soil), to test a sampling protocol for soil biodiversity. The campaign collected samples from 1 000 locations with diverse land cover and use. This is currently the most extensive EU assessment of soil biodiversity, based on DNA meta-barcoding (Orgiazzi et al., 2015).

The main aims of the survey are:

i to be able to develop a quantitative indicator of soil biodiversity, based on the genetic signatures,
ii to look for correlations between the DNA evidence and land cover/land use, especially in intensive agricultural areas,
iii to match the DNA data with residues of plant protection products. Information on the latter is available from a separate analysis of the concentrations of 70 active ingredients and metabolites in around 3000 LUCAS samples, and
iv eventually to expand the analysis to look at functional groups.

The DNA meta-barcode analysis will cover different types of soil-living organisms, from micro-organisms to macrofauna, and precisely (Orgiazzi et al., 2018):

- bacteria and archaea – target region 16S ribosomal DNA (rDNA),
- fungi – target region internal transcribed spacer (ITS), and
- eukaryotes other than fungi – target region 18S ribosomal DNA (rDNA).

Possibly, nematodes, arthropod mesofauna, and earthworms will be included. The protocol defined by the Earth Macrobiome Project (EMP, 2017) will be applied.

Soil samples should be frozen as soon as possible after collection. This requires precise logistical arrangements (Fernandez-Ugalde et al., 2017): the surveyors need to prepare freezer packs well in advance of the sampling and place them in a polystyrene box the day of the survey, wash the sampling material before each sampling with alcohol and water and wear plastic gloves during the collection of the sample. Once sealed in the polystyrene box, the sample, wherever the sampling point is located in Europe, should reach the final storage location preferably within 48 hours.

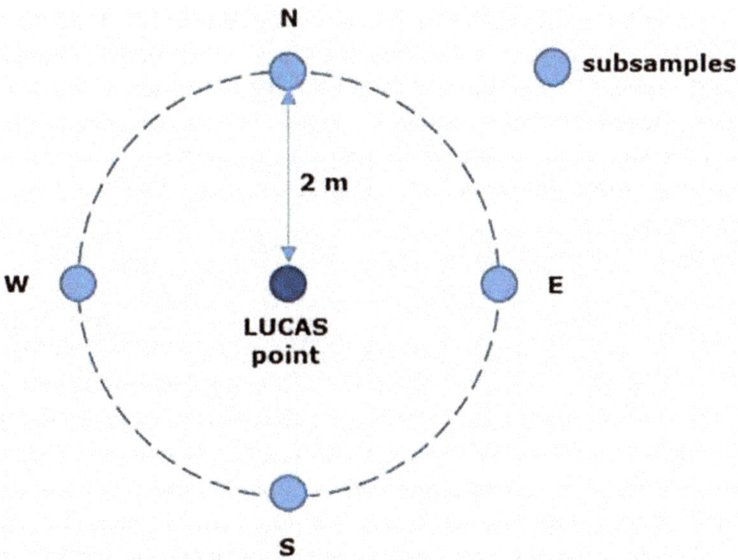

Figure 3 Spatial distribution of samples to be collected for soil biodiversity analysis per each visited LUCAS point.

The reference for locating the sampling point is the LUCAS grid. Soil samples should be taken within a maximum of 100 m distance from the LUCAS point and in the same field where the LUCAS point is located. Surveyors use a metallic ring that they press in the soil with the help of a mallet to extract the soil sample. Ideally, a sample is taken at the LUCAS point, and mixed with four other samples collected 2 m away from the point in each of the main cardinal directions (Fig. 3). A subsample of soil is extracted and put in a plastic jar, labelled and sealed; the jar is put in the polystyrene box together with the freezer packs, labelled and sealed with tape.

Initial results should be available during 2020. All information will be hosted by the European Soil Data Centre (ESDAC). This will include options for the online generation of maps as well as the entire methodology. Other parties (e.g. national organisations, NGOs) who wish to make the database grow through their own contribution may adopt the same protocol (Orgiazzi et al., 2018).

6 Monitoring under the Habitats Directive 92/43/EEC

The Habitats Directive 92/43/EEC is aimed at maintaining or restoring 'at favourable conservation status, natural habitats and species of wild fauna and flora of Community interest', which are listed in the Annexes of the Directive. Such requirements involve (i) defining 'conservation status'; (ii) knowing where the habitat/species is found and (iii) what its conservation status is.

Conservation status is defined as 'the overall assessment of the status of a habitat type or a species at the scale of a Member State's biogeographical or marine region' (DG Environment, 2017) and more precisely Article 1 of the Directive links the term to the extent of the area in which the habitat/species is found, the surface of the habitat area, its structure and functions (in case of habitat), the size of the population, its age structure, mortality and reproduction (of species) (EC, 2009).

Article 17 of the Habitats Directive (CEC, 1992) requires the Commission to draw up, every six years, a composite report based on the national reports delivered by the Member States of the European Union, including 'in particular information concerning the conservation measures referred to in Article 6(1) as well as evaluation of the impact of those measures on the conservation status of the natural habitat types of Annex I and the species in Annex II and the main results of the surveillance referred to in Article 11'. The report should be available for the other EU institutions and the public in general.

Three of these reports are available (the fourth to be released in 2020), two of which focus on the conservation status of the habitat types and species included in the Annexes to the Directive for the periods 2001–2006 and 2007–2012. The assessment of conservation status takes as a reference the concept of favourable conservation status, defined as 'a situation where a habitat type or species is prospering (in both quality and extent/population) and with good prospects to continue to do so in the future', which is the overall objective of the Directive.

The conservation status of a species in the Habitats Directive (Article 1(i)) will be taken as 'favourable' when:

- population dynamics data on the species concerned indicate that it is maintaining itself on a long-term basis as a viable component of its natural habitats,
- the natural range of the species is neither being reduced nor is likely to be reduced for the foreseeable future, and
- there is, and will probably continue to be, a sufficiently large habitat to maintain its populations on a long-term basis.

The conservation status of a habitat in the Habitats Directive (Article 1(e)) will be taken as 'favourable' when:

- its natural range and areas it covers within that range are stable or increasing,
- the specific structure and functions which are necessary for its long-term maintenance exist and are likely to continue to exist for the foreseeable future, and
- the conservation status of its typical species is favourable as defined in (i).

Table 3 Pressure categories in the list of pressures and threats (from the explanatory notes and guidelines for the period 2013-2018, http://cdr.eionet.europa.eu/help/habitats_art17)

Pressure code	Pressure category	Note
A	Agriculture	Includes pressures and threats caused by agricultural practice.
B	Forestry	Includes pressures and threats caused by forestry activities, including thinning, wood harvesting, pest control in trees.
C	Extraction of resources (minerals, peat, non-renewable energy resources)	Includes pressures related to extraction of materials, such as mining or quarrying, pollution or waste disposal.
D	Energy production processes and related infrastructure development	Includes pressures related to production of energy, for example, the construction and operation of power plants, water-use for energy production, waste from energy production, activities and infrastructure related to renewable energy.
E	Development and operation of transportation and service corridors	Includes pressures related to transportation of materials or energy, such as construction of infrastructure, pollution and disturbances or increased mortality due to traffic.
F	Development, construction and use of residential, commercial, industrial and recreational infrastructure and areas	Includes pressures related to development, construction and use of residential, commercial, industrial and recreational infrastructure, for example, infrastructural changes on existing built areas, expansion of built areas, land use and hydrological changes for urban or industrial development, disturbances or pollution due to residential, commercial, industrial or recreational infrastructure. Includes also pressures related to sport, tourism and leisure activities and infrastructure.
G	Extraction and cultivation of biological living resources (other than agriculture and forestry)	Includes pressures linked to uses of biological resources other than agriculture and forestry.
H	Military action, public safety measures, and other human intrusions	Includes pressures related to public safety and other human intrusions.
I	Invasive and problematic species	Includes pressures related to problematic inter-specific relationships with non-native species which cannot be associated with other pressure categories. Includes also problematic relationships with native species, which came out of balance due to human activities.
J	Mixed source pollution	Includes pollution which cannot be associated with other pressure categories.
K	Human-induced changes in hydraulic conditions	Includes hydrological and physical modifications of water bodies, which cannot be associated with other pressures categories.
L	Natural processes (excluding catastrophes and processes induced by human activity or climate change)	Includes natural processes, such as natural succession, competition, trophic interaction and erosion.
M	Geological events, natural catastrophes	Includes pressures such as natural fires, storms and tsunamis.
N	Climate change	Includes pressures related to climate change.

Four parameters have been identified for evaluating the conservation status, which, combined, provide the overall assessment. The parameters are:

- species: range, population, habitat for the species, future prospects, and
- habitat types: range, area, structure and functions, future prospects.

These four parameters are combined to assess conservation status for individual species and habitats using the following four categories: 'favourable' (FV), 'unfavourable-inadequate' (U1), 'unfavourable-bad' (U2) and 'unknown' (XX). Trends are equally assessed on the basis of a combination of the individual trends of the four parameters listed above, classified as: increasing, decreasing, stable and unknown.

Of particular interest for this chapter is the fact that in addition to the respective conservation status, trends and presence per biogeographical and MS provided in tabular format, distribution maps for species and habitats should be provided as geospatial information with a 10 km cell resolution. This allows using the geospatial layers for EU-wide assessments of the status of agroecosystems (Masante et al., 2015). Interestingly, pressures and/or threats are recorded as well, including a ranking of its impact on the conservation status of species for each pressure/threat (Table 3).

7 European Monitoring of Biodiversity in Agricultural Landscapes (EMBAL)

European Monitoring of Biodiversity in Agricultural Landscapes (EMBAL) is a rapid approach based on an in situ survey of plots with a size of 500 m × 500 m (25 ha). This plot size yields an optimum ratio between survey effort and detailedness, and effectively represents an agricultural landscape. The development of this survey is part of the monitoring activities coordinated by the European Commission (DG Environment).

The survey aims at recording information on characteristics of agricultural land in each plot, and namely on grassland, cropland and landscape elements. The survey is carried out during optimum vegetation conditions (peak of the growing season). Land cover is mapped, including landscape elements, surveyed parameters include the description of the site (e.g. habitat coding, type of land cover), cropland (e.g. coverage of crops, wild plants, bare soil, number of flowering forbs, key species list) and grasslands (parameters harmonized with LUCAS Grasslands protocol). A nature value is assigned to all mapped elements.

Four 20 m transects are walked, with an observation frame of 1.25 m to the left and 1.25 m to the right of the surveyor. During the transect walks, the surveyors will record the presence of potential key plant species given in a list as well as information on the structure of the fields. This information will serve as an additional qualification of the land-use intensity, which is estimated in the

Published by Burleigh Dodds Science Publishing Limited, 2020.

main sheet for each parcel and as biodiversity record which enables further data interpretation.

The parameters are to be assessed in the field and deliver results at the area level (area and types of land cover/landscape elements, change in landscape structure, nature values for the types of land cover/landscape elements, area per EUNIS habitat type etc.) and at the transect level (coverage of wild plants in arable land, number of key species in grassland, arable land and fallow land, species composition, flower density etc. and respective changes in these parameters). As in the case of the LUCAS grassland survey, training of field surveyors is of utmost importance for ensuring data quality.

EMBAL is set to be the first large-scale survey to provide information on biodiversity values in arable land and a detailed mapping and identification of landscape elements. It is also set to be the first survey applying the landscape approach. By including as well the qualitative estimate of the natural value of the area, once the survey is established and repeated it will provide evidence of the trajectory of evolution of agricultural landscapes with regard to some aspects of biodiversity (increase or decrease), as shown by previous experiences on which the survey is based (IFAB, 2017; Hünig and Benzler, 2017). The EMBAL methodology[14] was piloted in 2020, and is going to be fine-tuned as necessary. The pilot is also used for testing potential integration of soil sampling into the EMBAL framework. Sampling squares are aligned with the LUCAS grid. Overall, the two surveys are harmonized, and complement each other in depicting the state of biodiversity in agricultural areas.

8 Alien species

'Alien species' are defined as live specimens of a species, subspecies or lower taxon of animals, plants, fungi or micro-organisms introduced outside their natural range, including any part, gametes, seeds, eggs or propagules of such species, as well as any hybrids, varieties or breeds that might survive and subsequently reproduce[15].

Alien species causing damage to agriculture are harmful organisms (insects, mites, nematodes, bacteria, parasitic plants, fungi, viruses and virus-like organisms) regulated under the European phytosanitary legislation, and stem from Council Directive 2000/29/EC[16] and related decisions laying down

14 https://ec.europa.eu/environment/nature/knowledge/pdf/embal_report.pdf https://ec.europa.eu/environment/nature/knowledge/pdf/embal_survey_manual.pdf

15 Art. 3 of Regulation (EU) No 1143/2014 of the European Parliament and of the Council of 22 October 2014 on the prevention and management of the introduction and spread of invasive alien species. OJ L 317, 4.11.2014, p. 35–55

16 Council Directive 2000/29/EC of 8 May 2000 on protective measures against the introduction into the community of organisms harmful to plants or plant products and against their spread within the community. OJ L 169, 10.7.2000, p. 1–112

emergency or control measures for specific pests and diseases under this specific remit (e.g. *Xylella fastidiosa*, *Dryocosmus kuriphilus*, *Anoplophora chinensis* and *A. glabripennis*). This Directive has been repealed by Regulation (EU) 2016/2031[17], whose requirements entered into force on 14 December 2019. The compulsory monitoring for agriculture-related alien species (open field crops, nurseries, and greenhouses) is laid down in technical annexes and must be implemented by official controls performed by Member States National Plant Protection Organizations. Producers are subject to specific obligations in relation to the health of their crops, and shall immediately notify the responsible official body of the Member State concerned of any unusual occurrence of harmful organisms, symptoms or any other plant abnormality, and collaborate with official authorities. Official surveillance and systematic inspections are conducted at the place of production in the most appropriate periods considering the organisms' biology, supplemented by laboratory analysis. The organization of the phytosanitary surveillance and organisms involved varies and reflects Member States' administrative organization.

Invasive Alien Species (IAS) constitute one of the most important threats to biodiversity, causing worldwide severe ecological and socio-economic impacts. Target 2.2.10 of the the EU Biodiversity Strategy for 2030 (EC, 2020) puts particular effort on addressing invasive alien species (IAS) and the protection of native biodiversity. Recognizing the need for a coordinated set of actions to prevent, control and mitigate IAS impact, the European Parliament and the Council adopted the EU Regulation 1143/2014 (the IAS Regulation), which gives priority to a list of species, named as IAS of Union concern, selected by scientific and legislative bodies after an evaluation procedure and the analysis of specific pest risk assessments. Species are included in this list, among others, because they can cause a significant damage to biodiversity justifying the adoption of dedicated measures at Union level. Three Implementing Commission Regulations[18] have listed 66 alien species (30 animals and 36 plants) as of Union concern

17 Regulation (EU) 2016/2031 of the European Parliament of the Council of 26 October 2016 on protective measures against pests of plants, amending Regulations (EU) No 228/2013, (EU) No 652/2014 and (EU) No 1143/2014 of the European Parliament and of the Council and repealing Council Directives 69/464/EEC, 74/647/EEC, 93/85/EEC, 98/57/EC, 2000/29/EC, 2006/91/EC and 2007/33/EC.
OJ L 317, 23.11.2016, p. 4-104

18 a Commission Implementing Regulation (EU) 2016/1141 of 13 July 2016 adopting a list of invasive alien species of Union concern pursuant to Regulation (EU) No 1143/2014 of the European Parliament and of the Council C/2016/4295.
OJ L 189, 14.7.2016, p. 4-8

b Commission Implementing Regulation (EU) 2017/1263 of 12 July 2017 updating the list of invasive alien species of Union concern established by Implementing Regulation (EU) 2016/1141 pursuant to Regulation (EU) No 1143/2014 of the European Parliament and of the Council C/2017/4755.
OJ L 182, 13.7.2017, p. 37-39

c Commission Implementing Regulation (EU) 2019/1262 of 25 July 2019 amending Implementing Regulation (EU) 2016/1141 to update the list of invasive alien species of Union concern C/2019/5360.
OJ L 199, 26.7.2019, p. 1-4

by 2019, which are affecting terrestrial, freshwaters and marine (2 species) environments. Under the IAS Regulation, EU Member States competent authorities must prevent the introduction and spread of IAS of Union concern, enforce effective early detection and rapid eradication mechanisms for new introductions and adopt management measures for species already widely spread.

Article 16 of the IAS Regulation requires the mandatory notification of early detections of listed species to the European Commission and to the other Member States. Detailed and up-to-date spatial information (baselines) on the distribution of 49 IAS of Union concern in Europe have been prepared by the European Alien Species Information Network of JRC (https://easin.jrc.ec.europa.eu/easin/) (baseline for the 17 species listed by Regulation (EU) 2019/1262 is under preparation) working in close collaboration with EU Member States competent authorities, setting a geographic baseline of their current distribution[19]. The baselines allow a shared knowledge of the IAS distribution, a better planning of the surveillance activity, the official notification of new detections, and foster the collaboration among MS. These baselines are also available in electronic format as shapefiles (https://easin.jrc.ec.europa.eu/easin/Documentation/Baseline, Fig. 4).

IAS are also considered by the EU Framework Directive for the evaluation of 'good ecological status' of freshwaters (Directive 2000/60/EC)[20] and by the Marine Strategy Framework Directive under Descriptor 2 (non-indigenous species) (Directive 2008/56/EC)[21]. Farmers are sometimes actively involved in monitoring and management of IAS, even if not regulated by law, such as *Fallopia japonica*, Japanese knotweed, like it happens, for example, in Ireland (Teagasc Invasive Alien Species Working Group).

Specific professional groups of farmers are more directly involved in detection and control of IAS. This is the case of the Asian hornet, *Vespa velutina*, which sees an active role of beekeepers in many EU countries in detecting the presence of the invasive insect, which preys on honeybees, deploying a network of sentinel hives provided with hornet-traps (e.g. interregional network in northern Italy: http://www.stopvelutina.it/nord-italia-nasce-lover-la-rete-di-sorveglianza-per-la-velutina/).

Many LIFE and Interreg projects in Europe (e.g. LIFE STopVespa, LIFE Invasaqua, LIFE Asap, Interreg Invalis, Interreg ACECA, http://ec.europa.eu/

19 Tsiamis K, Gervasini E, Deriu I, D'Amico F, Katsanevakis S., Cardoso AC, 2019. Baseline distribution of species listed in the first update of Invasive Alien Species of Union concern. Publications Office of the European Union, Ispra. https://doi.org/10.2760/75328
Tsiamis K, Gervasini E, Deriu I, D'Amico F, Nunes A, Addamo A, De Jesus Cardoso A, 2017. Baseline Distribution of Invasive Alien Species of Union concern. Publications Office of the European Union, Ispra. https://doi.org/10.2760/772692

20 https://eur-lex.europa.eu/legal-content/EN/TXT/?uri=celex:32000L0060

21 https://eur-lex.europa.eu/legal-content/EN/TXT/?uri=CELEX:02008L0056-20170607

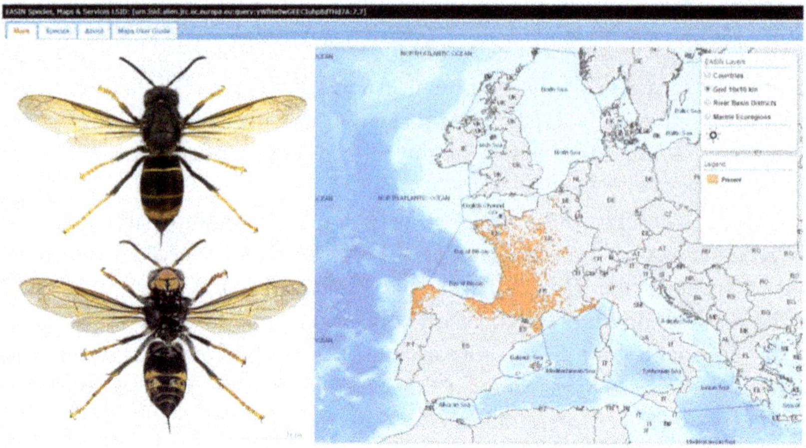

Figure 4 Distribution map of Vespa velutina nigrithorax in Europe at 10 km ×10 km grid level, from EASIN data. Picture: Didier Descouens – Own work, CC BY-SA 3.0, https://commons.wikimedia.org/w/index.php?curid=26722179.

environment/life/project/Projects/index. Cfm, https://www.interregeurope.eu/invalis/, https://camaloteaceca.eu/?page_id=347) are also dedicated to the monitoring, eradication and management of invasive alien species in different environments, or restoration activities which entail the removal of IAS, for example, weeds or invasive plants, preserving natural areas and protecting agriculture from their further spread.

The environmental DNA (e-DNA) is progressively being used to detect the presence of IAS, for example, *Procambarus fallax* f. *virginalis*, *Pseudorasbora parva*, *Vespa velutina* (Takahashi et al., 2018), while many sequenced genomes are already available in gene banks (https://www.ncbi.nlm.nih.gov/), including for *Lithobates catesbeianus* and *Heracleum mantegazzianum*.

A great potential for early detection of IAS and provision of massive data is offered by the contribution of volunteer citizens through Citizen Science, nowadays a consolidated activity, which favours citizens' awareness and active engagement in IAS management. A repository of Citizen Science projects in Europe which target alien species is available in EASIN (https://easin.jrc.ec.europa.eu/easin/CitizenScience/Projects). Following this line, JRC has developed a smartphone application called 'Invasive Alien Species Europe' allowing citizens to report species of Union concern across Europe, where species list, identification factsheets, and guidelines are kept up-to-date (http://digitalearthlab.jrc.ec.europa.eu/app/invasive-alien-species-europe). Data gathered via the App, after validation, are uploaded and available through EASIN geodatabase.

9 Other monitoring approaches

The monitoring frameworks presented in this chapter do not represent the only sources of information on biodiversity currently available, but those for which regular updating is either in place or planned.

The Biodiversity Information System for Europe (BISE, https://biodiversity.europa.eu/info) provides information at the European level in relation to the Biodiversity targets for the European Union, and as regards data it collects and makes available data sources, statistics and maps related to land, water, soil, air, marine, agriculture, forestry, fisheries, tourism, energy, land use and transport. Most of these data are made available under projects or programmes that do not foresee a regular update.

A structured approach to farmland monitoring is a main outcome of the BioBio Project[22]. The authors Geijzendorffer et al. (2015) and Herzog et al. (2012), starting from the assumption that no single all-inclusive index for biodiversity can be devised[23], proposed a framework composed by 15 indicators that describe genetic, species and habitat diversity in farmland. The indicators set is a result of thorough scientific screening and testing in 12 case-study regions with various farm types and farming systems across Europe, as well as regular stakeholder consultation (Herzog et al., 2012).

Such framework can be used as a reference versus the EU-wide monitoring efforts to check what data gaps are and what should still be improved. Table 4 briefly summarizes BioBio indicators, with the exclusion of farm management indicators.

Species diversity is the category in which efforts at the EU level are more advanced and going in the right direction. Plants, pollinators and soil organisms are covered by monitoring efforts either already in place or under development. To these, birds and butterflies must be added, and the progress on the European bat indicator is worth mentioning (EEA, 2013b; Battersby, 2010). Compared to other taxa, though, it must be noted that soil biodiversity is still the less known.

Indicators of habitat diversity can be partially derived from the EMBAL monitoring system (e.g. habitat richness, habitat diversity, patch size). Moreover, the Copernicus Land Monitoring Service, part of the European Union's Earth Observation Programme[24], is providing geographical information on land cover and its changes, land use and vegetation state which can be used to derive indicators on habitat diversity and structural habitat characteristics, such as the presence of landscape elements (tree rows, hedges and groups of trees). The LUCAS module on landscape features in 2022 will collect in-situ information on

[22] EU 7th Framework Programme, Grant no. KBBE 227161
[23] BioBio webpage http://www.biobio-indicator.org/indicators.php?l=1
[24] https://www.copernicus.eu/en/services/land

Table 4 BioBio indicators (source: Herzog et al., 2012)

Genetic diversity	Species diversity	Habitat diversity
Number and amount of different breeds	Vascular plants	Habitat richness
Cultivar diversity	Wild bees and bumblebees	Habitat diversity
Origin of crops	Spiders	Patch size
	Earthworms	Linear habitats
		Crop richness
		Shrub habitats
		Tree habitats
		Semi-natural habitats

smaller landscape elements in agricultural areas to complement information from Copernicus. Data (small woody features and hedgerows, grass fringes, ditches, small ponds, stone walls and terraces) will be collected on 93.000 points with the purpose to provide quantitative information at EU and Member State level. The potential of the new Sentinel imagery applied to habitat monitoring has still to be fully exploited. It is important to note that the mentioned approaches have their limitations in terms of resolution or density of sampling points, and provide results that are mostly applicable at country level or macro-regions (e.g. LUCAS grasslands, LUCAS soil, EMBAL), and satellite images are constrained by the resolution of the images (both geometric and spectral).

Despite the efforts, the first column in Fig. 2 remains substantially empty. Though projects aiming at the conservation of genetic material do exist, at global (e.g. Svalbard Global Seed Vault, https:// www.seedvault.no/), national (RIBES, the Italian network of seed banks, http://www.reteribes.it/) and local (https:// www.communityseedbanks.org/the-csb-map/) levels, an overview at EU level of the number, amount and geographical distribution of traditional breeds, cultivars, landraces, wild crop relatives, traditional and ancient varieties is not yet available. The FAO reports that 'although crop wild relatives represent about 13 percent of the world's gene bank holdings, about 70 percent of such species are still missing' (FAO, 2018). To a great degree, the quantification of that part of biodiversity that is directly embedded into agriculture and constitutes its core, as well as an insurance for agricultural ecosystems to remain resilient and adaptive, still has to happen.

What analysed so far concerns initiatives in place or planned. Before reaching some conclusions, it is worth looking into the near future and understand which substantial improvements in biodiversity monitoring can be expected from the operational use of new technologies.

10 New approaches and technologies

Recent and ongoing technological developments are reshaping our capacity to monitor biodiversity. Digital connectedness, cataloguing, and storage, combined with advanced analytical techniques and fast computation, provide frameworks that can gather observations across large areas. In Europe, LifeWatch ERIC was established as a European Research Infrastructure Consortium[25] to provide continued support to the scientific community studying biodiversity. LifeWatch ERIC is building virtual, instead of physical, e-laboratories supplied by the most advanced facilities to capture, standardize, integrate, analyse and model biodiversity (Basset and Los, 2012).

Flexibility and scalability allow designing of monitoring networks for specific tailor-made purposes. Importantly, these frameworks allow expert biologists, citizen scientists, as well as relative non-experts to make valuable contributions (e.g. Chandler et al., 2017). Indeed, these developments facilitate contributions by expert volunteers, who have contributed to biodiversity monitoring for centuries (McKinley et al., 2017). Several tools exist. Functionality varies from geo-locating ones' own observations to sharing and mapping across the globe. The most well-known citizen-science platform in this context is probably iNaturalist (www.inaturalist.org). iNaturalist provides a generic platform to catalogue species. Another example, which started in 2010, is Pl@ntNet (https://plantnet.org).

Pl@ntNet allows a user to identify a flower, plant or tree species by photographing them with a smartphone (Joly et al., 2014). Specific examples relate to identifying weeds, cultivated and/or ornamental plants, invasive species (Botella et al., 2018) or a focus on specific geographic regions.

Using computer-vision-based algorithms relies on massive amounts of good-quality training data. Pl@ntNet for instance is able to identify a wide range of species with increasing accuracy through a novel collection and validation approach. The Pl@ntNet algorithm continuously learns and improves its accuracy. Each new picture that is submitted provides new data to train the algorithm, benefitting from user feedback stating whether the correct species was identified (or not). These developments highlight that 'automated plant identification systems are now mature enough for several routine tasks, and can offer very promising tools for autonomous ecological surveillance systems' (Bonnet et al., 2018). This will drastically reduce the time needed to generate significant biodiversity data flows.

Clearly, there are also limitations. For example, essential taxonomic details may (currently) not be visible on a picture. Resembling a Turing test, Pl@ntNet pitted its algorithms against botanical experts (Bonnet et al., 2018).

25 By European Commission Implementing Decision (EU) 2017/499 of 17 March 2017.

One of the conclusions was the need for details to make certain taxonomic distinctions.

Ethical considerations also arise. How best to safeguard the geo-location of a protected red-listed species - say an orchid notoriously difficult to find? Surveying is thus undergoing changes. There is still a huge potential in Citizen Science and crowdsourcing. Maintaining such monitoring over longer time periods is one of the challenges. Furthermore, tapping this potential requires having procedures to set standards and for thorough quality. This starts at the collection of observations. The detection and monitoring of pollinator communities, for instance, provides guidance and training to guarantee minimum identification skills of contributing surveyors (LeBuhn et al., 2016). Besides building inventories of species occurrence, near real-time observations underpin better process understanding of why and how species move through landscapes and interact with their environment. Voluntary contributions tracking birds, for instance, have revealed that migratory tracks have been shifting due to climate change (Cooper et al., 2014). New technologies also enable fast and efficient ways to gather data. One example is the efficient collection and extraction of information from pictures with street-level cameras (e.g. d'Andrimont et al., 2018).

Green infrastructures critical for biodiversity can now be mapped at relevant temporal and spatial scales. For example, Lucas et al. (2019) used LiDAR to identify linear vegetation elements in a rural landscape. While new approaches have been developed to retrieve in situ data, the satellite remote sensing community, via the Group on Earth Observations Biodiversity Observation Network (GEO BON, https://geobon.org/ebvs) has developed the concept of Essential Biodiversity Variables (EBVs) (Pereira et al., 2013). They provide the first level of abstraction between low-level primary observations and high-level indicators of biodiversity. So far, there are six EBV classes (genetic composition, species populations, species traits, community composition, ecosystem function and ecosystem structure) with 21 EBV candidates. The EBVs are defined as the derived measurements required to study, report and manage biodiversity change. The EBVs should broker between monitoring initiatives and decision makers.

11 Conclusions

Some lessons can be learnt from the various experiences of running monitoring schemes and surveys at the supranational level. The first is that setting up a monitoring scheme takes time. Setting up large-scale surveys from planning to execution takes a few years, and this is the case of all surveys described in this chapter. Preparing the sampling scheme, the survey protocol, identifying the funding bodies, running contracts, finding surveyors, actually running the survey and finally preparing the final database and a first analysis of the data easily takes three to five years, and more if, as in the case of birds and butterflies, national schemes have to be set up individually. Therefore, the operational phase should start as soon as the knowledge gap is identified.

Secondly, the role of volunteers is essential. Whether skilled ornithologists and entomologists participating in counts, or people willing to use apps to communicate species and location of plants and animals, biodiversity monitoring cannot occur without volunteers. The cost for monitoring would simply be too high if tens of thousands of volunteers needed to be paid. Moreover, Schmeller et al. (2009) show that volunteer-based schemes can yield unbiased results, if surveys are appropriately designed and the number of volunteers is sufficiently high. In this regard, the raising awareness that biodiversity is under threat will hopefully convey growing interest into active participation in surveys.

The potential of new technologies is high and not yet fully exploited. We can expect in a near future much higher capacity of automatic recognition of plants and animals, improved algorithms for processing images sent by smartphones, higher availability of centralized databases, platforms to store and retrieve information. This will greatly reduce the time needed to be spent on the ground (e.g. taking photos of plants rather than recognizing in the field each single species), but planning, post-processing and maintenance of the infrastructure will still take a considerable share of resources.

Lastly, there is a point that is often undervalued: taxonomy. In order to correctly identify a specimen, a reference must exist, and species must have been named and classified. Moreover experts must be sufficiently trained to correctly classify specimens, and research on this topic should be reported correctly. It is recognized that taxonomy especially for insects and micro-organisms is not complete (FAO, 2016; Audisio, 2017), that there is a lack of taxonomists and experts with 10–15 years of experience in recognizing species (FAO, 2016), that taxonomic research is often lacking accurate and replicable taxonomic identification (Packer et al., 2018). Also in this case the important role of amateur taxonomists is recognized.

In conclusion, three main pillars can be identified in the complex architecture of (farmland) biodiversity monitoring: the experts, who hold the scientific knowledge of biodiversity and tools to measure it; the decision makers and funding bodies, who take decisions about what to monitor and support monitoring activities and monitoring architecture; and the citizens, essential in information recording.

12 Where to look for further information

It is clear that the effort for reaching a sufficient level of biodiversity monitoring is a collective and specialised effort, that requires coordination among many actors at different levels. The improvement of scientific knowledge is supported by research funding, for example, under the EU Research and Innovation programmes (Horizon 2020, Horizon Europe). Moreover, bridging science and technology eases active citizenship involvement. Monitoring schemes can be set up on the basis of existing scientific knowledge, and improved at each survey round. It is a process that takes time, for example, the LUCAS grassland survey developed in approximately seven years from the discussions about the approach to take in 2015, to a first test survey in 2018, to an enlarged survey on 20 000 transects planned for 2022. The involvement of scientific communities (e.g. birds, lepidoptera, pollinators) holding the necessary knowledge about individual taxa or habitats is key. Coordination and support provided by governance bodies (national ministries, EU bodies) is needed to provide the framework for the implementation of the surveys (e.g. funding, coordination) and to secure their repetition through time.

13 References

A European Union Strategy for Sustainable Development (Commission's proposal to the Gothenburg European Council), COM(2001)264 final https://eur-lex.europa.eu/legal-content/EN/TXT/PDF/?uri=CELEX:52001DC0264&from=EN.

Alliance Environnement (2017). Literature reviews on the effects of farming practices associated with the CAP greening measures on climate and the environment, carried out for the Evaluation study of the payment for agricultural practices beneficial for the climate and the environment ("greening" of direct payments), ISBN978-92-79-73406-9.EEIG, 266 p. Available at: https://ec.europa.eu/agriculture/sites/agriculture/files/fullrep_en.pdf (Accessed on 31st October 2019).

Audisio, P. (2017). Insect taxonomy, biodiversity research and the new taxonomic impediments. *Fragmenta Entomologica* 49(2): 121–124.

Barrios, E. (2007). Soil biota, ecosystem services and land productivity. *Ecological Economics* 64(2): 269–285.

Basset, A. and Los, W. (2012). Biodiversity e-Science: LifeWatch, the European infrastructure on biodiversity and ecosystem research. *Plant Biosystems-An International Journal Dealing with All Aspects of Plant Biology* 146(4): 780–782.

Batáry, P., Dicks, L. V., Kleijn, D. and Sutherland, W. J. (2015). The role of agri-environment schemes in conservation and environmental management. *Conservation Biology* 29(4): 1006–1016. https://doi.org/10.1111/cobi.12536.

Battersby, J. (comp.) (2010). *Guidelines for Surveillance and Monitoring of European Bats. EUROBATS Publication Series No. 5*. UNEP / EUROBATS Secretariat, Bonn, Germany, 95 pp.

Bonnet, P., Goëau, H., Thye Hang, S., Lasseck, M., Sulc, M., Malécot, V., Jauzein, P., Melet, J.-C., You, C. and Joly, A. (2018). Plant identification: experts vs. machines in the era of deep learning, p. 131–149. In: *Multimedia Tools and Applications for Environmental & Biodiversity Informatics*. Springer International Publishing AG, part of Springer Nature 2018.

Botella, A. J., Bonnet, P., Monestiez, P. and Munoz, F. (2018). Species distribution modeling based on the automated identification of citizen observations. *Applications in Plant Sciences* 6: 1–11.

Brereton, T., van Swaay, C. and van Strien, A. (2009). Developing a butterfly indicator to assess changes in Europe's biodiversity. *Avocetta* 33: 19–27.

Bumblebee Conservation Trust (2017). The step-by-step guide to setting up a BeeWalk transect. Available at: https://www.bumblebeeconservation.org/wp-content/uploads/2017/08/BeeWalkGuidelines2017.pdf (Accessed on 5th August 2019).

Chandler, M., See, L., Copas, K., Bonde, A. M. Z., López, B. C., Danielsen, F., Legind, J. K., Masinde, S., Miller-Rushing, A. J., Newman, G., Rosemartin, A. and Turak, E. (2017). Contribution of citizen science towards international biodiversity monitoring. *Biological Conservation* 213(Part B): 280–294.

Communication from the Commission to the European Parliament, the Council, the Economic and Social Committee and the Committee of the Regions: our life insurance, our natural capital: an EU biodiversity Strategy to 2020, COM/2011/0244 final.

Comont, R. F. and Dickinson, H. (2018). *BeeWalk Annual Report 2018*. Bumblebee Conservation Trust, Stirling, Scotland, UK. Available at: https://www.bumblebeeconservation.org/wp-content/uploads/2017/05/BBCT096-BeeWalk-Annual-Report-2018-04.18-LOW-RES-1.pdf (Accessed on 8th August 2019).

Cooper, C. B., Shirk, J. and Zuckerberg, B. (2014). The invisible prevalence of citizen science in global research: migratory birds and climate change. *PLOS ONE* 9(9): e106508. https://doi.org/10.1371/journal.pone.0106508.

Council of the European Communities (1979). Council Directive 79/409/EEC of 2 April 1979 on the conservation of wild birds, OJ L 103, 25.4.1979, p. 1–18.

Council of the European Communities (1992). Council Directive 92/43/EEC of 21 May 1992 on the conservation of natural habitats and of wild fauna and flora, OJ L 206, 22.7.1992, p. 7.

d'Andrimont, R., Yordanov, M., Lemoine, G .,Yoong, J., Nikel, K. and Van der Velde, M. (2018). Crowdsourced street-level imagery as a potential source of in-situ data for crop monitoring, *Land* 7 (4), 127.

DG Environment (2017). Reporting under Article 17 of the habitats directive: explanatory notes and guidelines for the period 2013–2018. Brussels, p. 188.

Dudley, N. and Alexander, S. (2017). Agriculture and biodiversity: a review. *Biodiversity* 18(2-3): 45–49.

EMP (Earth Microbiome Project) (2017). Protocols and standards. Earth Microbiome Project, La Jolla, CA (WWW document). Available at: http://www.earthmicrobiome.org/protocols-and-standards/ (Accessed on 2nd August 2019).

European Bird Census Council (2019). What is Pan-European Common Bird Monitoring Scheme? Available at: http://www.ebcc.info/pan-european-common-bird-monitoring-scheme-pecbms/what-is-pan-european-common-bird-monitoring-scheme/ (Accessed on 23rd July 2019).

European Commission (2001). Communication from the Commission - A Sustainable Europe for a Better World: A European Union Strategy for Sustainable Development. COM(2001)264 final.

European Commission (2009). *Report from the Commission to the Council and the European Parliament Composite – Report on the Conservation Status of Habitat Types and Species as Required under*: Article 17 of the Habitats Directive COM(2009)0358 final.

European Commission (2015). Report from the Commission to the Council and the European Parliament: the State of Nature in the European Union, Report on the status of and trends for habitat types and species covered by the Birds and Habitats Directives for the 2007–2012 period as required under Article 17 of the Habitats Directive and Article 12 of the Birds Directive. COM(2015) 219 final.

European Commission (2018). EU pollinators initiative. COM(2018) 395 final.

European Commission (2020). EU Biodiversity Strategy for 2030 – Bringing nature back into our lives. COM(2020) 380 final.

European Environment Agency (2013a). The European Grassland Butterfly Indicator: 1990–2011. EEA Technical report No 11/2013, 35 pp.

European Environment Agency (2013b). European bat population trend – a prototype biodiversity indicator, EEA Technical report No 19/2013, 61 pp.

European Parliament and the Council of the European Union (2009). Directive 2009/147/EC on the conservation of wild birds, OJ L 20, 26.1.2010: 7–25.

EUROSTAT (2018). LUCAS 2018 (Land Use / Cover Area Frame Survey) Technical reference document C2 Field Form and Ground Document (template) (Online). Available at: https://ec.europa.eu/eurostat/documents/205002/8072634/LUCAS2018-C2-FieldForm-GD- Template.pdf (Accessed on 23rd July 2019).

EUROSTAT (2019). Available at: https://ec.europa.eu/eurostat/statistics-explained/index.php/LUCAS_-_Land_use_and_land_cover_survey#The_LUCAS_survey (Accessed on 19th July 2019).

FAO (2016). Protocol to detect and monitor pollinator communities. FAO report 19367e/1/01.16. Available at: http://www.fao.org/3/a-i5367e.pdf (Accessed on 5th August 2019).

FAO (2018). Sustainable agriculture for biodiversity- biodiversity for sustainable agriculture. FAO Brochure I: 9577EN/1/05.2018. Available at: http://www.fao.org/3/a-i6602e.pdf (Accessed on 30th October 2019).

FAO (2019). The state of the world's biodiversity for food and agriculture Bélanger, J. and Pilling, D. (Eds). *FAO Commission on Genetic Resources for Food and Agriculture Assessments*. Rome. 572 pp. Available at: http://www.fao.org/3/CA3129EN/CA3129EN.pdf Licence: CC BY-NC-SA 3.0 IGO (Accessed on 5th August 2019).

Fernández-Ugalde, O., Orgiazzi, A., Jones, A., Lugato, E. and Panagos, P. (2017). LUCAS 2018 – *Soil Component: Sampling Instructions for Surveyors*, EUR 28501EN, doi 10.2760/023673.

Geijzendorffer, I. R., Targetti, S., Schneider, M. K., Brus, D. J., Jeanneret, P., Balázs, K., Báldi, A. and Kovács-Hostyánszki, A. (2015). How much would it cost to monitor farmland biodiversity in Europe? *Journal of Applied Ecology* 53(1): 140–149.

Geisen, S., Briones, M. J. I., Gan, H., Behan-Pelletier,V. M., Friman, V.-P., de Groot, G. A., Hannula, S. E., Lindo, Z., Philippot, L., Tiunov, A. V.. and Wall, D. H. (2019). A methodological framework to embrace soil biodiversity. *Soil Biology and Biochemistry* 136:107536.

Goulson, D., Nicholls, E., Botías, C. and Rotheray, E. L. (2015). Bee declines driven by combined Stress from parasites, pesticides, and lack of flowers. *Science* 347(6229): 1255957.

Gregory, R. D., Van Strien, A. J., Vorisek, P., Gmelig Meyling, A. W., Noble, D. G., Foppen, R. P. B. and Gibbons, D. W. (2005). Developing indicators for European birds. *Philosophical Transactions of the Royal Society of London. Series B, Biological Sciences* 360(1454): 269–288.

Gregory, R. D., Willis, S. G., Jiguet, F., Voříšek, P., Klvaňová, A., van Strien, A., Huntley, B., Collingham, Y. C., Couvet, D. and Green, R. E. (2009). An indicator of the impact of climatic change on European bird populations. *PLoS ONE* 4(3): e4678.

Hallmann, C. A., Sorg, M., Jongejans, E., Siepel, H., Hofland, N., Schwan, H., Stenmans, W., Müller, A., Sumser, H., Hörren, T., Goulson, D. and de Kroon, H. (2017). More than 75 percent decline over 27 years in total flying insect biomass in protected areas. *PLoS ONE* 12(10): e0185809.

Herrando, S., Keller, V., Bauer, H., Brotons, L., Eaton, M., Kalyakin, M., Voltzit, O., Lehikoinen, A., Milanesi, P., Noble, D., Ramírez, I., Voříšek, P. and Foppen, R. (2019). Using the first European Breeding bird Atlas for science and perspectives for the new Atlas. *Bird Study* 66(2): 149–158. doi: 10.1080/00063657.2019.1618242.

Herzog, F., Balázs, K., Dennis, P., Friedel, J., Geijzendorffer, I. R., Jeanneret, P., Kainz, M. and Pointereau, P. (2012). *Biodiversity Indicators for European Farming Systems. A Guidebook, ART-Schriftenreihe* (vol. 17). Forschungsanstalt Agroscope Reckenholz-Tänikon ART Reckenholz, Zürich.

Herzog, F., Jeanneret, P., Ammari, Y., Angelova, S., Arndorfer, M., Bailey, D., Balázs, K., Báldi, A., Bogers, M., Bunce, R., Choisis, J.-P., Cuming, D., Dennis, P., Dyman, T., Eiter, S., Elek, Z., Falusi, E., Fjellstad, W., Frank, T., Friedel, J., Garchi, S., Geijzendorffer, I., Gomiero, T., Jerkovich, G., Jongman, R., Kainz, M., Kakudidi, E., Kelemen, E., Kölliker, R., Kwikiriza, N., Kovács-Hostyánszki, A., Last, L., Lüscher, G., Moreno, G., Nkwiine, C., Opio, J., Oschatz, M.-L., Paoletti, M. G., Penksza, K., Pointereau, P., Riedel, S., Sarthou, J.-P., Schneider, M., Siebrecht, N., Sommaggio, D., Stoyanova, D., Szerencsits, E., Szalkovski, O., Targetti, S., Viaggi, D., Wilkes-Allemann, J., Wolfrum, S., Yashchenko, S. and Zanetti, T. (2013). Measuring farmland biodiversity. *Solutions* 4(4): 52–58.

Hünig, C. and Benzler, A. (2017). Das Monitoring der Landwirtschaftsflächen mit hohem Naturwert in Deutschland. – BfN-Skripten 476, 40 S. u. Anhänge. Available at: https://www.bfn.de/en/activities/monitoring/hnv-farmland-indicator.html (Accessed on 31st October 2019).

IEEP and IUCN (2018). Consultation workshop on an EU Pollinators Initiative. Available on line at https://ec.europa.eu/environment/nature/conservation/species/pollinators/documents/consultation_ workshop_report.pdf (Accessed on 5th August 2019).

IFAB (2017). Landscape Infrastructure and Sustainable Agriculture (LISA) – Report on the investigation in 2016. Available at: http://www.ifab-mannheim.de/pdf/LISA%20report%202016-final-30oct2017.pdf (Accessed on 31st October 2019).

IPBES (2016). The assessment report of the Intergovernmental Science-Policy Platform on Biodiversity and Ecosystem Services on pollinators, pollination and food production Potts, S. G., Imperatriz-Fonseca, V. L. and Ngo, H. T. (Eds). Secretariat of the Intergovernmental Science-Policy Platform on Biodiversity and Ecosystem Services, Bonn, Germany. 552 pp.

IPBES (2019). Summary for policymakers of the global assessment report on biodiversity and ecosystem services of the Intergovernmental Science-Policy Platform on biodiversity and Ecosystem Services Díaz, S., Settele, J., Brondizio, E. S., Ngo, H. T., Guèze, M., et al. (Eds). IPBES Secretariat, Bonn, Germany. XX pages.

Joly, A., Goëau, H., Bonnet, P., Bakić, V., Barbe, J., Selmi, S., Yahiaoui, I., Carré, J., Mouysset, E., Molino, J., Boujemaa, N. and Barthélémy, D. (2014). Interactive plant identification based on social image data. *Ecological Informatics* 23: 22–34.

Jeffery, S., Gardi, C., Jones, A., Montanarella, L., Marmo, L., Miko, L., Ritz, K., Peres, G., Römbke, J. and van der Putten, W. H. (Eds). (2010). *European Atlas of Soil Biodiversity*. European Commission, Publications Office of the European Union, Luxembourg.

Kettunen, M. and ten Brink, P. (2006). Value of biodiversity - documenting EU examples where biodiversity loss has led to the loss of ecosystem services. Final report for the European Commission. Institute for European Environmental Policy (IEEP), Brussels, Belgium. 131 pp. Available at: https://ieep.eu/uploads/articles/attachments/18218 6d5-18d5-4c4b-b009- 54005ddd73de/Value_of_biodiversity-June_06.pdf?v=6 3664509703 (Accessed on 31st October 2019).

Kleijn, D., Baquero, R. A., Clough, Y., Díaz, M., Esteban, J. De, Fernández, F., Gabriel, D., Herzog, F., Holzschuh, A., Jöhl, R., Knop, E., Kruess, A., Marshall, E. J. P., Steffan-Dewenter, I., Tscharntke, T., Verhulst, J., West, T.M. and Yela, J. L., (2006), Mixed biodiversity benefits of agri-environment schemes in five European countries, *Ecology Letters,* 9(3): 243-254.

Kok, M. T. J., Alkemade, R., Bakkenes, M., van Eerdt, M., Janse, J., Mandryk, M., Kram, T., Lazarova, T., Meijer, J., van Oorschot, M., Westhoek, H., van der Zagt, R., van der Berg, M., van der Esch, S., Prins, A. and van Vuuren, D. P. (2018). Pathways for agriculture and forestry to contribute to terrestrial biodiversity conservation: a global scenario-study. *Biological Conservation* 221: 137-150.

LeBuhn, G., Droege, S., Connor, E., Gemmill-Herren, B. and Azzu, N. (2016). *Protocol to Detect and Monitor Pollinator Communities - Guidance for Practitioners*. FAO. Avilable at: http://www.fao.org/3/a-i5367e.pdf (Accessed on 31st October 2019).

Lucas, C., Bouten, W., Koma, Z., Kissling, W. and Seijmonsbergen, A. (2019). Identification of linear vegetation elements in a rural landscape using LiDAR point clouds. *Remote Sensing* 11(3): 292.

Gianpiero, P., Thomas, C., Hans, S., Sirkiä, P. M., Teufelbauer, N. and Trautmann, S. (2018). Declining population trends of European mountain birds. *Global Change Biology* 25(2): 577-588.

Masante, D., Rega, C., Cottam, A., Dubois, G. and Paracchini, M. L. (2015). Indicators of biodiversity in agroecosystems: insights from Article 17 of the Habitats Directive and IUCN Red List of Threatened Species. *JRC Technical Report JRC97800*. European Union, 112 pp.

McKinley, D. C., Miller-Rushing, A. J., Ballard, H. L., Bonney, R., Brown, H., Cook-Patton, S. C., Evans, D. M., French, R. A., Parrish, J. K., Phillips, T. B., Ryan, S. F., Shanley, L. A., Shirk, J. L., Stepenuck, K. F., Weltzin, J. F., Wiggins, A., Boyle, O. D., Briggs, R. D., Chapin, S. F., Hewitt, D. A., Preuss, P. W. and Soukup, M. A. (2017). Citizen science can improve conservation science, natural resource management, and environmental protection. *Biological Conservation* 208: 15-28.

McLaughlin, A. and Mineau, P. (1995). The impact of agricultural practices on biodiversity. *Agriculture, Ecosystems and Environment* 55(3): 201-212.

Orgiazzi, A., Ballabio, C., Panagos, P., Jones, A. and Fernández-Ugalde, O. (2018). LUCAS soil, the largest expandable soil dataset for Europe: a review. *European Journal of Soil Science* 69(1): 140-153.

Orgiazzi, A., Dunbar, M. B., Panagos, P., de Groot, G. A. and Lemanceau, P. (2015). Soil biodiversity and DNA barcodes: opportunities and challenges. *Soil Biology and Biochemistry* 80: 244-250.

Packer, L., Monckton, S. K., Onuferko, T. M. and Ferrari, R. R. (2018). Validating taxonomic identifications in entomological research. *Insect Conservation and Diversity* 11(1): 1-12.

Pereira, H. M., Ferrier, S., Walters, M., Geller, G. N., Jongman, R. H. G., Scholes, R. J., Bruford, M. W., Brummitt, N., Butchart, S. H. M., Cardoso, A. C., Coops, N. C., Dulloo, E., Faith, D. P., Freyhof, J., Gregory, R. D., Heip, C., Höft, R., Hurtt, G., Jetz, W., Karp, D. S., McGeoch, M. A., Obura, D., Onoda, Y., Pettorelli, N., Reyers, B., Sayre, R., Scharlemann, J. P., Stuart, S. N., Turak, E., Walpole, M. and Wegmann, M. (2013). Essential biodiversity variables. *Science* 339(6117): 277–278.

Potts, S. G., Roberts, S. P. M., Dean, R., Marris, G., Brown, M., Jones, R. and Settele, J. (2009). Declines of managed honey bees and beekeepers in Europe. *Journal of Apicultural Research* 49(1): 15–22.

Potts, S. G., Biesmeijer, J. C., Kremen, C., Neumann, P., Schweiger, O. and Kunin, W. E. (2010) Global pollinator declines: trends, impacts and drivers. *Trends in Ecology & Evolution*, 25(6): 345–353. ISSN 0169-5347. doi: 10.1016/j.tree.2010.01.007.

Sanchez-Bayo, F. and Wyckhuys, K. A. G. (2019). Worldwide decline of the entomofauna: a review of its drivers. *Biological Conservation* 232: 8–27.

Schmeller, D. S., Henry, P. Y., Julliard, R., Gruber, B., Clobert, J., Dziock, F., Lengyel, S., Nowicki, P., Déri, E., Budrys, E., Kull, T., Tali, K., Bauch, B., Settele, J., Van Swaay, C., Kobler, A., Babij, V., Papastergiadou, E. and Henle, K. (2009). Advantages of volunteer-based biodiversity monitoring in Europe. *Conservation Biology* 23(2): 307–316.

Sevilleja, C. G., van Swaay, C. A. M., Bourn, N., Collins, S., Settele, J., Warren, M. S., Wynhoff, I. and Roy, D. B. (2019). Butterfly Transect Counts: Manual to Monitor Butterflies. Report VS2019.016. Butterfly Conservation Europe & De Vlinderstichting/Dutch Butterfly Conservation, Wageningen.

Soldaat, L. L., Pannekoek, J., Verweij, R. J. T., van Turnhout, C. A. M. and van Strien, A. J. (2017). A Monte Carlo method to account for sampling error in multi-species indicators. *Ecological Indicators* 81: 340–347.

Statistics Netherlands (2017). A tool for calculating multi-species indices and trends therein by Monte Carlo simulation (Accessed on 29th July 2019). Available at: https ://www.cbs.nl/en-gb/society/nature-and- environment/indices-and-trends--trim--/m si-tool.

Steffen, W., Richardson, K., Rockström, J., Cornell, S. E., Fetzer, I., Bennett, E. M., Biggs, R., Carpenter, S. R., De Vries, W., De Wit, C. A., Folke, C., Gerten, D., Heinke, J., Mace, G. M., Persson, L. M., Ramanathan, V., Reyers, B. and Sörlin, S. (2015). Planetary boundaries: guiding human development on a changing planet. *Science* 347(6223): 1259855.

Sutcliffe, L. M. E., Schraml, A., Eiselt, B. and Oppermann, R. (2019). The LUCAS Grassland Module Pilot – qualitative monitoring of grassland in Europe. *Palearctic Grasslands* 40: 27–31.

Takahashi, R., Okuyama, H., Minoshima, Y. N. and Takahashi, J. I. (2018). Complete mitochondrial DNA sequence of the alien hornet *Vespa velutina* (Insecta: hymenoptera) invading Kyushu Island, Japan. *Mitochondrial DNA Part B* 3(1): 179–181.

United Nations (1993). Convention on Biological Diversity (CBD). Available at: https://www.cbd.int/convention/text/default.shtml (Accessed on 31st October 2019).

United Nations Environment Programme World Conservation Monitoring Centre (UNEP WCMC) (2017). Workshop on understanding the knowledge on pollinators and pollination in the EU. Available at https://ec.europa.eu/environment/nature/conservation/species/pollinators/documents/summary_pollinator_knowledge_workshop_23_october_2017.pdf (Accessed on 5th August 2019).

Van Strien, A. J., Pannekoek, J. and Gibbons, D. W. (2001). Indexing European bird population trends using results of national monitoring schemes: a trial of a new method. *Bird Study* 48(2): 200–213.

Van Swaay, C. A. M., Warren, M. and Loïs, G. (2006). Biotope use and trends of European butterflies. *Journal of Insect Conservation* 10(2): 189–209.

Van Swaay, C. A. M. and van Strien, A. J. (2005). Using butterfly monitoring data to develop a European grassland butterfly indicator. In: Kuehn, E., Thomas, J. and Settele, J. (Eds). *Studies on the Ecology and Conservation of Butterflies in Europe*. Proceedings of the Conference held in UFZ Leipzig, 5–9 December 2005.

Van Swaay, C. A. M., Dennis, E. B., Schmucki, R., Sevilleja, C. G., Balalaikins, M., Botham, M., Bourn, N., Brereton, T., Cancela, J. P., Carlisle, B., Chambers, P., Collins, S., Dopagne, C., Escobés, R., Feldmann, R., Fernández-García, J. M., Fontaine, B., Gracianteparaluceta, A., Harrower, C., Harpke, A., Heliölä, J., Komac, B., Kühn, E., Lang, A., Maes, D., Mestdagh, X., Middlebrook, I., Monasterio, Y., Munguira, M. L., Murray, T. E., Musche, M., Õunap, E., Paramo, F., Pettersson, L. B,, Piqueray, J., Settele, J., Stefanescu, C., Švitra, G., Tiitsaar, A., Verovnik, R., Warren, M. S., Wynhoff, I. and Roy, D. B. (2019). The EU Butterfly Indicator for Grassland species: 1990-2017: Technical Report. Butterfly Conservation Europe & ABLE/eBMS (www.butterfly-monitoring.net). https://butterfly-monitoring.net/sites/default/files/Publications/Technical%20report%20EU%20Grassland%20indicator%201990-2017%20June%202019%20v4%20(3).pdf.

Chapter 2

Modelling biodiversity in agriculture

G. R. Squire, James Hutton Institute, UK

1 Introduction

Modelling biodiversity in agriculture lies within the broader concepts and methods of ecological modelling but can be differentiated from modelling climate or biogeochemical cycles, for example, because it explicitly operates within a managed ecosystem that delivers products and has life forms as the mediator or end point of the process. Biodiversity modelling has a long and diverse history that includes population dynamics, invasions and spread, vegetation succession, genetic differentiation, evolution of traits and environmental risk assessment (ERA). In the present context, the topic can be narrowed to cover the coexistence and coevolution of biota with the managed culture of crops, grass, trees and livestock.

Agricultural intensification in the 20th century moved agriculture towards reliance on chemicals, while globalisation opened markets internationally to a degree that some forms of agriculture are largely export driven, producing little for the local population. Not all agriculture intensified, and there were also subsequent moves to deintensify and diversify production methods. Today, therefore, agriculture covers a wide range of practices that have been arranged on two axes (Therond et al., 2017), one defined by the type and intensity of agronomic inputs and the other the degree to which products go to global or local markets. At one end is the high-input, monocultured cereal intended for global trade; at the other is the mixed field or intercropped orchard grown without chemical additives, collectively, for local consumption. Wherever a form

http://dx.doi.org/10.19103/AS.2020.0071.05

of agriculture lies along these axes, it inevitably relies on and in turn influences biodiversity, in terms of the crops grown, the soil organisms in the field and the plants invertebrates and animals that coexist with and, in many cases, support production.

There is no single agriculture therefore, and, gradually over the past 60 years, modelling has come to cover many of its diverse forms. Crop simulation was initiated in the 1960s, mainly to help maximise output (Jones et al., 2016) but has broadened to include economic appraisal. The overall scope of modelling has also widened to include many ecological processes integral to managed habitat and the biota that coexists with it. The representation of biota in models differs in the unit of biodiversity, scale, trophic level and degree of coupling between organisms and their environment. Models are constructed for both theoretical and practical purposes, a prime example of the latter being decision aids for managing complex agro-ecosystems. The resulting methodologies are similarly wide ranging (Chopin et al., 2019). This chapter therefore illustrates several possibilities rather than a single recommended approach, but first it sets out a number of concepts that the author believes should form the background to any approach to modelling biodiversity. The chapter is written from the standpoint of a quantitative biologist who works and interacts with mathematical modellers and programmers.

1.1 Why model biodiversity?

At the core of many questions on biodiversity in agriculture is the need to achieve a balance between a range of desired states or outputs, for example, feeding a human population without extinguishing wildlife, exporting for profit without denying local consumption or adopting new technology without risk to the wider environment. In such questions, the biota is influenced by a range of interacting factors whose combined effects are difficult to predict. The trade-offs between multiple factors are likely to vary with temporal (annual vs long-term) and spatial (field, farm, landscape, region) scales. Moreover, the effects of a change in a single factor such as tillage practice or livestock density are highly dependent on the context as defined by the other components of the agroecosystem, including biotic and abiotic factors, pedoclimatic conditions, policy support and political will. This complexity offers a serious challenge to those who have to design new systems or give consent to innovations.

How then can sustainable agricultural systems be designed and managed? Formal experiments testing an innovation against a current state or control are standard practice at scales of the plot or part-field. Yet many of the important questions in agriculture today are set at scales well above the field, from many fields operating in a landscape, emitting or losing genetic material or chemical pollutant, to whole regions and even continents over which energy and materials are exchanged. Even when the field remains the scale of interest, its boundaries are porous, not just to materials including the biota but to market demand and prices and to decisions of policy and politics often made in other parts of the world. Moreover, the current state against which an innovation is compared in an experiment may not be one that is ecologically safe or stable. An ideal, 'safe' state would then need to be identified in terms of biota and ecological processes. Experimentation may be able to discover ways to move the current state towards the ideal, but it may still not be possible to 'create' the ideal state given present conditions. In the case of a degraded system, for example, an ideal state may require a higher soil carbon content or pollinator population size than currently exists. In such cases, modelling could, in principle, create the ideal state, assess its dynamics and devise the most effective way to transform the current system.

Before considering and comparing models as such, it is necessary to examine two fundamentals: the role of the biota in the chain of effect linking the basic conditions in a region to desired high-level outputs which might be expressed in terms of ecosystem services for example, and the forms, interactions and units in which biodiversity is defined in models.

1.2 Chain of effect - which direction to take?

Whatever the modelling approach taken, the biota assumes a specific position in a chain or net of effect within an ecosystem (Fig. 1). Any field, farm or region has a pedoclimatic base which determines what can be grown - for example, rough grazing, managed grassland, mixed farming or arable cropping. The type of biota that once inhabited or now coexists with the cultivated species is similarly determined by that base. The biota, including native and introduced species, mediates most of the ecological processes in the ecosystem: for example, nitrogen fixation, denitrification, carbon acquisition and carbon breakdown, soil formation and soil loss, crop growth and grazing intake. Human interventions can act through the biota or ecological processes independently, but in agriculture they mostly act through the biota (Fig. 1). If, for example, agricultural policy leads to the removal of legume crops, weeds and forages, the process of biological nitrogen fixation diminishes. If it is to deploy excessive tillage and so destroy soil fungi and mycorrhizae, the process of soil binding diminishes and soil erosion rises. The rates and durations of ecological processes together then determine which high-level states and outputs are satisfied. These states and outputs have been variously expressed as ecosystem services, the four 'spheres' or tetrahedron of sustainability (O'Connor, 2006; Maxim et al., 2009) or the United Nations' Sustainable Development Goals (UN SDG).

Questions may be asked of the chain in Fig. 1 in both directions. Conventionally, in ERA, for example, questions are asked left to right - what

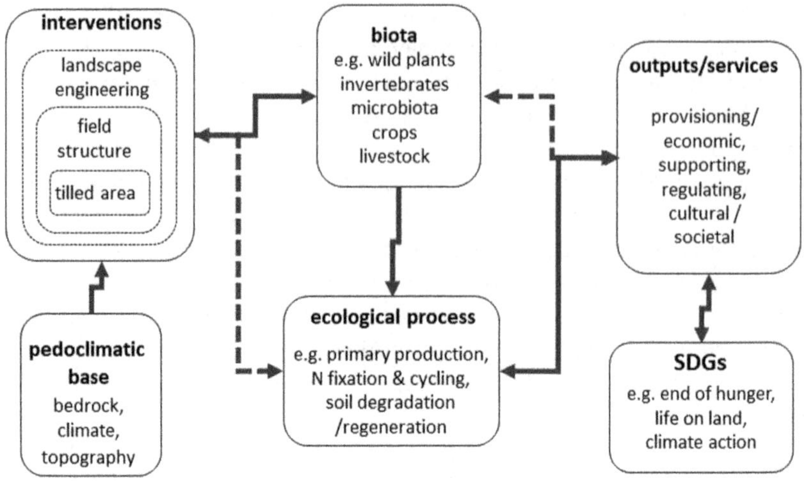

Figure 1 Chain of effect linking the pedoclimatic base of a region with agricultural outputs and high-level aims such as Sustainable Development Goals (UN SDG) via interventions, biota and ecological processes (modified after Squire, 2017). Solid arrows indicate the main linkages in agricultural systems.

is the effect of an intervention such as a new chemical or biotech crop on the biota or ecological processes. The more valuable direction in ecological design is to define the balance of required outputs first and then to see what combination of processes, biota and hence interventions would achieve those outputs. Whichever way the chain is worked, it is useful to classify the biota in relation to whether primarily it benefits agriculture or benefits from agriculture (Bàrberi et al., 2010; Bàrberi, 2015). If biodiversity (or the biota) is B and agriculture A, then a condition in which organisms explicitly support agriculture may be termed 'B for A'. For example, if interventions provide local habitat for predatory beetles that consume insect pests and allow crops to bulk faster, then the state is B for A. Alternatively, if management desires to create habitat for rare mammals and birds, then it is A for B. The two are not mutually exclusive, and both are relevant in many systems but are differentiated here because much of the modelling in the topic of agricultural biodiversity is A for B, yet a sustainable future depends on achieving B for A.

1.3 Units, trajectory, scale and boundary

Commonly, the state and dynamics of plant communities are recorded by broad-scale descriptors that condense into simple terms large batches of information on the number, size and mass of taxa in a population or community. Species-area curves, species-abundance relations, the family of diversity indices and demographic classifications are typically used to make comparisons between and within types of vegetation (Hill, 1973; Kempton, 1979; Rozenweig, 1995). These community-scale features are valuable comparators and aids in management but commonly remain empirical descriptors.

Modelling populations and communities first assumed, in a comparable manner, that taxa and resource were homogeneous, yet, to be realistic, modelling needs to account for the individuals differing by several orders of magnitude in attributes such as basal area and longevity (Crawley and May, 1987). The assumption of homogeneity was further criticised by Huston et al. (1988) on the grounds that it 'violates two of the basic tenets of biology', that each individual is different and exists in a different 'place'. Ecological modelling as a discipline made great strides in the late 1980s towards understanding how interactions between individuals and their environment determined ecological states and functions (Huston et al., 1988; Lomnicki, 1988; DeAngelis and Gross, 1992). Subsequent arguments emphasised the importance of individual behaviour in experiment, observation and modelling (Bolnick et al., 2003; Breckling et al., 2006).

If individuals are taken as the basic functional unit, they can be viewed as proceeding through life along a unique trajectory (e.g. Sumida et al., 1997; Squire, 2000). The trajectory may be through time as it develops and grows, geographical space as in the movement of a gene or organism across a landscape or a cascade of interactions such as a trophic web. Comparably, plants can be considered to progress through evolution along a complex trajectory or 'fitness landscape' (Niklas, 1999). More generally, a trajectory may be envisaged as a movement through 'trait space' (Pachepsky et al., 2001) – the multidimensional arena defined by the properties or phenological traits in which organisms interact with each other and their surroundings. The idea of individual trajectories is relevant in agriculture because an external event, such as a storm, drought or chemical application, may affect individuals differently, depending on their position in the trajectory.

Back therefore to the question – 'why model biodiversity'. An ecological process is usually experimented at a restricted scale and on the assumptions that its boundary is definite and impermeable (i.e. the trajectories of individuals being contained within the boundary). Yet modelling is repeatedly asked to go beyond the known boundary. Over two decades ago, problems in scaling-up were actively pursued in a range of biophysical research topics (Van Gardingen et al., 1997), the main question being whether 'behaviour' measured at one spatial or temporal scale could be extrapolated. Marshall et al. (1997) concluded that in many instances, the 'scaling laws' that allowed such extrapolation to higher scales and across experimental or observational boundaries were yet to be discovered. In effect, many ecosystem processes were too complex to be modelled at scales of time and space beyond those for which knowledge was available. At least one reason for the uncertainty was that research programmes generally fund research at scales that can be manipulated and which lead to definitive results and conclusions, yet these scales are well below those that determine many phenomena in agriculture (Squire and Gibson, 1997). These authors noted a related mismatch in the 'applicability' of research, much of it too reductionist to be directly useful in guiding decisions in agriculture and biodiversity. Moreover, research in agriculture was mostly in the biophysical sphere, whereas many of the factors influencing biodiversity were from the economic and social spheres.

In the following sections (Sections 3 to 5), approaches of modelling are grouped in several main categories (the white boxes in Fig. 2), each located in a space defined by physical scale and the four 'spheres' of sustainability (O'Connor, 2006; Maxim et al., 2009) - biophysical, economic, social and political (Section 3.2). Individual-based models (Sections 2.1 and 2.2) are mainly within the biophysical but extend from patch to landscape. Process-based models (Sections 3.1, 3.2 and 4.2) are mainly biophysical but some extend into the economic, while integrated system approaches (Section 4.3) encompass also the societal sphere. Global frameworks (Section 4.1) and multi-attribute decision models (MADMs) (Sections 5.1 and 5.2) can, in principle, explore the entire space. Notably, the rise of MADMs in the last 10 to 15 years has made biodiversity modelling accessible to land managers and other stakeholders in a way that no other models have been able to do.

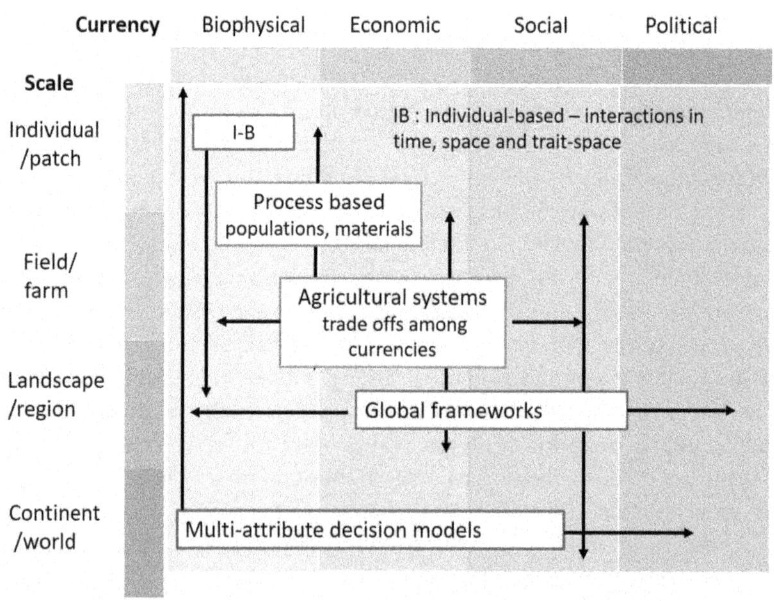

Figure 2 Modelling approaches described in this chapter (white boxes), with arrows showing the extent of coverage across (top) the biophysical, economic, social and political spheres and (left) increasing scales.

2 Interaction among individuals for place, space or resource

In nearly all individual-based approaches to modelling biodiversity, the entities involved interact with the result that they gain or lose something. The entities typically interact over a resource base. Those that lose might die or reproduce less, and those that gain dominate and spread. In consequence the composition of the group changes. The way entities interact in models has been defined in many different ways. In the following brief synopsis (Section 2.1), modelling evolves from simple, rule-based automata where the resource is not defined but exists in terms of one type changing at the expense of another to complex simulated 'fields' in which millions of individuals grow and reproduce over an explicit resource base (Section 2.2).

2.1 Towards incorporating variability among individuals in vegetation

Spatial models began to adopt the use of a heterogeneous resource (usually on a spatial lattice), but most models were still inflexible in the way the plants physiologically interact with the resource. A move to more realistic plant types was made in the cellular automaton model of Silvertown et al. (1992), while a step to incorporate the resource as well as functional type was taken by Colasanti and Grime (1993). Individual-based modelling continued to progress, as summarised by Bown et al. (2007) but remained limited by several conditions, one being 'non-overlap', in which an individual could occupy only one cell in a lattice and could not extend into surrounding cells as it expanded.

In contrast, most modelling of agricultural production did not adopt individuality or spatial interactions. Its early aims and emphases (1960s to 1990s) were on growth and yield, and hence the main biota modelled was crops and grass, treated as homogeneous. Agricultural modelling did, however, take the lead on making explicit that the flux of resource such as solar radiation, water or nutrients was the way to link environment, land management and productivity (Monteith, 1972). Subsequent developments also accounted for individuals and their interactions over space and time. Evers et al. (2007), for example, constructed and validated a model of wheat growth in which canopy closure and solar energy interception were related to fine mechanisms of foliar development and expansion. As argued by Bown et al. (2007), such models needed detailed parameterisation from experiments and were not intended to be used as scaling tools to examine the consequences of individual variation in complex communities.

It was nevertheless imperative that models should accommodate variation between individuals and their growth into an explicit resource base. Situations had to be captured in which external events such as drought, soil cultivation or other agronomic interventions acted on a group whose individuals were in different stages of development or growth (Squire, 2000). The divergence of trajectories within a group has major implications for defining risk in agriculture: for example, in a large-scale study of the effects of weed management on plants and trophic groups, the larger individual weeds produced most of the seed that replenished the seedbank and provided food for certain invertebrates and birds (Heard et al., 2003). A model that does not allow for such individual variation would be missing a large part of the behaviour of such weed populations. Modelling in agriculture was therefore faced with an explicit challenge - arising from arguments on the introduction of GM cropping to Europe in the late 1990s - that it had to incorporate differences in traits between individuals operating at a range of interconnecting scales from patches and fields to complex landscapes (SIGMEA, 2007).

2.2 Modelling individuals in trait space across an explicit resource

As a case study, the approach described here first arose from the need to understand questions about diversity and function in plant communities of semi-natural grassland and arable land but was then extended to cover issues around dispersal and geneflow. In relation to Figs. 1 and 2, the models were designed to explore how biota and ecological process were affected by interventions. For illustration (Fig. 3) two individuals are first shown in isolation, one covering a larger base than the other and each taking up a resource (Fig. 3a). If they grow or extend the area from which they take resource, they come to overlap (Fig. 3b), each occupying a portion of the resource base in which they have to share or compete. In many situations, the individuals differ so much in size that some individuals live within the resourcing range of others (Fig. 3c). These groupings of individuals might occupy land discontinuously, one grouping here, another over there, due to fragmentation of the landscape, and exist as a metapopulation in which turnover in local populations is influenced by geneflow between populations (Fig. 3d). Some or all the individuals can move around the various domains, for example, at seeding, and some might exchange pollen and genetic material. The requirements for modelling such scenarios are therefore demanding, both in the biophysical knowledge and in the supporting mathematical and computational needs.

As a means of incorporating the interactions in Fig. 3, a generic approach was developed in which the community dynamics are determined by resource flux and the sharing of resource among individuals rather than, say, species rules (Pachepsky et al., 2001, 2007; Bown et al., 2007). Individuals are characterised by four groups of traits - development through time, resource

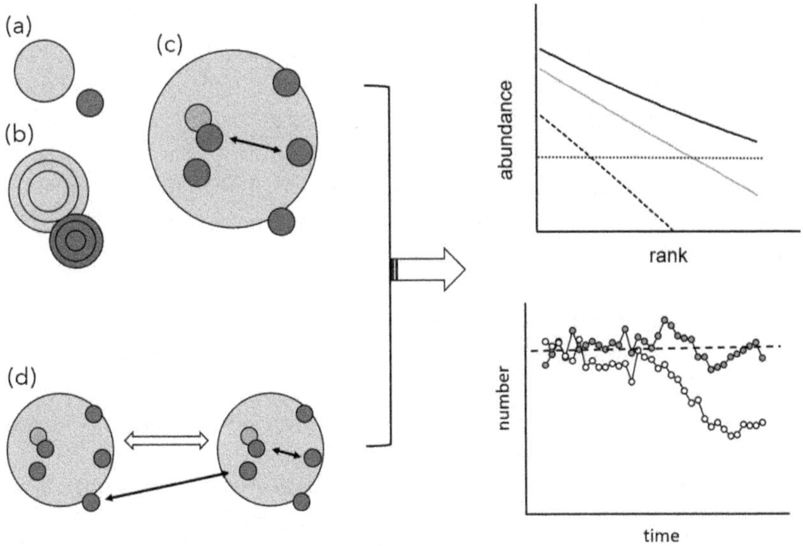

Figure 3 Representation of interactions (left) between individuals of different types in individual-based models (a) not in contact, (b) growing into each other's resource base, (c) living entirely within another's base and sometimes exchanging pollen (arrow), and (d) groups occurring as non-contiguous populations that interact via seed (block arrow) and geneflow (arrows); and (right) the use of such configurations to explore summary attributes such as rank-abundance curves and coexistence time series (text gives details).

acquisition, resource allocation and reproduction, all of which can be derived from experimentation and observation. Plants are located at a point in a lattice, but the areas from which they acquire resource may grow over time. Progression through development occurs when certain states are reached. The resource (e.g. representing soil nutrients) may be distributed heterogeneously and characterised by a release rate (e.g. by mineralisation) and replenishment rate (e.g. as individuals die).

One of the aims of this modelling was to explore how combinations of individuals defined by traits evolve higher-level attributes such as a species-area and rank-abundance relations and ultimately which combinations promote coexistence over time. For example, a community might be judged as sustainable if the number of component species or their rank-abundance curve remain above a certain level (indicated by the horizontal dashed line in Fig. 3 right, upper and lower), but circumstances might drive the number of species or less abundant members to fall below the line. The models in question - based not on species but on individuals differing in trait-values - were able to recreate community-scale features typically derived where species is the unit of diversity. For example, the number of types coexisting on a domain was strongly governed by the slope of the trade-off between traits - for example, between time to reproduction and fecundity (Pachepsky et al., 2001, 2007). The greater the differences in this parameter among individuals on the domain, the more types coexist. In terms of currency (Fig. 2), the modelling remained within the biophysical but extended to the landscape scale.

The above-mentioned case study is one example of the way modelling biodiversity in agriculture evolved from the late 1990s to link individual behaviour to characteristics of 'the group' or the emergent properties of the system. More generally, developments included:

- treating the individual as the unit of biodiversity;
- defining individuals by eco-physiological traits which linked them either to each other or to their environment;
- use of an explicit resource base and resource flux (solar radiation, nutrients) through which traits interact; and
- expanding the scale to heterogeneous landscapes over which the biota disperses, forages and exchanges genetic material.

Models did not necessarily incorporate all of the above-mentioned developments, but the capacity to model biodiversity in agriculture had moved into a new phase (Breckling et al., 2006). Some examples, including within-field effects on the weed flora and landscape-scale mammal behaviour, are covered under the topic of process-based modelling described next.

3 Process-based modelling

The principal form of modelling relevant to agricultural biodiversity had come to be 'process-based', which here means that the model works through biophysical mechanisms in which units of biota interact with each other or with their surroundings by means of an ecological process in which flows of energy or matter occur. Biodiversity has also been examined by statistical models in which the processes are implied but not directly featured (e.g. Bohan et al., 2011). Among the most widely studied processes are those based around the seeding plant life cycle - seed persistence, germination, growth, competition, and so on. Life cycle stages form the core of the many weed-crop models developed for various systems and regions of the world (Holst et al., 2007). These authors noted that the emphasis in most models is on the harm to a crop and how to reduce that harm but that few deal with interspecific competition or coexistence among weed communities (see Section 3.1 for more recent developments).

Those models summarised here typically define a pedoclimatic base and a set of interventions that act on biota and ecological process through transparent mechanisms (Fig. 1). A few process-based models take account of individual performance as described in Sections 2.1 and 2.2, but others rely on a mean-field approach in which all entities of one type have the same properties, though the types are defined by traits or functional forms (e.g. Hawes et al., 2009; Meziere et al., 2015). The models variously operate at scales from patch to landscape and through mainly biophysical but also economic currencies (Fig. 2). Typically, equations in the models define the rate of transfer of materials, growth of organisms, movements and sometimes genetic change.

3.1 The stimulus of environmental risk assessment

Most process-based models can be paralleled by measurements in real ecosystems. Restating the earlier arguments, modelling is advantageous where measurement and experimentation are not feasible, usually because the systems under study are too large or complex for some aspect of the system to be controlled, or where an outcome has to be predicted resulting from an intended or unintended effect on the system, but that outcome might be deleterious. Environmental risk analysis, specifically to address concerns over whether it is ecologically 'safe' to release an organism or a chemical, has been a great stimulus to process-based modelling and in particular the concepts and methods of up-scaling (Landis, 2003, 2010). Models may be used to devise a strategy which might include mitigation, should deleterious effects occur. Life forms may be the end points that the models are designed to assess, but they may also mediate the processes.

A few examples of the many process-based models are given in Table 1, with brief descriptions of the purpose, the biota, the processes being modelled and the scale at which the model operates. For example, the US Environmental Protection Agency (US EPA) has formally incorporated a suite of models into its approach to risk assessment. Its Centre for Exposure Assessment Modeling (CEAM) groups models under the headings of groundwater, surface water and food chain. An example is AQUATOX (Table 1), which predicts the movement of pollutants and their effects on fish, invertebrates and plants. It simulates the transfer of pollutants between ecosystem compartments and the effect of their concentration on a range of biota. The UK Environment Agency has also reviewed a wide range of ecological models for their potential in risk assessment (Roast et al., 2007).

The scale of activity is a particularly important feature when choosing models for ERA. Most information gathered from initial experiments following the release of a new crop or first trial of a chemical is typically at a small spatial unit such as a field plot (10 × 10 m, 100 × 100 m). If the questions are about persistence and effect at that scale, then adequate information can be gained to formulate, populate and test models. The main in-field models of crop-weed interactions in which the beneficial and harmful effects of weed

Table 1 Process-based models that include biodiversity either by mediating a process or as an end point for assessment. Examples are selected from a large field of work to demonstrate the range of activity from effects of management on mainly sedentary plants within or close to fields, through general system-level impacts of human activity on several types of organism to effects at the landscape scale due to changing habitat or multiple exposure to hazards. In column 4, S is the spatial scale, T temporal scale and LF the scale of the life forms (e.g. individual, population average, species). ERA refers to a formal Environmental Risk Assessment. The models listed are a very small selection from many available

Purpose/source	Biodiversity/life forms	Processes modelled	Spatial/temporal scales
Balance between crop production and biodiversity: upscaling of earlier FLORSYS model. Colbach et al. (2018).	Organisms: crops, weed species in modelled maize-based, GM crop systems. (Model can be applied to any crop system.)	Processes: life-cycle biology from germination to seeding, seed bank, plant-plant competition for resource, dispersal across areas grown with and without crops.	S: field or other management unit (e.g. uncropped strip) aggregated into a small landscape, Aquitaine France. T: runs over many years of crop sequence. LF: 'mean' or representative trait values for species or population.
Optimum cropping systems balancing positive and negative effects of weed flora. Meziere et al. (2015).	Weed flora divided into 5 weed-harmfulness and 5 biota-positive groups.	Life cycle of weeds in relation to functional group properties, management and environment; link to fungal pest model.	S: twenty-six cropping systems in two French regions (simulated but based on actual) T: 30 year-time span. LF: broad weed functional groups.
ERA for water: (AQUATOX) US Environment Protection Agency (US EPA)	Plants (phytoplankton, macrophytes etc.), fish, invertebrates in various water bodies and water courses; generalised model for impact and scenario assessment.	Transfer and exchange of pollutants (e.g. N, P) between ecosystem compartments; transformation of (e.g. organic) matter; exposure of biota to pollutant concentration; oxygen demand; impacts on functioning and survival of the biota.	S: variable, depending on application, for example, river systems, lakes. T: as spatial, varies with application. LF: generally broad functional types or main biotic groups.
Exposure of butterflies and moths to crop derived toxin. Perry et al. (2010).	Maize crop modified to produce insecticidal protein; larvae of three non-target Lepidoptera.	Pollen movement from crop; pollen deposition on larval food plants, exposure of larvae, mortality.	S: field and field margin for 11 maize cultivation regions in Europe differing in exposure. T: growing season. LF: larvae parameterised at species level based on range of experimental data.

Geneflow across agricultural landscapes (variations on the GENESYS model) Colbach (2008); Colbach et al. (2009).	Oilseed rape crops, crop-derived weeds and ferals (potentially other species).	Pollen flow from fields, genetic exchange, seed bank dynamics, transmission of impurity to crops.	S: simulated landscape patterns, e.g. 100 fields. T: years. LF: crop and crop derived varying in genetic locus, for example, GM and non-GM.
Multiple effects on honey bee colonies, including toxicity from agrochemicals. Becher et al. (2014); Rortais et al. (2017).	Organisms: honey bees, varroa mite, virus transmission.	Processes: population dynamics and reproduction of different bee cohorts, within-hive food storage; foraging, dispersal; exposure to toxins, damaging effects of toxins.	S: within hive to foraging range across landscape. T: within a year to several years. LF: individuals, cohorts.
To assist ERA for European brown hare exposed to ecotoxins. Topping et al. (2016); based on ALMaSS platform in Topping et al. (2003).	Brown hare (*Lepus europeaus*). Based on ALMaSS modelling platform.	Distribution of hares across the landscape, the latter defined by biophysical and management attributes; effect of hypothetical pesticide on hare performance, movement and survival, mainly through chronic effects on litter size.	S: ten landscapes in Denmark, each 10 × 10 km on which were placed a range of pesticide treatment scenarios. T: over several years. LF: individual animals (agents).
Defining regional conservation corridors in response to change in land use. Cushman et al. (2009).	Black bear (*Ursus americanus*).	Resistance-to-movement of the bears; modelling based on a range of factors leading to definition of least-cost migration routes. Model potentially testable via GPS collars.	S: extensive area of US Rocky Mountains. T: not directly modelled. LF: individuals.
Estimation of regional scale cropland carbon budgets. Zhang et al. (2015).	Biota not explicitly included except crops, but soil % carbon is a useful indicator of soil biodiversity status; included here as an example of combining models with a range of data to estimate status of systems.	Carbon exchange, crop carbon acquisition, soil carbon storage, carbon decline.	Instructive example of up-scaling combining geospatial databases (land use, soil, etc.) and management (fertiliser, tillage) and process-based modelling to examine state and change in land parcels covering 250 000 km^2, validated against crop survey, USA.

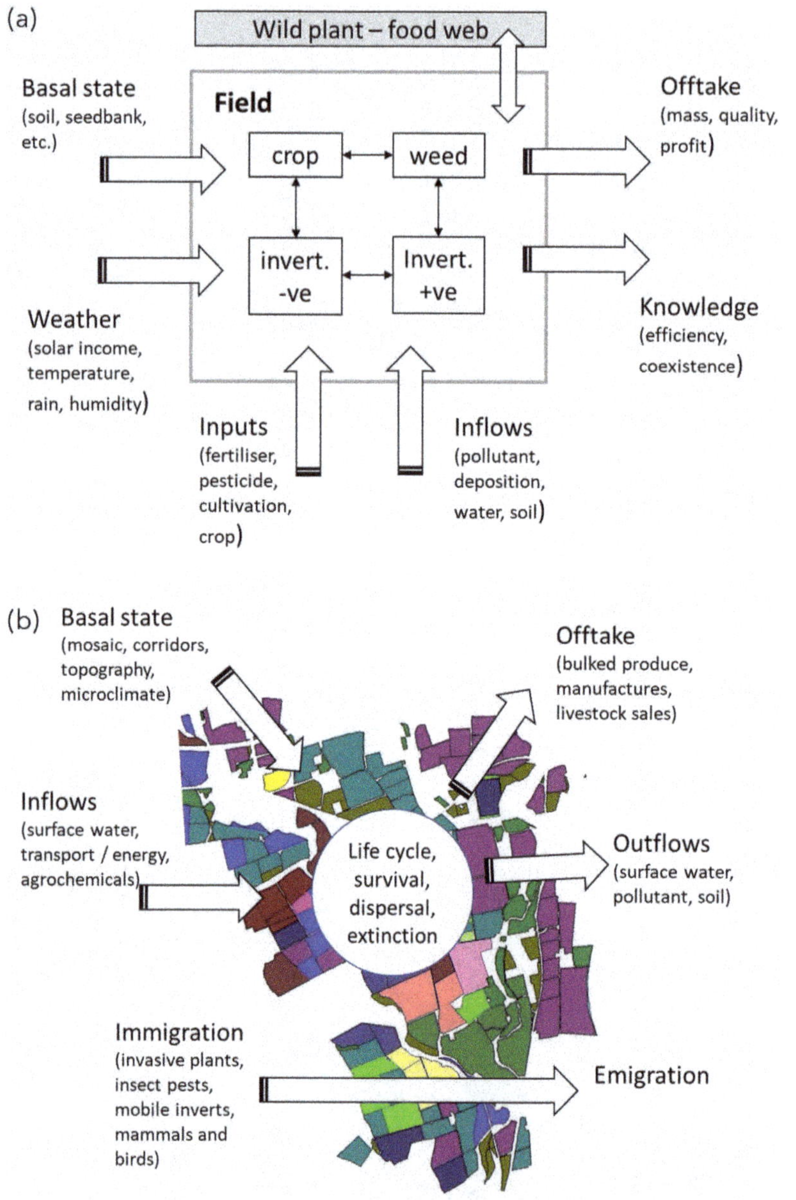

Figure 4 Examples of processes modelled when (a) considering an individual field and its immediate margin and (b) an agricultural landscape (text gives explanation).

species are assessed in relation to soil and management (Fig. 4a) are well-developed examples of this type (e.g. Colbach et al., 2018, Table 1). Basal conditions, weather, agronomic inputs and sometimes unmanaged inflows such as atmospheric deposition are generally well described by data, such that agricultural offtake can be predicted accurately. Interactions within the field between crops, weeds and the invertebrates of the food web determine the status of the biota, comprising, for instance, seedbank, plants, herbivores, predators and parasitoids (Hawes et al., 2009). A limited interaction may be modelled outside the field, for example, exchanges of biota between a field and its immediate margin. Other than mobile invertebrates that contribute to the in-field food web, the biota that traverses the field is usually difficult to account for at this scale of modelling. However both in ERA and in agricultural modelling, generally the extent of a contained field is too limited for a complete assessment.

3.2 Scaling-up

The problems of scaling-up in agricultural and environmental science (Section 1.3) became a major topic of research in the 1990s (Van Gardingen et al., 1997). Some processes were readily scalable - for example, the measured photosynthesis rate of many individual leaves could be scaled to the crop canopy using known physiological and physical relations. Others, such as scaling across heterogeneous landscapes or over decades of time, were often intractable due to complex couplings and emergent properties (Marshall et al., 1997). Despite improvements in computing power, up-scaling of process-based modelling still has its challenges. A modelled landscape comprises basal states such as topography and landscape structure, the movement into the landscape of physical and biological entities, the biotic interactions inside the landscape which are the main object of interest and the movements of biota and materials out of the landscape (Fig. 4b). With developments in remote sensing and mapping, many of the basal attributes may be well documented, but modelling at this scale is commonly limited by uneven data on the states of the various entities and poor understanding of the biotic interactions. The greatest limitations in the 'input' data are typically of the management of the landscape, which needs to take account of the variation in management among individual fields and non-agricultural areas. While the basal conditions determine which life forms can inhabit the area, their dynamics and fate depends very largely on how the habitat is treated (e.g. Topping et al., 2016). If the object of modelling is ERA, then the feasibility and success of modelling may rely on prior knowledge of biota or agrochemicals that are similar to those under assessment.

Where such information is available, landscape-scale modelling is feasible, given appropriate understanding and technology (Landis, 2003, 2010). Models of geneflow and persistence mediated by GM pollen and seed relied for their success on prior studies in various European environments since the late 1990s (SIGMEA, 2007). Information existed on similar biota, including non-GM crops of the same species which emitted pollen, gave rise to weedy 'volunteers' in fields and feral populations outside fields, all of which came into contact with wild relatives, some sexually compatible with the crop, others not. Such a body of information could be used to enact scenarios that would predict how GM crop traits would spread across landscapes following cultivation (Colbach, 2008). Processes such as pollen movement by wind or insects were modelled over a landscape populated by a heterogeneous distribution of receptor plants. The genetic constitution of the receptor population changed as a result and had consequential effects on non-receptor organisms, which ranged from a non-GM crop that was rendered unsaleable if contaminated to a weed community that developed herbicide resistance (SIGMEA, 2007).

The coming on-stream of large-scale data gave a major stimulus to landscape-scale modelling, demonstrated in EU countries by the Integrated Administration and Control System (IACS), which records the crops and grass grown on each parcel of land registered for agricultural subsidy. Before this, basal attributes had to be derived from measurements at points on a grid, for example, the Land Use and Coverage Area Survey or LUCAS, from which data for most individual fields could not be interpolated. As an example, Topping et al. (2003, 2016, Table 1.2) combined IACS data with biophysical survey and mapping to determine the fate of model organisms such as the vole *Microtus agrestis* and European brown hare *Lepus europaeus*. The life cycle biology and behaviour of these organisms was to a degree known from prior studies. They moved across the landscape as individuals, grew and reproduced and were affected by agrochemicals and other interventions. Methodologies for up-scaling to estimate biophysical states across thousands of square kilometres have now been demonstrated, for example, by combining a range of basal data and process models to estimate cropland carbon budgets (Zhang et al., 2015).

Landscape-scale modelling of such biota is therefore now possible to a degree that could hardly be envisaged in the 1990s. Many advanced methodologies are in use for analysing landscape structure and the way it affects biodiversity (Schroder and Seppelt, 2006). Uncertainties still remain, however, in the emergent properties that arise as processes interact across complex landscapes. There are so many possible interactions in reality that they cannot be understood from the level of observation possible in most research projects. In the case of the brown hare, the modelling of the effect of an endocrine disruptor pesticide on reproduction and survival resulted in some unintuitive findings; for example, the locations of the greatest pressure on the hare did not always coincide with the greatest pesticide loading (Topping et al., 2016). Similarly, when comparing a range of land-sparing and land-sharing scenarios, the fate of arable plant species was highly sensitive to the basal conditions and landscape structure (Colbach et al., 2018). In part to cope with the uncertainties presented by 'context', process-based models tend to be channelled towards specific aspects of the system or specific life forms for which there is already extensive knowledge. Process-based models may help determine policy but rarely assist in making routine decisions about how to manage a sector of land.

4 Trading biophysical, economic, social and political currencies

A range of frameworks and modelling approaches have come to place biodiversity in a much broader context. If agricultural biodiversity is to be supported or protected, then action is needed through two groups of factors (Squire and Gibson, 1997). One is that of the management of the actual habitat itself, where decisions are made of what is grown, how it is grown and how the biota can coexist with production – the scale here typically ranges from the management unit to the configuration of structures and human activity in a landscape. The other group of factors operates from well outside the area being managed and comprises international trade and markets, local and international politics and ultimately land policy. To manage these factors and scales properly requires trading between different 'currencies' and defining exchange rates, for example, between secure populations of a rare animal, an economic loss, or a societal gain.

Trading across currencies has been stimulated by a range of approaches. Global policy frameworks (Section 4.1) have encouraged potential conflicts and trade-offs to be appreciated, even where they do not include any explicit modelling activity. Trade-offs can be examined more explicitly in system models that evolved partly from crop simulation but that now integrate production with economics and some wider biophysical processes (Section 4.2). More recently, biodiversity has been incorporated explicitly into whole-system models deployed to quantify trade-offs and seek optimal states in which biodiversity and other ecosystem services and functions coexist (Section 4.3). The literature in Sections 4.1 to 4.3 is substantial, and only a few studies are selected here to illustrate scope and capability. However, most frameworks and models are still, to a large degree, remote from practitioners, and hence the final part of this chapter (Section 5) turns to more accessible decision models.

4.1 The context of global, multisectoral policy frameworks

One of the most comprehensive of multisectoral frameworks is *The Economics of Ecosystems and Biodiversity* (TEEB, 2010), hosted by the United Nations Environment Programme. It aims to bring biodiversity and ecosystem services into decision-making at all levels, in part by helping to recognize the wide range of benefits provided by biodiversity. In one of their studies – 'The Visible and Invisible Flows of Agricultural Production' – 'human systems' are linked with 'agriculture and food systems' and 'biodiversity and ecosystems'. The latter comprises what they term 'positive flows' that include soil formation, erosion control, moderation of extreme events and provision of water, and 'negative flows' such as habitat encroachment, species reduction and greenhouse gas emissions. Other global frameworks include the *Intergovernmental Science-Policy Platform on Biodiversity and Ecosystem Services* (IPBES) and the *System of Environmental-Economic Accounting* (SEEA), the latter of which promotes agreed standards and accounting rules for internationally comparable statistics on the environment and its relationship with the economy.

Increasingly the political realm is considered in ecosystem studies along with the environmental, economic and societal spheres. This 'tetrahedron of sustainability' (O'Connor, 2006; Maxim et al., 2009) is represented very simply in Fig. 5a, where an agricultural system (the circular symbol) is located hypothetically within the multi-dimensional space of the four spheres (the corners of the square). In the last decade, a widely used framework known

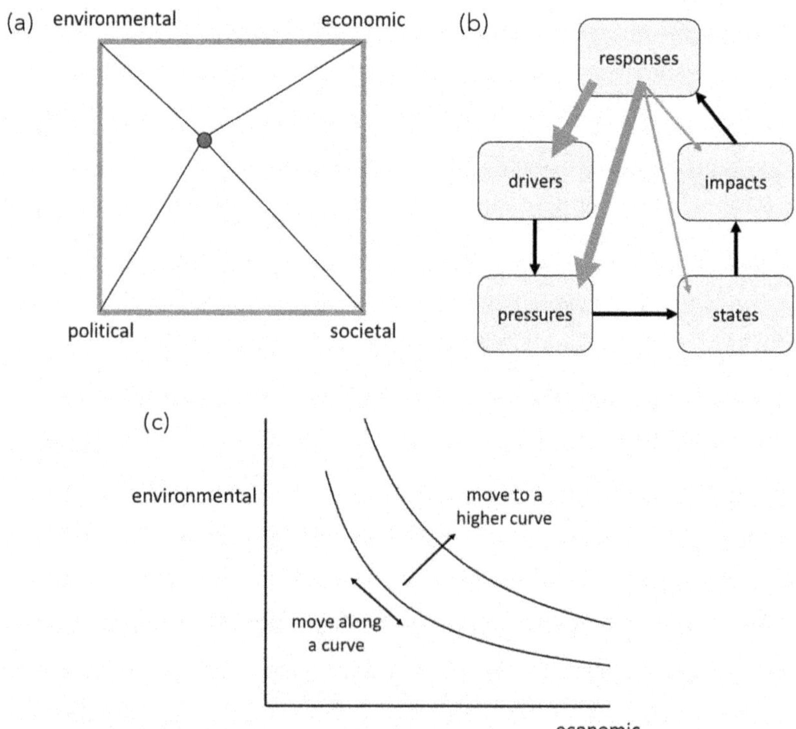

Figure 5 The context of integrative modelling: (a) locating a system (circle) within the tetrahedron of sustainability; (b) DPSIR framework indicating greater response to drivers and pressures than to states and impacts may be needed; and (c) optimisation curves, arrows indicating the movement between environmental and economic outcomes or preferably moving to a higher curve.

as *Driver-Pressure-State-Impact-Response* or DPSIR (OECD, 1993; Gari et al., 2015) has been integrated with the tetrahedron to emphasise that the future of biodiversity depends as much on the political as the other three (Maxim et al., 2009). The approach is applied, for example, to chemicals in the environment (Maxim and Spangenberg, 2009) and biological invasions (Rodriguez-Labajos et al., 2009). In such studies, *Drivers* are changes in social and economic conditions, such as developments in the petrochemical industry in the 1900s, *Pressures* might be the release of chemicals to the environment, *States* the populations or organisms and their trophic interactions, *Impacts* the degradation and loss of species or ecological function and *Responses* those actions by society, industry and governments to limit fossil fuel usage, pollutants to water and greenhouse gas emissions (Omann et al., 2009). Analysis via DPSIR has concluded that if biodiversity loss needs an immediate remedy, then systems may be too complex and knowledge too uncertain for *Responses* to target *States* or *Impacts* directly and should more effectively ease *Drivers* and *Pressures* (Fig. 5b). This major conclusion accepts that while human effects on agricultural ecosystems may be too complex to understand in full, they should be regulated and scaled back if the frameworks or models indicate that harm is occurring.

Global frameworks such as TEEB and DPSIR - which can be classed as conceptual models - sometimes use modelling to quantify states and explore scenarios, as in the references to DPSIR mentioned earlier (see also Table 2). They encourage work across sectors (e.g. biophysical, economic, social, political) and across ecosystems (land-based, freshwater, marine) using common terms, language and concepts and tend to reach high-level conclusions that may be influential to policy.

4.2 Extending crop simulation modelling - trading the biophysical and economic spheres

Flux-based simulation modelling - a form of process-based modelling as defined previously (Sections 3.1 and 3.2) - typically links attributes of the pedoclimatic base through agronomic interventions to a few selected processes driving agricultural productivity. In the 1960s, the combination of environmental physics and physiology laid the foundation for the study of global fluxes such as evaporation and carbon assimilation (Monteith, 1965). The physiological processes were defined that lead to yield through climatic and agronomic variables such as temperature, solar radiation, atmospheric humidity, rainfall and nutrients (Squire, 1990). Operating through these processes, flux models were able to estimate the biomass and yield of the world's major crops and high-input pastures (Ritchie, 1991; France and Thornley, 1984). Crop models are testable because the end points (yield)

Table 2 Models of agricultural systems that aim for optimisation of multiple ecosystem services (ES) and include biodiversity. Here, both agricultural production and economic output are considered ES. A for B and B for A are used to facilitate comparison but are not necessarily used in the studies referred to: A for B, organising agricultural land (including non-agricultural habitats) to support species of conservation concern; B for A, supporting life forms that in turn mediate ecological processes that aid production

Purpose/optimisation of	Biodiversity and other factors examined	Scales/methods	Main findings
Land use and ES in Brazil Cerrado, Kennedy et al. (2016).	Three groups of ES: bird and mammal species/water quality/sugarcane and cattle ranching – combining biophysical and economic models. Biodiversity modelled through land required for breeding, habitat suitability and dispersion.	Catchment (watershed) 400 000 ha, search algorithms to find optimum states among the three ES at scale of 90 m^2 units.	Possible to achieve win-win solutions – increase economic production and biodiversity – in many areas.
Biodiversity and production in a peri-urban region of The Netherlands via agri-environment measures, Verhagen et al. (2018).	ES include habitat for endangered species (e.g. amphibians, birds), landscape aesthetics, fruit and dairy farming – various models, for example pollination model to estimate fruit tree production; social media uploads of landscape photographs; suitability for great crested newt based on ponds and surrounding landscapes.	Netherlands, 219 km^2 agricultural land – compared introduction of on-farm measured (organic practices and linear features) and off-farm (taking land out of agriculture), estimated trade-offs, for example, between newts and pasture production.	Most ES could be improved by better management within both categories of agri-environment measures, but combination of on- farm and off-farm measures achieves the optimum trade-off, spatial allocation important.
Agri-environment schemes in Europe for multiple ES – conceptual study, Ekroos et al. (2014).	Distinguishes species of conservation concern (A for B) from life forms that support more general ES, for example, contributing to agriculture (e.g. beetle banks, B for A)	Conceptual (no specific location), examining the contribution and spatial arrangement of non-agricultural areas of high habitat quality (A for B) as distinct from more frequent but smaller habitat patches (B for A).	Agri-environment schemes should distinguish whether they target A for B or B for A and be designed accordingly.
Drivers-Pressures-States-Impacts-Response (DPSIR) framework for assessing effects of agrochemicals on biodiversity in Europe. Maxim & Spangenberg (2009).	Biota considered in total and not explicit, Argument: the pressure on biodiversity of the thousands of agrichemicals in past and current use is not assessable by formal ERA due to lack of data on long term effects of multiple exposure.	The 'tetrahedral DPSIR' (see text), conceptual model combining the four spheres of sustainability with DPSIR. Primarily examines drivers – social-economic-political, policy/regulatory, and management with respect to agrochemical usage.	Impacts of multiple chemical exposure in ecosystems are complex, uncertain and uncontrollable; need to progress primarily via reducing *Pressures*.

are readily measured on the ground: a generic model of growth and yield may be 'trained' on one set of data then tested on another. Crop simulation has followed an expanding trajectory since the 1960s (Belcher et al., 2004; Boote et al., 2010; Jones et al., 2016). Given data on climate and soil, from remote sensing, for example, models can be used to predict output in any part of the world. They are now widely applied in comparing management options, designing systems and predicting the effects of weather (Boote et al., 2010). Some have been developed into comprehensive research platforms, combining modelling across a range of crop and animal systems with software networks and extensive databases (e.g. APSIM, Holzworth et al., 2014).

Crop simulation, based for long in the biophysical sphere, has also been extended to incorporate economics and some other biophysical processes, but the main body of crop simulation has tended to restrict itself to the world's major crops as the main form of biota. Belcher et al. (2004), for example, combined crop and economic models to assess the sustainability of agricultural systems, but while attributes such as soil organic matter and greenhouse gas emissions were used as indicators, their model does not include biodiversity explicitly. In principle, however, flux-based crop modelling could be applied to multi-species communities. For example, flux modelling was soon adapted to include mixed cropping (Marshall and Willey, 1983) and then agro-forestry (Atangana et al., 2014; Magcale-Macandog, 2014). In tropical agro-forestry, in particular, flux modelling began to fuse with economic and social considerations. When comparing systems or attempting to improve existing practices for the benefit of the human population, agro-forestry modelling took on the form of integrative ecosystem models (Section 4.3).

4.3 Integrative ecosystem models: optimisation of biodiversity and other system features

Integrative system models that include the biophysical, economic and societal spheres, and in some cases, the political sphere (Figs. 1 and 5a), have developed rapidly in the last two decades, some including biodiversity as a major part. Several separate trends contributed to that shift in opinion and practice. Taking the example of pest (and more specifically weed) control in agriculture, Doyle (1997) concluded that weed models up to that time had limited relevance to integrated pest management (IPM) or sustainable agriculture. They concentrated on chemical control, took cost-effectiveness as the criterion of success and targeted one species or weed type rather multiple pests under the influence of changing crops and weather. There was little, if any, view or appreciation of 'weeds' as biodiversity or of any damaging effects of agrochemicals on the wider biota. Part of the problem was that research funding had a narrow focus on the pesticide and had little appreciation of the importance of many other attributes of the system.

Dent et al. (1995) went further to argue that any agricultural policy is likely to fail if directed only by economic and biophysical considerations and without the inclusion of social factors such as the livelihood and attitude of people who manage land. They proposed that models of farming should include economic and social (or societal) components along with production of livestock and crop and could be extended to include wider environmental effects on farming systems. The main extant ecological models that could be 'clipped on' to system models at that time were limited to habitat dynamics and pollutant exposure, but the principle was established. There was a push therefore from within farming systems modelling to include multiple factors, long time scales and influences from societal and political spheres.

Also in the latter part of the 20th century, the environmental risks posed by many agricultural practices gained increasing attention in public, scientific and political discourse. The need for agricultural land to conserve and enhance biota of rare or conservation status became widely appreciated. This was mostly A for B, agriculture supporting biodiversity (Section 1.1). But the concept of B for A, biodiversity supporting agricultural output, also came to the fore. For example, major GM crop trials that examined the effects of herbicide-tolerant crops on non-noxious weed flora and non-target invertebrates explicitly considered these forms of the biota as essential B for A (Squire et al., 2003; Hawes et al., 2003). There has been a substantial shift therefore in the scope of modelling over the past two decades to broaden the emphasis well beyond agricultural production.

Environmental considerations therefore became integral to the assessment of sustainability in agriculture. Initially, the biota was not defined explicitly, but subsumed within broad indicators such as 'agroecological practice' (e.g. Dantsis et al., 2010). However, several recent studies use modelling to attempt optimisation of systems in which named biota, as well as other outputs or services, including economic output, are in some form of balance that as near as possible satisfies all needs (Table 2). The trade-offs in such work can be envisaged through the hypothetical optimisation curves in Fig. 5c. The aim is to define (by various techniques) the envelope along which the system might move to realise various balances between environmental and economic attributes (societal and political also, not shown). A further aim is to define whether 'higher' envelopes exist to which the system might be moved if it was managed more efficiently for multiple gains.

Two examples, from very different ecosystems, by Kennedy et al. (2016) and Verhagen et al. (2018) examine how species of conservation concern can exist within agricultural landscapes. Land is assessed in terms of suitability for the target species (e.g. newts) based on known habitat preferences. The process relies not only on there being extensive data on land classification for the life forms but also on the economic and other ecosystem services such as control of pollution. The optimisation modelling (described in the articles cited) in both of these studies is consistent in showing that the provision of biodiversity and other services can all be increased beyond present levels in each type of land parcel but that the highest level of optimisation is achieved across a heterogeneous landscape in which life forms do best in certain habitats, whereas, say, the economic output is best served in other land parcels. Modelling suggested that the systems had scope to move towards a higher optimisation curve (Fig. 5c).

Optimisation modelling has also shown that ecological designs that promote A for B, which involves engineering landscapes for rare or endangered species, require different landscape configurations and within-landscape management than optimising B for A to generate habitat for predatory beetles or pollinators (Ekroos et al., 2014). The practical potential of recent integrative modelling of this type has been to point out that agricultural landscapes may be better able to support multiple functions than is generally realised. Questions remain as to whether the outputs from most such models can be validated and whether they can be used as tools by land managers. They suffer problems of scale and complexity similar to process-based landscape models (Section 3.2). If systems are too complex to model, or as Marshall et al. (1997) contend, their scaling laws have yet to be understood, the safest option to secure a future for biodiversity would be to reduce the pressures and not simply try to redress the negative impacts directly (Maxim and Spangenberg, 2009).

5 Whole-system multi-attribute decision modelling

The process-based models referred to so far deal with parts of ecosystems, generally those for which there are already much data and some understanding of processes (Sections 3.1 and 3.2). Where they scale up, they remain largely in the biophysical domain, where the currencies are typically numbers and mass of biota. The integrative models in Sections 4.1, 4.2 and 4.3 work across currencies from the biophysical to the social and sometimes to the political. In most instances, however, modelling of biodiversity tends to be remote and inaccessible to those stakeholders who would be affected, and often most interested, such as land managers and consumers of natural products. Moreover, when operating across currencies, large gaps in information appear: one process might be supported by adequate data and knowledge, but in order to to reach a conclusion on what to do, that process has to be judged against another for which there is almost no formal understanding. To proceed in the face of such imbalances, whole-system frameworks are being developed that allow processes to be combined through semi-quantitative or qualitative descriptors and at the same time encourage direct engagement of stakeholders. The most commonly used of such models in biodiversity studies are termed MADM or decision trees.

5.1 The general structure of a multi-attribute model (MADM)

Many of the current MADMs used in agriculture consist of 'hierarchically structured variables called attributes', which are 'qualitative rather than numeric', 'take only discrete symbolic values' and are aggregated by decision rules (Bohanec et al., 2007). The typical software is open-source DEXi (Bohanec, 2014), in which decision trees of varying complexity are constructed from measured or inferred attributes (Bohanec et al., 2007; Bohanec and Znidarsic, 2013; Sadok et al., 2009). For example, a hypothetical tree in Fig. 6a is constructed so as to define the status of a specific invertebrate population named 'omnivore type A', whose survival depends on two attributes, 'food supply' and 'predation', each of which in turn depends on other attributes. Those that are measured are usually termed 'input attributes', the rest aggregate attributes.

Irrespective of how much is known about them, each attribute is quantified by qualitative values termed 'classes', which might be categories such as

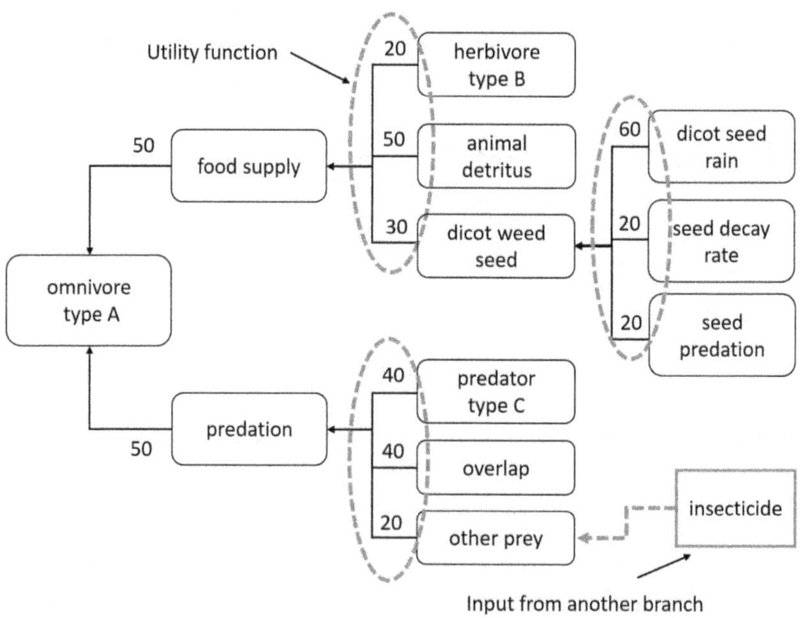

Figure 6 Hypothetical example of a DEXi multi-attribute decision model (MADM) showing several attributes combining to determine the state of an invertebrate population (omnivore type A). Text gives description.

'high', 'medium', 'low' or more simply 'yes' or 'no'. For simplicity and ease of operation classes tend to number between 5 and 2. The class boundaries of input attributes are pre-defined often based on the stakeholders' experience. The classes of the aggregated attributes are then defined by utility functions comprising 'IF ... THEN' decision rules. Rules have to be made for every possible combination. To allow greater flexibility, attributes can be weighted so one has a greater downstream effect than another.

The primary use of MADMs is to compare two or more agricultural systems or environments or a current environment with a potential future one. Biodiversity in one form or another is a component of several MADMs (Table 1). Those by Bohanec et al. (2007), Bohanec et al. (2008) and Griffiths et al. (2010) dealt primarily with soil biodiversity as might be affected by a new technology or carbon additions to soil. The remainder (Pelzer et al., 2012; Sadok et al., 2009; Hawes et al., 2019) increase the emphasis on above-ground biodiversity. The MADMs differ in the degree to which they assess agriculture as a means to support wider biodiversity (A for B, for example, farmland birds) or agriculture as dependent on the functions provided by biodiversity (B for A, for example, supporting pest-predators).

The application of MADMs often leads to firm conclusions on the questions asked: those on soils in Table 3 showed unequivocally that soil organisms and processes have inertia enough to be hardly affected by a change in crop variety (e.g. non-GM to GM insect resistant). While most MADMs used to date are based on crops, they could readily be extended to grass and grazing. The strength of these models is their holistic inclusion of many types of attributes and their accessibility that allows stakeholders to run the model to show how a change in class or the weighting of an attribute propagates along the tree. Moreover, the requirement that all attributes have to be assigned categories exposes where there is lack of knowledge.

5.2 Biota as integral to the working of a multi-attribute model

In the models in Table 3, elements of biodiversity (e.g. weeds, collembola, predators) are generally treated as aggregate indicators of environmental sustainability, comparable to, for instance, pollutant losses or soil quality. Moreover, the biota does not in general occupy a consistent position in the hierarchy of these models. In Fig. 1, the biota was placed so as mediate the effect of human interventions on ecological processes. Based on these linkages, Squire et al. (2016) devised a MADM in which the biota is set at a formal position within a chain or net of cause and effect. Human interventions in terms of ecological engineering or field management act with basal states (soil, climate) to affect the numbers or activity of the biota. These in turn mediate ecological processes that lead to the delivery of ecosystem services or

Table 3 Multi-attribute decision models (MADM) or decision trees including some aspect of biodiversity. Column 3 assesses the extent to which the biodiversity component offers functional support to agriculture (Biodiversity for Agriculture, B for A) or whether agriculture provides enhanced opportunity for in-field and landscape-scale biodiversity (Agriculture for Biodiversity, A for B)

Purpose	Biodiversity profile in the model/ life forms incorporated	A for B *versus* B for A	Scales covered
Fields and field plots: comparison of experimental plots in two countries with and without GM varieties (ECOGEN). Bohanec et al. (2007).	Soil biodiversity (no above-ground) is a main component in two major branches: (1) 'soil diversity' comprising bacteria, mesofauna and earthworms; (2) 'soil function' in which many life forms influence processes such as comminution and mineralisation.	Both: A for B in the soil diversity branch, B for A in the 'soil function' branch.	Tested for field plot but relevant to and could be extended to larger scales.
Cropping systems, GM versus conventional: maize, with and without herbicide and insect-resistant varieties. Bohanec et al. (2008).	Biodiversity constitutes two of four sub-branches in the 'Ecology' branch: (1) 'soil biodiversity', components not distinguished, characterised implicitly through farm management; (2) '(above-ground) biodiversity' including beneficial invertebrate functional groups (e.g. pollinators, parasitoids) characterised through weed flora and pest control.	Mainly B for A: control of pest organisms to maintain yield potential; support via crop management of organisms beneficial for crop production.	Field, both current and hypothetical systems.
Soil quality, following organic matter addition and de-intensified rotation. Griffiths et al. (2010).	Soil biodiversity (no above ground) in both main branches: (1) 'structure', as influenced by bacterial (e.g. denitrification) and faunal (e.g. nematode, earthworm) activity; (2) 'function', through comminution and mineralisation, involves a wide range of indicators of bacterial-fungal ratio, nematodes, protozoa, arthropods, enchytraeids, earthworms.	Mainly B for A: soil organisms that mediate processes relevant to soil quality.	Field, cropping system, rotation.
Cropping systems, assessment and comparison in terms of environmental, social and economic, with emphasis on pest management (DEXiPM). Pelzer et al. (2012).	Above ground biodiversity forms one of three sub-branches of the 'Environment' branch then divides into: (1) flora comprising weeds and semi-natural vegetation, for example, field margin; (2) fauna concentrating on natural enemies and pollinators as influenced by management and surrounding habitat.	Both, but emphasis on B for A, since the life forms assessed are mainly beneficial to pest management.	Field but incorporates wider influences such as 'habitat network'.

(*Continued*)

Table 3 (*Continued*)

Purpose	Biodiversity profile in the model/ life forms incorporated	A for B *versus* B for A	Scales covered
Cropping systems, designed for 'before the event' assessment, for example, before a change in crop type or agronomy (MASC). Sadok et al. (2009).	'Biodiversity conservation' is one of three sub-branches of the 'Environmental sustainability' branch, but its dependent attributes are mainly pesticide use, crop type and other management inputs.	Organisms not explicitly identified.	Field, cropping system.
Cropping systems, assessment and comparison in terms of environmental and economic sustainability with fuller inclusion of biodiversity (DEXiCSC). Hawes et al. (2019).	Biodiversity occupies one of the main branches in environmental sustainability: weed and margin vegetation, invertebrates, soil organisms.	Both: biodiversity as supporting agroecological function (e.g. weed coexistence with crops supporting a beneficial food web), and the quality of agriculture as habitat for uncommon plants, pollinators and vertebrates.	Primary scale is the field or unit of management defined as a 'cropping system'. Landscape (habitat, intensity) influences a range of attributes, for example, mobile organisms.

assessment of sustainability. In this scheme, very few operations occur without affecting or being affected by the biota.

An indication of the MADM's operation is given in Fig. 7. Attributes are shown by boxes joined by lines. Each input attribute, mostly one or more interventions, links through biota and processes to determine the status of one or more ecosystem services, provisioning, regulating or supporting (cultural are not included in this depiction). The boxes superimposed on the tree offer broad descriptions (e.g. new crop traits) of groups of attributes in the MADM beneath. Note that the regulating ecosystem services shown to the right are not linked to any processes, biota or interventions to the left of them because they originate in another part of the tree, not shown here.

The structure of the tree in Fig. 7 emphasises that nearly all processes and services depend on the biota in one form or another. Exceptions might include, for example, land shaping directly by machinery, but even here soil biota would quickly assume importance by determining whether the newly shaped soil will be stable or not. The structure in Fig. 7 can be worked both ways, from interventions to ecosystem services or as a planning tool to define the options that would achieve desired outputs.

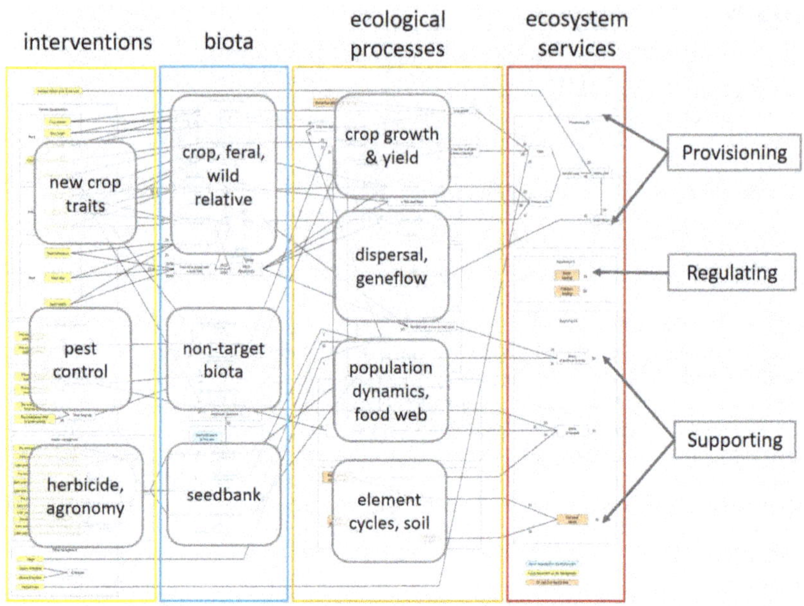

Figure 7 Representation of part of a MADM (DEXiES) to show scope and linkages in an arable system from interventions through the biota and ecological processes to ecosystem services. Superimposed boxes indicate general topics in the underlying tree. Adapted from the EU AMIGA project (Squire et al., 2016).

6 Future trends and conclusion

Biodiversity in agriculture is determined not only by what happens in fields and hillsides. Its future will depend as much on national, regional and global policies and politics. Models based on ecological processes operating in agricultural land will therefore need interpretation in relation to much wider influences. Scaling is particularly challenging because most of the globally important questions need action at a large scale - landscape, region, continent - at which experimentation is not possible and to which up-scaling presents difficulties of unpredictable emergent behaviour (Landis, 2010). Admittedly, data on landscape-scale biodiversity is incomparably greater than it was even two decades ago, but the landscape-scale models illustrated here still have to assume to varying degrees that what happens at small scale still happens in a complex ecosystem. Questions, for example, as to the combined effects of many pesticides acting on the biota of a field or region cannot be answered from a series of independent, laboratory-based toxicity tests on 'model' organisms. Modelling should therefore be wary of outstripping knowledge. Looking broadly at the trajectory of crop simulation, Boote et al. (2010) consider that the capacity to model has run far ahead of the knowledge needed to synthesise the complexity of biophysical processes. And that is for the form of process modelling with the longest pedigree and with a limited focus on crops and grass.

The cautious tone of Squire and Gibson's (1997) recommendations on scaling-up are still relevant, in that modelling needs to be conscious of 'applicability' - that is, the knowledge gained from modelling biodiversity in agriculture should aim to enable policy and management to agree a definitive solution. The advent of accessible decision models or MADMs, based on simple to use, open-source software, brought a major shift in appreciation of applicability (Bohanec et al., 2007, 2008). The process of constructing them forces those involved to confront major deficiencies or inequalities in knowledge. MADMs are a major development in linking science to practice.

Modelling biodiversity in agriculture therefore needs to centre in the subject and not the technology. If an investigator wishes to use modelling in order to understand how biodiversity should be best managed, the following steps might be followed:

- The question: the first step is to decide what cannot be understood from experiment and observation and therefore what might be understood by modelling.
- The target: is the biota in question operating as A for B or B for A, or a bit of both; and can it be represented as individuals or functional groups, defined by traits and located on an explicit base.
- MADM: constructing a decision tree using open-source software will likely confirm the question and define the important attributes that most influence the respective biota.
- Framework: in parallel, the investigator might look at all the factors that might affect the biota through one of the frameworks mentioned earlier, for example, DPSIR applied though the tetrahedron of biophysical, economic, societal and political factors. (There is no point simply opting for the biophysical or that with some economic added - the drivers and pressures on biodiversity will be much wider!)
- Up to this point (i.e. constructing a MADM and DPSIR framework) no specialised mathematical or programming ability is needed. When the questions, data and context are fully appreciated, process-based or integrated system modelling might be attempted as a route to guiding decisions on how to manage the biota.

Most modelling effort today in the topic of biodiversity is highly multidisciplinary and uses many techniques and tools (Chopin et al., 2019). However, the modellers, programmers, geospatial analysts and statisticians that the author has worked with all agree that the question should come first before any discussion on the type of modelling or the expertise needed.

7 Where to look for further information

The scope and practical utility of biodiversity modelling can be examined through the following links and downloads:

- US Environmental Protection Agency hosts probably the best single repository of environmental models and associated data, many of which include impacts on biodiversity. The following link gives a general introduction from which the reader can move to specific models and data sources: https://www.epa.gov/measurements-modeling/learn-about-models-epa.
- DEXi (Bohanec, 2014), a web site and base for multi-attribute decision modelling, offers downloadable free software that can be used without any specific training to explore a problem and define input and output variables. Mendoza and Martins (2006) give a general analysis of multi-attribute modelling in natural resource management.

Models linking biodiversity and agricultural are published in a wide variety of journals, but the following two should give the interested beginner an idea of the coverage of the subject:

- Ecological Modelling: https://www.journals.elsevier.com/ecological-modelling.
- Environmental modelling and software: https://www.journals.elsevier.com/environmental-modelling-and-software/.

The central principles and methodologies examined in the 1980s and 1990s are still highly relevant today and many have not been resolved. The following give an introduction to concepts explored two to three decades ago and still debated today:

- Individual-based approaches to biodiversity modelling: Huston et al. (1988), then the symposium volume authored by DeAngelis and Gross (1992).
- Up-scaling in ERA: Landis (2003) considers the challenges faced since the 1980s in using limited experimental data with few variables to model processes in catchments and landscapes.
- Up-scaling generally: the symposium volume edited by Van Gardingen et al. (1997) includes chapters on a wide range of habitats, processes, concepts and methods.

Chopin, P., Bergkvist, G. and Hossard, L. 2019. Modelling biodiversity change in agricultural landscape scenarios - a review and prospects for future research. *Biological Conservation* 235, 1-17. doi: 10.1016/j.biocon.2019.03.046.

Colasanti, R. L. and Grime, J. P. 1993. Resource dynamics and vegetation processes: a deterministic model using two-dimensional cellular automata. *Functional Ecology* 7, 169-176.

Colbach, N. 2008. How to model and simulate the effects of cropping systems on population dynamics and gene flow at the landscape level: example of oilseed rape volunteers and their role for co-existence of GM and non-GM crops. *Environmental Science and Pollution Research* 16, 348-60. doi: 10.1007/s11356-008-0080-6.

Colbach, N., Monod, H. and Lavigne, C. 2009. A simulation study of the medium-term effects of field patterns on cross-pollination rates in oilseed rape (*Brassica napus* L.). *Ecological Modelling* 220, 662-72. doi: 10.1016/j.ecolmodel.2008.11.020.

Colbach, N., Cordeau, S., Garrido, A., Granger, S., Laughlin, D., Ricci, B., Thomson, F. and Messéan, A. 2018. Landsharing vs landsparing: how to reconcile crop production and biodiversity? A simulation study focusing on weed impacts. *Agriculture, Ecosystems and Environment* 251, 203-17.

Crawley, M. J. and May, R. M. 1987. Population dynamics and plant community structure: competition between annuals and perennials. *Journal of Theoretical Biology* 125, 475-89.

Cushman, S. A., McKelvey, K. S. and Schwartz, M. K. 2009. Use of empirically derived source-destination models to map regional conservation corridors. *Conservation Biology* 23, 368-76. doi: 10.1111/j.1523-1739.2008.01111.x.

Dantsis, T., Douma, C., Giourga, C., Loumou, A. and Polychronaki, E. A. 2010. A methodological approach to assess and compare the sustainability level of agricultural plant production systems. *Ecological Indicators* 10, 256-63.

DeAngelis, D. L. and Gross, L. J. (Eds). 1992. *Individual-based Models and Approaches in Ecology: Populations, Communities and Ecosystems*. Chapman and Hall, New York, NY.

Dent, J. B., Edwards-Jones, G. and McGregor, M. J. 1995. Simulation of ecological, social and economic factors in agricultural systems. *Agricultural Systems* 49, 337-51.

Doyle, C. J. 1997. A review of the use of models of weed control in Integrated Crop Protection. *Agriculture, Ecosystems and Environment* 64, 165-72.

Ekroos, J., Olsson, O., Rundlof, M., Watzold, F. and Smith, H. G. 2014. Optimising agri-environment schemes for biodiversity, ecosystem services or both? *Biological Conservation* 172, 65-71.

Evers, J. B., Vos, J., Fournier, C., Andrieu, B., Chelle, M. and Struik, P. C. 2007. An architectural model of spring wheat: evaluation of the effects of population density and shading on model parameterization and performance. *Ecological Modelling* 200, 308-20.

France, J. and Thornley, J. H. M. 1984. *Mathematical Models in Agriculture*. Butterworths, London, UK.

Gari, D. K., Newton, A. and Icely, J. D. 2015. A review of the application and evolution of the DPSIR framework with an emphasis on coastal socio-ecological systems. *Ocean and Coastal Management* 103, 63-77.

Griffiths, B. S., Ball, B. C., Daniell, T. J., Hallett, P., Neilson, R., Wheatley, R. E., Oslerc, G. and Bohanec, M. 2010. Integrating soil quality changes to arable agricultural systems following organic matter addition, or adoption of a ley-arable rotation. *Applied Soil Ecology* 46, 43-53.

8 References

Atangana, A., Khasa, D., Chang, S. and Degrande, A. 2014. Agroforestry modeli In: *Tropical Agroforestry*. Springer, Dordrecht, the Netherlands. c 10.1007/978-94-007-7723-1_21.

Bàrberi, P. 2015. Functional biodiversity in organic systems: the way forward? *Sustaina Agriculture Research* 4, 26–31.

Bàrberi, P., Burgio, G., Dinelli, G., Moonen, A. C., Otto, S., Vazzana, C. and Zanin, G. 20 Functional biodiversity in the agricultural landscape: relationships between we and arthropod fauna. *Weed Research* 50, 388–401.

Becher, M. A., Grimm, V., Thorbek, P., Horn, J., Kennedy, P. J. and Osborne, J. L. 20 BEEHAVE: a systems model of honeybee colony dynamics and foraging to expl multifactorial causes of colony failure. *Journal of Applied Ecology* 51, 470–82. c 10.1111/1365-2664.12222.

Belcher, K. W., Boehm, M. M. and Fulton, M. E. 2004. Agroecosystem sustainabilit system simulation model approach. *Agricultural Systems* 79, 225–41.

Bohan, D. A., Powers, S. J, Champion, G., Haughton, A. J., Hawes, C., Squire, Cussans, J. and Mertens, S. K. 2011. Modelling rotations: can crop sequen explain arable weed seedbank abundance? *Weed Research* 54, 422–32. c 10.1111/j.1365-3180.2011.00860.x.

Bohanec, M. 2014. DEXi: program for multi-attribute decision making user's man Institut Josef Stefan, Slovénie. http://kt.ijs.si/MarkoBohanec/dexi.html.

Bohanec, M. and Znidarsic, M. 2013. DEX methodology: three decades of qualitat multi-attribute modelling. *Informatica* 37, 49–54.

Bohanec, M., Cortet, J., Griffiths, B., Znidarsic, M., Debeljak, M., Caul, S., Thompson and Krogh, P. H. 2007. A qualitative multi-attribute model for assessing the impac cropping systems on soil quality. *Pedobiologia* 51, 239–50.

Bohanec, M., Messean, A., Scatasta, S., Angevin, F., Griffiths, B., Krogh, P. H., Znidarsic, and Dzeroski, S. 2008. A qualitative multi-attribute model for economic and ecologi assessment of genetically modified crops. *Ecological Modelling* 215, 247–61.

Bolnick, D. I., Svanbäck, R., Fordyce, J. A., Yang, L. H., Davis, J. M., Hulsey, C. D. and Foris M. L. 2003. The ecology of individuals: incidence and implications of individ specialization. *American Naturalist* 161, 1–28.

Boote, K. J., Jones, J. W., Hoogenboom, G. and White, J. W. 2010. The role of crop syste simulation in agriculture and environment. *International Journal of Agriculture a Environmental Information Systems* 1, 41–54.

Bown, J. L., Pachepsky, E., Eberst, A., Bausenwein, U., Millard, P., Squire, G. R. and Crawfo J. W. 2007. Consequences of intraspecific variation for the structure and functi of ecological communities. Part 1: model development and predicted patterns diversity. *Ecological Modelling* 207, 264–76.

Breckling, B., Middelhoff, U. and Reuter, H. 2006. Individual based models as tools f ecological theory and application: understanding the emergence of organisatior properties in ecological systems. *Ecological Modelling* 194, 102–13. doi: 10.1016 ecolmodel.2005.10.005.

CEAM. Web site for the US EPA Centre for Exposure Assessment Modeling CEAM. htt ://www.epa.gov/exposure-assessment-models (accessed August 2019).

Hawes, C., Haughton, A. J., Osborne, J. L., Roy, D. B., Clark, S. J., Perry, J. N., Rothery, P., Bohan, D. A., Brooks, D. R., Champion, G. T., Dewar, A. M., Heard, M. S., Woiwod, I. P., Daniels, R. E., Young, M. W., Parish, A. M., Scott, R. J., Firbank, L. G. and Squire, G. R. 2003. Responses of plant and invertebrate trophic groups to contrasting herbicide regimes in the Farm Scale Evaluations of genetically-modified herbicide-tolerant crops. *Philosophical Transactions of the Royal Society of London B* 358(1439), 1899–913.

Hawes, C., Haughton, A. J., Bohan, D. A. and Squire, G. R. 2009. Functional approaches for assessing plant and invertebrate abundance patterns in arable systems. *Basic and Applied Ecology* 10, 34–42.

Hawes, C., Young, M., Banks, G., Begg, G. S., Christie, A., Iannetta, P. P. M., Karley, A. J. and Squire, G. R. 2019. Whole-systems analysis of environmental and economic sustainability in arable cropping systems: a case study. *Agronomy* 9, 438. doi: 10.3390/agronomy9080438.

Heard, M. S., Hawes, C., Champion, G. T., Clark, S. J., Firbank, L. G., Haughton, A. J., Parish, A. M., Perry, J. N., Rothery, P., Roy, D. B., Scott, R. J., Skellern, M. P., Squire, G. R. and Hill, M. O. 2003. Non-crop plants in fields with contrasting conventional and genetically modified herbicide-tolerant crops II. Species abundance and seasonal dynamics. *Philosophical Transactions of the Royal Society of London B* 358(1439), 1833–46.

Hill, M. O. 1973. Diversity and evenness: a unifying notation and its consequences. *Ecology* 54, 427–32.

Holst, N., Rasmussen, I. A. and Bastiaans, L. 2007. Field weed population dynamics: a review of model approaches and applications. *Weed Research* 47, 1–14. doi: 10.1111/j.1365-3180.2007.00534.x.

Holzworth, D. P., Huth, N. I., deVoil, P. G., Zurcher, E. J., Herrmann, N. I., et al. 2014. APSIM – evolution towards a new generation of agricultural systems simulation. *Environmental Modelling and Software* 62, 327–50.

Huston, M., DeAngelis, D. and Post, W. 1988. New computer models unify ecological theory: computer simulations show that many ecological patterns can be explained by interactions among individual organisms. *BioScience* 38, 692-1. doi: 10.2307/1310870.

IACS. Integrated Administrative and Control System. European Commission. https://ec.europa.eu/agriculture/direct-support/iacs_en (accessed August 2019).

IBPES. Intergovernmental Science-Policy Platform on Biodiversity and Ecosystem Services. http://www.ipbes.net/(accessed August 2019).

Jones, J. W., Antle, J. M., Basso, B., Boote, K. J., Conant, R. T., Foster, I., Godfray, C. J., Herrero, M., Howitt, R. E., Janssen, S., Keating, B. A., Munoz-Carpena, R., Porter, C. H., Rosenweig, C. and Wheeler, T. R. 2016. Brief history of agricultural systems modelling. *Agricultural Systems* 155, 240–54.

Kempton, R. A. 1979. The structure of species abundance and measurement of diversity. *Biometrics* 35, 307–21.

Kennedy, C. M., Hawthorne, P. L., Miteva, D. A., Baumgarten, L., Sochi, L., Matsumoto, M., Evans, J. S., Polasky, S., Hamel, P., Vieira, E. M., Ferreira Develey, P., Sekercioglu, C. H., Davisdon, A. D., Uhlhorn, E. M. and Kiesecker, J. 2016. Optimising land use decision-making to sustain Brazilian agricultural profits, biodiversity and ecosystem services. *Biological Conservation* 204, 221–30.

Landis, W. G. 2003. Twenty years before and hence; ecological risk assessment at multiple scales with multiple stressors and multiple endpoints. *Human and Ecological Risk Assessment* 9(5), 1317–26. doi: 10.1080/10807030390248500.

Landis, W. G. 2010. The frontiers in ecological risk assessment at expanding spatial and temporal scales. *Human and Ecological Risk Assessment* 9(6), 1415-24. doi: 10.1080/10807030390250912.

Lomnicki, A. 1988. *Population Ecology of Individuals*. Princeton University Press, Princeton, NJ.

LUCAS. Land Use and Coverage Area frame Survey. Joint Research Centre, European Commission. https://esdac.jrc.ec.europa.eu/projects/lucas (accessed August 2019).

Magcale-Macandog, D. B. 2014. Agroforestry models for promoting effective risk management and building sustainable communities. In: Kaneko, N., Yoshiura, S. and Kobayashi, M. (Eds), *Sustainable Living with Environmental Risks*. Springer, Tokyo, Japan, pp. 57-71.

Marshall, B. and Willey, R. W. 1983. Radiation interception and growth in an intercrop of pearl millet/groundnut. *Field Crops Research* 7, 141-60.

Marshall, B., Crawford, J. W. and Porter, J. R. 1997. Variability and Scaling: matching methods and phenomena. In: Van Gardingen, P. R., Foody G. M. and Curran P. J. (Eds), *Scaling-up: From Cell To Landscape*. Society for Experimental Biology Seminar Series 63. Cambridge University Press, Cambridge, UK, pp. 253-72.

Maxim, L. and Spangenberg, J. H. 2009. Driving forces of chemical risks for the European biodiversity. *Ecological Economics* 69, 43-54.

Maxim, L., Spangenberg, J. H. and O'Connor, M. 2009. An analysis of risk for biodiversity under DPSIR framework. *Ecological Economics* 69, 12-23.

Mendoza, G. A. and Martins, H. 2006. Multi-criteria decision analysis in natural resource management: a critical review of methods and new modelling paradigms. *Forest Ecology and Management* 230, 1-22.

Meziere, D., Colbach, N., Dessaint, F. and Granger, S. 2015. Which cropping systems to reconcile weed-related biodiversity and crop production in arable crops? An approach with simulation-based indicators. *European Journal of Agronomy* 68, 22-37.

Monteith, J. L. 1965. Evaporation and environment. *Symposium of the Society for Experimental Biology* 19, 205-24.

Monteith, J. L. 1972. Solar radiation and productivity in tropical ecosystems. *Journal of Applied Ecology* 9, 747-66.

Niklas, K. J. 1999. Evolutionary walks through a land plant morphospace. *Journal of Experimental Botany* 50, 39-52.

O'Connor, M. 2006. The 'four spheres' framework for sustainability. *Ecological Complexity* 3, 285-92.

OECD. 1993. *Core Set of Indicators for Environmental Performance Reviews*. Organization for Economic Cooperation and Development, Paris, France.

Omann, I., Stocker, J. and Jager, J. 2009. Climate change as a threat to biodiversity: an application of the DPSIR approach. *Ecological Economics* 69, 24-31.

Pachepsky, E., Crawford, J. W., Bown, J. and Squire, G. R. 2001. Towards a general theory of biodiversity. *Nature* 410, 923-6.

Pachepsky, E., Bown, J. L., Eberst, A., Bausenwein, U., Millard, P., Squire, G. R. and Crawford, J. W. 2007. Consequences of intraspecific variation for the structure and function of ecological communities. Part 2: linking diversity and function. *Ecological Modelling* 207, 277-85.

Pelzer, E., Fortino, G., Bockstaller, C., Angevin, F., Lamine, C., Moonen, C., Vasileiadis, V., Guerin, D., Guichard, L., Reau, R. and Messean, A. 2012. Assessing innovative cropping systems with DEXiPM, a qualitative multi-criteria assessment tool derived from DEXi. *Ecological Indicators* 18, 171-82.

Perry, J. N., Devos, Y., Arpaia, S., Bartsch, D., Gathmann, A., Hails, R. S., Kiss, J., Lheureux, K., Manachini, B., Mestdagh, S., Neemann, G., Ortego, F., Schiemann, J. and Sweet, J. B. 2010. A mathematical model of exposure of non-target Lepidoptera to Bt-maize pollen expressing Cry1Ab within Europe. *Proceedings of the Royal Society B* 277, 1417–25.

Ritchie, J. T. 1991. Specifications of the ideal model for predicting crop yields. In: Muchow, R. C. and Bellamy, J. A (Eds), *Climatic Risk in Crop Production: Models and Management*. CAB International, Wallingford, UK, pp. 97–122.

Roast, S., Northing, P., Smart R., Boxhall, A., Fishwick, S. and Ashton, D. 2007. A review of models and methods for ecological risk assessment. Science Report SCO30003/SR. Environment Agency, UK.

Rodriguez-Labajos, B., Binimelia, R. and Monterroso, I. 2009. Multi-level driving forces of biological invasions. *Ecological Economics* 69, 63–75.

Rortais, A., Arnold, G., Dorne, J-L., More, S. J., Sperandio, G., Streissl, F., Szentes, C. and Verdonck, F. 2017. Risk assessment of pesticides and other stressors in bees: principles, data gaps and perspectives from the European Food Safety Authority. *Science of the Total Environment* 587, 524–37. doi: 10.1016/j.scitotenv.2016.09.127.

Rozenweig, M. L. 1995. *Species Diversity in Space and Time*. Cambridge University Press, Cambridge, UK.

Sadok, W., Angevin, F., Bergez, J.-E., Bockstaller, C., Colomb, B., Guichard, L., Reau, R., Messean, A. and Dore, T. 2009. MASC, a qualitative multi-attribute decision model for *ex ante* assessment of the sustainability of cropping systems. *Agronomy for Sustainable Development* 29, 447–61.

Schroder, B. and Seppelt, R. 2006. Analysis of pattern-process interactions based on landscape models – overview, general concepts and methodological issues. *Ecological Modelling* 199, 505–16.

SEEA. The System of Environmental-Economic Accounting, United Nations. https://seea.un.org/(accessed August 2019).

SIGMEA. 2007. European Policy Oriented Research Project FP6-2002-SSP1. https://www6.inra.fr/sigmea (accessed August 2019).

Silvertown, J., Holtier, S., Johnson, J. and Dale, P. 1992. Cellular automaton models of interspecific competition for space – the effect of pattern on process. *Ecology* 80, 527–34.

Squire, G. R. 1990. *The Physiology of Tropical Crop Production*. CAB International, Wallingford, UK.

Squire, G. R. 2000. Plant and canopy diversity. In: Marshall, B. and Roberts, J. A. (Eds), *Leaf Development and Canopy Growth*. Sheffield Academic Press, Sheffield, UK, pp. 280–309.

Squire, G. R. 2017. Defining sustainable limits during and after intensification in a maritime agricultural ecosystem. *Ecosystem Health and Sustainability* 3. doi: 10.1080/20964129.2017.1368873.

Squire, G. R. and Gibson, G. J. 1997. Scaling-up and scaling-down: matching research with requirements in land management and policy. In: Van Gardingen, P. R., Foody G. M. and Curran P. J. (Eds), *Scaling-up: From Cell To Landscape*. Society for Experimental Biology Seminar Series 63. Cambridge University Press, Cambridge UK, pp. 17–34.

Squire, G. R., Brooks, D. R., Bohan, D. A., Champion, G. T., Daniels, R. E., Haughton, A. J., Hawes, C., Heard, M. S., Hill, M. O., May, M. J., Osborne, J. L., Perry, J. N., Roy, D. B., Woiwod, I. P. and Firbank, L. G. 2003. On the rationale and interpretation of the farm-scale evaluations of genetically-modified herbicide-tolerant crops. *Philosophical Transactions of the Royal Society of London B* 358(1439), 1779–800.

Squire, G. R., Young, M. W., Demade, M., Messean, A. and Angevin, F. 2016. DEXiES - a multi-attribute model of ecosystem functioning for use in environmental risk assessment. EU AMIGA Project, Final Report Deliverable 3.6 Annex A.

Sumida, A., Ito, H. and Isagi, Y. 1997. Trade-off between height growth and stem diameter growth for an evergreen oak, *Quercus glauca*, in a mixed hardwood forest. *Functional Ecology* 11, 300–9.

Therond, O., Duru, M., Roger-Estrade, J. and Richard, G. 2017. A new analytical framework of farming system and agriculture diversities. A review. *Agronomy for Sustainable Development* 37, 21. doi: 10.1007/s13593-017-0429-7.

TEEB. 2010. The Economics of Ecosystems and Biodiversity: Mainstreaming the Economics of Nature: a synthesis of the approach, conclusions and recommendations of TEEB. http://www.teebweb.org (accessed August 2019).

Topping, C. J., Hansen, T. S., Jensen, T. S., Jepson, J. U., Nikolajsen, F. and Odderskaer, P. 2003. ALMaSS, an agent-based model for animals in temperate European landscapes. *Ecological Modelling* 167, 65–82. doi:10.1016/S0304-3800(03)00173-X.

Topping, C. J., Dalby L. and Skov, F. 2016. Landscape structure an management alter the outcome of a pesticide ERA: evaluating impacts of endocrine disruption using ALMaSS European Brown Hare model. *Science of the Total Environment* 541, 1477–88. doi: 10.1016/j.scitotenv.2015.10.042.

UN SDG. Sustainable Development Goals, United Nations. https://sustainabledevelopment.un.org (accessed August 2019).

US EPA. United States Environmental Protection Agency. Source for AQUATOX exposure assessment model. https://www.epa.gov/exposure-assessment-models/aquatox (accessed August 2019).

Van Gardingen, P. R., Foody G. M. and Curran P. J (Eds). 1997. *Scaling-up: From Cell To Landscape*. Society for Experimental Biology Seminar Series 63. Cambridge University Press, Cambridge, UK.

Verhagen, W., Zanden, E. H. v., Strauch, M., Teeffelen, A. J. A. v. and Verburg, P. H. 2018. Optimizing the allocation of agri-environment measures to navigate the trade-offs between ecosystem services, biodiversity and agricultural production. *Environmental Science and Policy* 84, 186–96.

Zhang, X., Izaurralde, R. C., Manowitz, D. H., Sahajpal, R., West, T. O., Thomson, A. M., Xu, M., Zhao, K., LeDuc, S. D. and Williams, J. R. 2015. Regional scale cropland carbon budgets: evaluating a geospatial agricultural modelling system using inventory data. *Environmental Modelling and Software* 63, 199–216.

Chapter 3

Assessing the economic value of agricultural biodiversity: a critical perspective

Corrado Topi, Stockholm Environment Institute at York, Department of Environment and Geography and Interdisciplinary Global Development Centre, University of York, UK; and Leonie J. Pearson, Stockholm Environment Institute, Thailand

1 Introduction

Over the course of the last few decades, the attempt to give a pure economic value to biodiversity has been a contentious issue and rightly so. Most of the methods that are currently in use at policymaking and decision-making level tend to take a classical economic perspective on the issue of valuing diversity, whether it be biodiversity in general or, more specifically, agricultural biodiversity, that is, agrobiodiversity (terms we will use interchangeably in the rest of the chapter).

The aim of such approaches is to optimise classic economic indicators, such as ROI, which in turn require maximising productivity, that is, taking production per hectare towards the maximum technical potential (MTP), while reducing all costs to the minimum.

This has important consequences for the way in which agrobiodiversity and agriculture are viewed and brings with it a wealth of problems of both an economic and non-economic nature. They range from the ethical impossibility

http://dx.doi.org/10.19103/AS.2020.0071.06

of allocating an economic value to living beings (the main components of biodiversity which includes, of course, humans), to issues of ownership, that is, the fact that biodiversity is a collective good while profit from productivity is an individual good, as well as the error and uncertainty problem, that is, the fact that you cannot price something you do not know, as is the case with many processes that link biodiversity to agricultural productivity. These are but a few of a wealth of similar issues.

The first aim of our chapter is then to highlight some of the most problematic among these many issues.

One answer in applying economic thinking to agriculture is to choose a systemic perspective. A systemic perspective would take into account agriculture within the context of the planetary social, environmental and economic-financial systems, seen as a complex organism made of a series of components, with all their properties, all their relationships and interdependencies (and eventually, as we will see at the end, their rights). Even if a discussion of the properties, structures and theory of complex systems is well beyond the scope of the present chapter, agriculture can be interpreted as a subsystem (i.e. a component), and more specifically, a component of the subsystem that provides food to the human community. By doing so:

- we broaden the system boundary of agriculture to include all of its relationships with all the other subsystems within the planetary boundary,
- we make the constraints imposed by the planetary boundaries on agriculture explicit, and
- we position agrobiodiversity within a much larger map of relationships that may serve other needs (i.e. not agriculture) of the planetary system, accepting the fundamental and intrinsic multidimensionality of the contributions of living beings.

The need to use systemic approaches to make agriculture sustainable is not new to the scholarly community, and we can address readers interested in delving into its historical roots to some seminal work (Ikerd, 1993a,b; Kirschenmann, 1991). However, for a series of reasons, there has been strong resistance to accept that systemic approaches need to be used also in economic analysis of agriculture. One of these reasons is that economics is often confused with accounting which, with its long history dating back to the fifteenth century[1] (Dean et al., 2016; Sangster, 2018), is often considered a well-oiled, completely

1 Living and working in what corresponds to modern day Italy towards the end of the Renaissance, Franciscan friar, polymath and archetypal renaissance man Luca Pacioli is widely regarded in scholarly literature as the 'Father of Accounting' (Fischer, 2000) for his 1494 work *Summa de Arithmetica, Geometrica, Proportioni et Proportionalita* that included a fundamental chapter on theoretical and empirical accounting that de facto gave birth to the discipline (Dean et al., 2016).

settled approach to reality and cost benefits. One other important reason is that within contemporary mainstream neoliberal economic thinking, simplicity, that is, reduction of a system to a small set of easy-to-interpret and formally rigid rules, seems to be a value (inherited from accounting, Sangster, 2018), because it is taken to be formally well founded, and also because it means 'easy replicability', and this is assumed to be possible and true even in the case of complex systems. The rise of systemic science, and of its economic branch, latterly referred to as ecological economics (Costanza, 1991a), is developing instruments for successfully challenging this world view (Costanza, 1991b).

Ecological economics is 'a new transdisciplinary approach that looks at the full range of inter-relationships between ecological and economic systems' where 'by transdisciplinary we mean that ecological economics goes beyond the normal conceptions of scientific disciplines and tries to integrate and synthesize many different disciplinary perspectives' (Costanza, 1991a). As such, ecological economics seems to be the best candidate to deal with systemic phenomena such as biodiversity and agrobiodiversity. Ecological economics also seems to offer the advantages of being a 'post-normal science' (Castro e Silva and Teixeira, 2011), first and foremost its attention to the 'linkages between the social, political, economic, biological, physical, chemical, and geological systems', its explicit claim to ethical non-neutrality, and its 'not being divorced from the policymaking and decision-making processes'. Mainstream orthodox economics comes indeed with a set of implicit underlying ideological assumptions, which, by being implicit, are hard to elicit, challenge and manage. As a consequence, it provides statements with strong ethical implications under the appearance of ethical neutrality, regards linkages mostly as unnatural constraints, and it links policymaking and decision-making only in the form of providing *diktats*[2]. As with all things economic, battle is raging, and the dust is far from settled (Smale et al., 2002), because interests are at stake and world views are being challenged, with repercussions that may go well beyond the realm of agriculture.

The second aim of our chapter is then to critically review developments in methods to assess the economic value of agricultural biodiversity.

The chapter is structured as nine sections, starting with this introduction. In the second section we briefly introduce and discuss the different definitions of agrobiodiversity. In the third section, we discuss ways of evaluating agrobiodiversity. This is followed by Section 4 where we introduce and discuss the ecosystem services framework (ESF), which we extend in the fifth section through the ecosystem interactions framework. In Section 6 we discuss two important limitations, that is, uncertainty and the problem of ecosystem services

2 The assumption of economics, in all of its forms, being an autonomous self-contained 'normal' science at all is a matter of contentious and unresolved epistemological debate.

whose value cannot be derived from the market. We propose an innovative perspective that could radically change the way we evaluate agrobiodiversity in Section 7. Finally, in Section 8 we draw conclusions. The chapter is completed in Section 9 by a series of references for the readers interested in expanding their understanding.

2 The relationship between definitions and economic approaches

In order to clearly identify an economic value for agrobiodiversity, we first need to have a clear definition of agrobiodiversity itself. However, within the literature, there are several, often competing, definitions of agrobiodiversity, which in turn generate very different approaches to its economic evaluation where that is even possible.

Different definitions imply different extrapolation techniques for the different valuation methodologies and for the values themselves and for the valuation of the cost-effectiveness and the benefits of management interventions, when we try to apply them to different situations.

This is especially true for those economic valuation methodologies that focus on indirect estimation and that are, as such, highly context dependent. We use indirect estimation whenever we cannot give a direct value to the real inputs, outputs and inner workings of a process. This requires the use of proxy variables, which in turns increases complexity and uncertainty further. A typical example is using national averages to evaluate the costs of a resource, or of an impact, or the value of a product. Any such transfer of values can thus be unreliable and needs to be treated with caution and adapted on a case-by-case basis.

Several definitions of agrobiodiversity have been given, of which we selected three, representative of the evolution of definitions of the concept, listed in order of time of appearance in the literature:

Definition I (Academic): 'Agricultural biodiversity (or agrobiodiversity) is a component of biodiversity referring to all diversity within and among species found in crop and domesticated livestock systems, including wild relatives, interacting species of pollinators, pests, parasites, and other organisms' (Qualset et al., 1995; Schifino-Wittmann, 2000).

Definition II (FAO definition): 'The variety and variability of animals, plants and microorganisms that are used directly or indirectly for food and agriculture, including crops, livestock, forestry and fisheries. It comprises the diversity of genetic resources (varieties, breeds) and species used for food, fodder, fibre, fuel and pharmaceuticals. It also includes the diversity of non-harvested species that support production (soil microorganisms, predators, pollinators), and those in the wider environment that support agro-ecosystems (agricultural, pastoral, forest and aquatic) as well as the diversity of the agro-ecosystems' (FAO, 1999) (Fig. 1).

Yet, within the literature, and, worryingly, within the more specialised agrobiodiversity economic literature, a different definition is often used:

Definition III (TEEB definition): 'An essential component of the global eco-agri-food system is the genetic diversity of crops and livestock. These genetic resources, including both the diversity of cultivated varieties as well as the wild relatives of crops ("crop wild relatives") and livestock, are a key form of natural capital, and the conservation and use of agrobiodiversity is essential for the development of a more sustainable and resilient global food system' (TEEB, 2018).

In this definition, the system boundary around the agrobiodiversity seems to be essentially restricted to mean genetic diversity of agricultural resources, but misses the contribution of all the wild diversity that provides essential services to the agri-food value chain. Without the latter, the agri-food value chain would have an overall value that is much less, but it is impossible to estimate exactly how large the loss of value would be.

To use the most systemic approach available, we have decided to adopt definition II that is used by UN agencies and that incorporates definition I. We do not use definition III and we suggest this definition is not used in practice. This has important implications for our understanding of how to value biodiversity, as we will discuss in the next section.

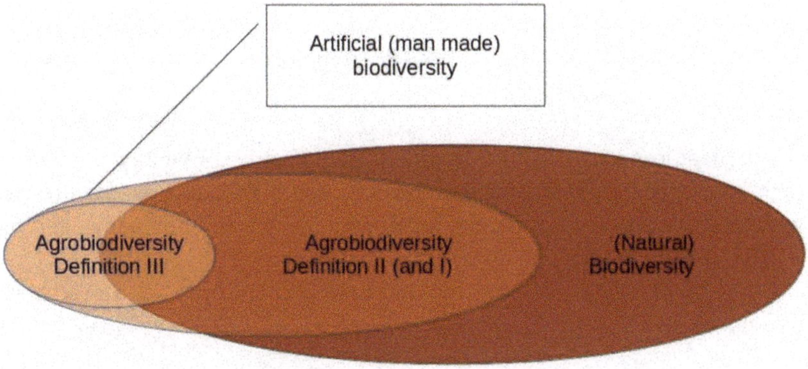

Figure 1 Relationships between the different definitions of agrobiodiversity and biodiversity (natural and artificial).

3 What does valuing agricultural biodiversity mean?

Putting an economic value on biodiversity and on agricultural biodiversity would mean that we can then apply economic abstraction, that can be followed by economic reasoning, on the basis of which decision-makers and policymakers would make (supposedly) rational decisions (Bateman et al., 2013).

Calculating this economic value would require a perfect understanding of the mechanisms by which biodiversity contributes to a human activity. In the case of agricultural biodiversity, decisions on management interventions concerning agricultural land use, including biodiversity, are often either made exclusively on the basis of economic indicators, or at least include economic considerations as a major component of the decision-making process. In the former case, economic abstraction is taken to its extreme, where the object, that is, the agricultural system, is not considered, and its economic value becomes the object of decision-making (as it is the case with some FDI in agricultural systems).

Research that specifically investigates the economic value of agricultural biodiversity is scarce. To validate the claim we conducted a rapid review adapting the methodology of Khangura et al. (2012), where the need was:

- to identify all papers explicitly referring to the economic or financial evaluation of agrobiodiversity, and
- to identify those papers that mentioned economic or financial evaluation of agrobiodiversity in the context of a systemic approach.

We focussed specifically on the Web of Knowledge, one of the most important databases of scientific references available, and, after synonymization and stemming, we formulated and used the query:

> *TS=((agr*biodiversity OR (agricultur* NEAR/0 biodiversit*)) AND (economic* OR financ*))*

Our search returned only 251 results on 2020/03/13 at 17:14, while a cumulative search for the last 5 years, and searches by each individual year for the last 5 years did not show a substantial movement away from the yearly average (Fig. 2).

Correspondingly, over the same database and time period, a search for the string:

> *TS=((agr*biodiversity OR (agricultur* NEAR/0 biodiversit*)) AND (economic* OR financ*) AND (system* OR integr* OR holist*))*

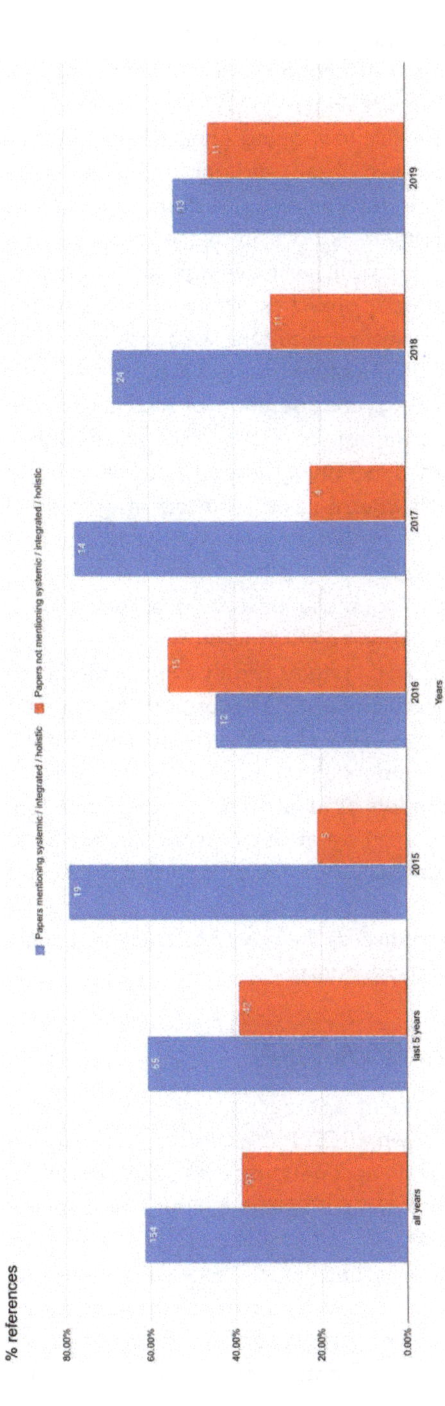

Figure 2 Chart of the % and number of papers on Web of Science mentioning agrobiodiversity economics versus papers that mention it in a systemic study context, 2015–2019, compared to overall average and last 5 years average, from Web of Science. % is given by the height of the column, while the number of papers is given by the label at the top of the individual column.

was run for all years - for the last five years cumulatively and then individually for each one of the last five years. Even this second run did not show again a substantial deviation from the yearly average (Fig. 2).

While this rapid assessment test is far from being conclusive, and must be taken *addito grano salis* (i.e. with a pinch of salt) and only in the context of a rapid review approach, it seems to suggest that using systemic approaches is not yet the norm. More worryingly there is no sign in the literature of a substantial increase or of movement from the year average in the number of papers making references to systemic approaches. While we see a steady number of papers considering biodiversity in agriculture, we see little change from a partial assessment (e.g. consideration of bees as pollinators to specific crops) to a more holistic assessment (e.g. consideration of mixed land use and management, use of native vegetation soil microbes to deliver enhanced agricultural production and ecosystem functioning services such as climate regulation). Without a broader approach, when a 'value' is placed on agricultural biodiversity it is partial, and specific for a given outcome, without considering the whole system.

Systemic farming practices, including those based on agroecological principles, have been delivering sustainable increases in production for many decades with regenerative and semi-closed systems widely seen as the future of agriculture (Pearson, 2007). Yet economics has not kept up with the demand for systemic assessments reflecting these new approaches. In placing a value on agricultural biodiversity, we need to understand the system of assessment - exactly what are we valuing (dollars, outputs or functioning climate), for whom (private landholders or public community), what spatial scale (paddock to landscape) and how (agriculture as a user of biodiversity, or agriculture as part of the biodiversity landscape).

In the next section, we are going to discuss the most important approach that has been devised so far to answer those questions.

4 The ecosystem services framework (ESF)

One of the most significant and important approaches used in numerous studies is the ecosystem services framework (ESF) as a basis to evaluate the economic value of biodiversity. Taking into account the caveats outlined in the previous sections, such an approach has some interesting strengths, even if it is marred by some significant drawbacks.

The idea behind ESF-based approaches is to consider the natural environment as a supplier of services and goods to human economic activities. Within this framework we subsume not only the classic ecosystem services approach, but also the discourse and the debate about natural capital and natural goods, as it is customary.

In theory, such an approach seems to answer directly and specifically the question we are trying to address. If the natural environment can be interpreted as a supplier of goods and services to human economic activities, we can choose to apply 'classic' value chain reasoning and the problem is solved. Unfortunately, this simplistic perspective does not hold true, as we will see in the next sections. As usual, when dealing with truly complex systems, the devil is in the details.

The history of the ESF is several decades long. The framework itself has become mainstream in decision-making and policymaking in the last 20 years, in part as a result of the global effort to produce the Millenium Ecosystems Assessment, also simply known as the MEA (Millennium Ecosystem Assessment, 2005a,b,c,d), where a specific approach to ecosystem services was introduced.

The idea behind this approach was relatively simple. An ecosystem is an extremely complex, intertwined system and the need was felt to capture such complexity within a simplified model. To do so, we can break down the complexity of services offered by the supplier (the ecosystem) into classes and classify all the functions (of which we are aware) into a limited number of homogeneous categories.

On the other hand, we also need to break down the demands of the client (human society) on the supplier (the ecosystem) into similarly homogeneous categories. Finally, we can establish supply and demand relationships between the suppliers' categories on the one hand and the client categories on the other. The results of this conceptual process is captured in Fig. 3 (taken from the MEA (Program), 2005).

Ecosystem services were classified into four major groups: supporting, provisioning, regulating and cultural. Each category or group was then broken down, within certain limits, into a list of services belonging to that group. Of course, a full discussion of the structure of the classification of services in the MEA is outside the scope of the present chapter, and we refer the interested reader to the relevant references about the MEA.

In the MEA approach to ecosystem services, the underlying complexity of functions and relationships relying on biodiversity within the living ecosystem was hidden within the generic category of supporting services rather than being made explicit.

In the early 2000s, policymaking and decision-making bodies started adopting the ESF, including the ESF contained in the MEA, and the language of the ecosystem services (and of natural goods and capital) became progressively more embedded in policy documents.

In parallel to, and following the MEA, several other attempts at classifying the ecosystem services were initiated with a similar world view.

Such classifications and assessments were undertaken or initiated at the global, macro-regional and even national levels, using classifications that

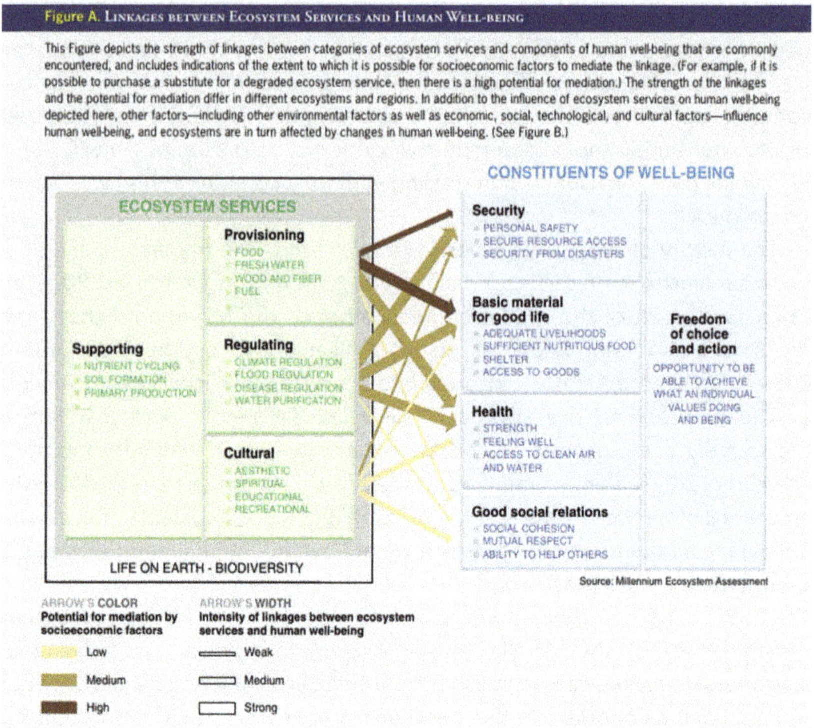

Figure 3 Ecosystem services and their contribution to well-being.

varied, resulting in considerable fragmentation in research, policymaking and decision-making.

While an exhaustive discussion of all the approaches is beyond the scope of this chapter, we think three initiatives are worth mentioning in this context:

- the Economics of Ecosystems and Biodiversity (TEEB) global initiative,
- the Common International Classification of Ecosystem Services (CICES) global initiative, and
- the Mapping and Assessment of Ecosystems and their Services (MAES).

In the following discussion we focus our attention specifically on the attempts at standardisation provided by the TEEB and the CICES initiatives. We leave it to the interested reader to explore the specific macro-regional approach developed by the MAES (https://biodiversity.europa.eu/maes) (Maes et al., 2018). A common feature of all these initiatives is that they consider biodiversity as the basis for the optimal functioning of the ecosystem as a provider of services and goods.

The TEEB global initiative was born out of a proposal of the environment ministers of the G8+5 group at a meeting in March 2007 in Potsdam, Germany.

The TEEB initiative was launched jointly by the German Federal Ministry for the Environment (BMUB) and the European Commission (EC) ('TEEB – The Initiative', n.d.). The aim of the initiative was to analyse the global economic benefits of biological diversity, the costs of the loss of biodiversity and the failure to take protective measures versus the costs of effective conservation. To frame this analysis, TEEB adopted three core principles:

- Recognizing value in all aspects of biodiversity and living ecosystems as a core value of all human societies,
- Demonstrating value in economic terms, considering full costs and benefits of an ecosystem, and
- Capturing value, that is, introducing mechanisms that incorporate the value of ecosystems into decision-making and policymaking.

The work of the TEEB initiative was split into three phases:

- the first phase concluding with the presentation of an interim report at the High-Level Segment of the Ninth Conference of the Parties to the Convention on Biological Diversity (CBD COP-9) in Bonn, Germany, in May 2008 (TEEB, 2008),
- the second phase, concluded with the presentation of four key reports addressed to four different groups of stakeholders in CBD COP-10 in Nagoya, Japan, in October 2010 (Kumar, 2010; TEEB, 2012; T. TEEB, 2009; Wittmer and Gundimeda, 2012), and
- the final current phase, where the findings of TEEB are expected to be embedded into policymaking at all scales following the publication of key studies and of guidelines, including on climate (TEEB, 2009) and the agri-food sector (Ecosystems and Biodiversity, 2018; TEEB, 2018, 2015).

The work of TEEB has been of particular importance for the evaluation of biodiversity and of agrobiodiversity.

The second major global initiative is the initiative by the European Environment Agency (EEA) to sponsor the development of a recognised framework, through the Common International Classification of Ecosystem Services (CICES, https://cices.eu/). As is the case for TEEB, the initiative is currently ongoing. The CICES is now at version 5.1, published in 2018 (Haines-Young and Potschin, 2018, 2013a,b, 2011, 2010).

Compared to other initiatives, CICES introduces two novel elements:

1. it eschews completely the debate around the underlying complexity of ecosystems, by only classifying final ecosystems services, that is, those that provide direct services to humans (Potschin and Haines-Young, 2011), and

2 it includes the abiotic environment in the classification (e.g. climate-related processes), in a provisional state in version 4.3 and in a more systematic way in version 5.1 (Haines-Young and Potschin, 2018)

While the former is concerning, particularly to those that attempt to use systemic approaches for the economic evaluation of (agro)biodiversity, the latter is a welcome addition to the mix.

A very important aspect of CICES is the attempt to harmonise and offer a correspondence between the different frameworks, namely CICES itself, the MEA and the TEEB, and two other classifications not discussed here:

- the Intergovernmental Science-Policy Platform on Biodiversity and Ecosystem Services (IPBES) Classifications, and
- the US-EPA Final Ecosystem Goods and Services Classification System (FEGS-CS).

The CICES makes available simple tools to compare frameworks (https://cices.eu/cices-structure/). The Operationalization of Natural Capital and Ecosystem Services: From Concepts to Real-World Applications (OpenNess) European project provides an online widget to evaluate equivalence between MEA, CICES and TEEB, as well as the UK National Ecosystems Assessment and CICES – Belgium, a nationally adapted version of CICES. While the OpenNess project finished in 2017, the equivalence widgets are available online at http://openness.hugin.com/example/cices and were tested by us on 2020-04-21 at 18:38.

In the next part of this section we review the most common ESF-based methods for valuing agricultural biodiversity based on a review of the literature. While, in the absence of a well-defined attribution protocol, the review makes no claim of being systematic (e.g. as proposed by Sutton et al., 2019 or Khangura et al., 2012), we believe it reflects the major research in the field.

As a starting point Power (2010) has outlined key ecosystem functional relationships within an agroecosystem, assuming that biodiversity is a core input to agroecosystems (Fig. 4). Importantly it recognises both services and disservices that originate from agrobiodiversity.

There are essentially four methods by which the value of agricultural biodiversity can be evaluated in terms of:

1 increasing productivity,
2 increasing resilience,
3 lowering risk, and
4 taking land out of production.

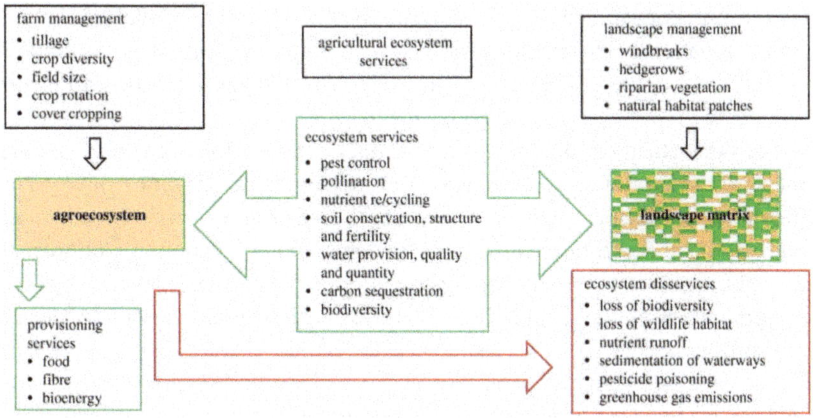

Figure 4 Impacts of agricultural farm management and landscape management on the flow of ecosystem services and disservices to and from agroecosystems. Source: Power, 2010.

These four methods are discussed below.

1 Role of agricultural biodiversity in increasing productivity

A standard economic production function for agriculture shows that, for an increasing production output (Y), there are essential factors of production. This output is only produced, through a functional form (f) of controlled factors like soil, water, seeds and so on (x) and the level of unknown or uncontrollable factors (u).

This can be written as $Y = f(x_1, x_2, x_3 \ldots x_n) + u$

Here, we can consider agriculture biodiversity (genetic components) as a factor of production, an (x) - which can have a positive or negative influence on the final outputs (Y). In this way, depending on the type of agricultural biodiversity and its relationship with the output, there are many functional forms to take, increasing linear, decreasing, asymptotic or exponential (see Seppelt et al. (2016) for varying functional forms).

2 Role of agricultural biodiversity in increasing resilience

A specific relationship between agricultural biodiversity and production has been found when biodiversity increases the resilience of the farming system. The value of resilience lies in its ability to maintain the provision of benefits under a given regime. A key assumption is that many farming systems are fragile and vulnerable to a range of intersecting and interacting stresses, for example, flooding, drought, salinization or soil nutrient deficiency. Once certain thresholds are crossed, or when the agroecosystem is 'flipped', a reduction in system resilience would occur, resulting in a decrease in farm productivity.

For example, there is clear evidence that increased soil microbial diversity, quality and quantity can increase farm resilience to drought. Increasing agricultural biodiversity therefore leads to increasing long-term productivity, resulting in increased farm resilience. However, these functional relationships are dependent on the agroecosystem context, level of biodiversity and management techniques (see TEEB 2018 for further references and work on valuation).

3 Role of agricultural diversity in lowering risk

A number of studies have established a relationship between the abundance of biodiversity and the lower risk of crop failure. In particular, there are numerous studies showing that high crop diversity reduces the variance of the yield under varying conditions while radically affecting the mean of the yield. This has been studied both as intraspecific diversity, in particular in cereal production, and in functional farm biodiversity, particularly in modern farms with poor biodiversity (Omer et al., 2007). This relationship seems to apply particularly in more challenging environments, such as more challenging climatic conditions in the developing world, or biodiversity-poor modern agricultural landscapes (see e.g. Di Falco (2012) for a more mathematical approach with a summary of empirical conclusions.

4 Taking land out of production

The relationship between biodiversity conservation and taking land out of production refers to the practice of setting aside a percentage of productive agricultural land in order to foster and promote biodiversity. This may be done in a permanent form, or by rotation, where the areas of the land that are devoted to biodiversity shift periodically, on a percentage of the land at a time, and over periods of time consistent with supporting diversity and with the required rest period needed by the soil. The latter is often one of the most important reasons behind the practice of rotation. Agricultural land that is devoted to biodiversity is taken out of production, which is then typically perceived as a cost of missing revenues. The owner perceives that they incur a pure cost without direct benefit and, for this reason, this is conventionally evaluated economically as a pure cost, that is a lack of productivity from those areas of agricultural landscape that are devoted to biodiversity conservation and cannot be used for direct production.

Additionally, if the owners have to directly manage the land for maximising biodiversity, there may be conversion costs. There are several caveats to this reasoning, in particular, in relation to the fact that productivity is estimated as if the same level of productivity was possible with and without biodiversity.

What are the drawbacks of using the ecosystem services approach in one of its many incarnations to evaluate the economic (and financial) value of agrobiodiversity? In the following discussion we outline four potential weaknesses that should be taken into account when producing any economic evaluation of agrobiodiversity.

The first weakness is that different approaches tend to set system boundaries of varying extensions and coverage, and decisions on what should be included or excluded from a calculation may radically impact the final value. A primary source of concern is the fact that even TEEB, an approach that is in all intentions systemic, puts a rather narrow system boundary around agrobiodiversity as can be seen by the definition of agrobiodiversity given by the initiative (see Definition III in Section 2). Another is that choosing an approach that evaluates only final direct services, while avoiding excessive complexity, does incur the risk of not evaluating correctly the underlying services that are the essential foundation of the final services.

Second, the complex causal and relational connections between the benefits provided by ecosystem services and biodiversity, and in particular agrobiodiversity, are not yet fully understood, particularly for those taxonomic groups that are yet not fully explored, and whose relationship with other taxonomic groups is still to be understood, which makes any overly simplified attempt at economic evaluation very problematic. This is a very important issue because, among those poorly-understood, underexplored taxonomic groups, there are several that are suggested to be of importance for the maintenance of the agri-food supply chain, such as some taxa of insects and of mycorrhizas. Any evaluation should then be used with extreme caution, understood as a provisional qualitative evaluation, and the principle of precaution should always be applied.

Moreover, the role of the abiotic environment in supporting ecosystem services remains unclear, and it is very often wrongly excluded by calculations. We have to remember that even the CICES and ESF did not integrate the abiotic environment until version 5.1, and that even the previous versions, up to version 4.3 that officially preceded the very last release, contained some of the abiotic services in a provisional state.

Fourth, but not least, many of the accounting methodologies based on the ESF appear to capture complexity, but in reality, do not do so, or do so only partially. In particular, they do not take into account five very important aspects of a complex system:

1 transversal interactions between these services,
2 feedback cycles within the ecosystem, within the sphere of the human client and between the two realms,
3 emerging properties, such as pseudo-homeostasis, that are essential to the maintenance of the system as a whole,
4 multi-stakeholder perspectives, conflicts of interests and trade-offs, and
5 tipping points.

While the most recent publications from the TEEB initiative have gone a long way in attempting to include some of these issues, at least conceptually, into consideration of evaluators (e.g. some of the feedback cycles), much is still to be done at a practical level. Discussing this in detail is outside the scope of the chapter. To illustrate our point and give the reader an overview of possible methodological approaches, in Section 5 we look specifically at one of the issues highlighted above: ecosystem interactions.

5 Ecosystem interactions

One of the drawbacks we have highlighted is that the goods and benefits delivered by the ecosystem services in a particular habitat depend heavily on interactions with the adjacent ecosystems and habitats. The current evaluation approaches, when applied tout court (i.e. as they are, without modifying or adapting them) and decontextualized from the actual agricultural landscape and layout of the land, do not take into account the fact that agrobiodiverse agricultural ecosystems behave very differently if interacting with different surrounding ecosystems. An area of farmland devoted to growing cereal crops in a highly agrobiodiverse farm would, for example, behave very differently if the adjacent ecosystems and habitats were rivers and wetlands or forest and shrubland, because the interactions among the ecosystems would be very different. This is particularly relevant for agricultural landscapes, in particular, those characterised by both intensive and extensive farming systems that split the landscape into (relatively) homogenous habitats, such as large adjacent parcels of land for arable for crops, for vegetable growing, tree plantations and so on as well as for those situations where interactions between agricultural habitats and adjacent areas are critical for resilience. For example, different habitats in a landscape can affect each other in the delivery of ecosystem services because there may be trade-offs (negative interactions) and synergies (positive interactions).

This issue is of primary importance and it is generally poorly understood and poorly explored. The statement holds true for all the dimensions: legislation, policymaking, decision-making, conservation, agricultural prices and, last but not at all least, economics.

Recent work for the British government by one of our research groups (Adriana Ford-Thompson et al., 2015) was among the first to explore these interactions using a multidimensional perspective to give practical recommendations for policymaking and decision-making. In the study, we looked at the case study of the complex interactions of agricultural systems, with a specific focus on lowland mixed-use agriculture, urban and fresh water systems.

For the agriculture ecosystems, the study focussed on the production of crops, livestock and fish through aquaculture, while for freshwater ecosystems, the focus was on the specific case of river and wetlands. The case study locations, where the evaluation framework was piloted, were two large multi-habitat regions in the North of England.

We focussed our attention on a subset of all processes and interactions, and more specifically;

1 Category 1: Interactions of agriculture ecosystems with river-wetlands ecosystems:
   1.1. Interaction 1.1: delivery of crops, livestock and fish from aquaculture caused a negative impact upon wild species biodiversity, water quality regulation, water supply, disease and pest regulations and environmental settings in river wetland habitats through pollution.
   1.2. Interaction 1.2: river unmanaged flooding onto agriculture ecosystems caused a negative impact on production of crops, livestock and fish from aquaculture, and on soil quality regulation.
   1.3. Interaction 1.3: river managed flooding for the creation of washlands caused positive impact on wild species diversity, hazard regulation and environmental settings.
   1.4. Interaction 1.4: drainage for agriculture caused positive impact of the production of crops, livestock and fish from aquaculture; negative impact on climate regulation from wetlands by increasing the release of CO_2 and N_2O; positive impact on climate regulation by this reducing emissions of methane; negative impact on trees, vegetation and peat, soil quality and water quality because of peat degradation and of subsidence; positive impact on wild species diversity; positive impact on environmental settings.

2 Category 2: Interactions of urban ecosystems and habitats with river-wetlands ecosystems:
   2.1. Interaction 2.1: pollution from urban areas has negative impacts upon water quality regulation, wild species diversity, disease and pest regulation, water supply and environmental settings in freshwater habitats.

When one looks at the economic evaluation of these interactions, or, more specifically, at the contribution of these interactions to the overall evaluation of the contribution of wild biodiversity, on the one hand, and of agricultural biodiversity, on the other, complexity tends to increase to a point where managing it is not feasible within the constraints of conventional decision-making. We need to adopt an approach that captures part of the complexity, while allowing for some degree of simplification.

Furthermore, we need to include both those processes and interactions whose economic value can be directly evaluated in a conventional economic sense, and those that do not. Among the former are those processes that have or produce or increase benefits and market value, or, on the other hand, destroy, damage or decrease market value and increase costs. Among the latter are those processes that do not have a direct market value.

In the specific case of this research project, we decided to do so by limiting the scope to a series of subprocesses, and then categorising these approaches into broader categories.

1 In the case of Category 1, Interaction 1.2 and, to some extent, Interaction 1.3, the approaches available to economic evaluation can be broadly categorised as 'damage costs avoided'.
2 In the case of Category 1, Interaction 1.4, costs are normally estimated using initial investment, that is, direct drainage costs, and operating costs. They are highly dependent on the stakeholder perspective, for example, for the private landowner the opportunity costs and benefits are a priority, for the community the damage to conservation costs are more important. Opportunity costs and benefits in this context relate to the private cost of agricultural output obtained, or lost if an area is to be destined for conservation for public use, and the operational costs linked to any protective measures adopted. In the latter case, we often need to use non-market value approaches for the benefits, with all the caveats of the case, such as willingness-to-pay approaches, hedonic pricing and travel cost method. We also need to include benefits to the overall community, such as carbon sequestration capacity and recreational spending, and in a limited number of cases, private revenues from activities, such as recreational activities, sports activities, hunting and so on.
3 In the case of both Category 1, Interaction 1.1, and Category 2, Interaction 2.1, the costs and benefits with direct market value can be estimated using replacement costs, substitute values, loss of opportunity costs or loss of earnings, loss of value (e.g. in the case of fisheries), avoided abatement costs, foregone (lost) farm income in relation to establishment of buffer zones and treatment of clean-up costs. In the

case of no market value, these can be estimated using willingness-to-pay and/or willingness-to-accept approaches. These can be based on discrete choice experiments, contingent evaluation and revealed preference methods.

These conclusions, and in particular the methodological framework used, can be easily extrapolated and/or upscaled to cover the cases of other ecosystems and habitats, other landscape scales and other interactions. By doing so, we add a further layer of information to the simpler direct approach offered by ecosystem services applied as such. The outcomes of taking account of interactions in economic assessments may change substantially the way we evaluate the costs and benefits of agricultural biodiversity,

6 Understanding the limitations of ecosystem service valuations

Ecosystem services are not directly traded in any market. Therefore, in undertaking any valuation process for agriculture biodiversity we need to acknowledge that there are some things that are known, but uncertain, others we know we do not know (known unknowns), and others that are unknown unknowns. How we manage the knowns and unknowns is key to determining how robust and transparent agrobiodiversity valuations will be.

As we have seen, there are some things we think we know, for example, how much production a farm typically achieves, what ratio of water, seed, fertilizer and so on is required to make a profit, or what amount of farmland could can be set aside to promote biodiversity (e.g. pollinating species) that would enhance agricultural production. Yet, within these knowns there are unknown factors, which we may label as 'uncertain knowns'. Examples include: How much rain will be delivered this year? What will be the fuel or fertiliser price? Will there be a pest or disease outbreak or locust plague? These are all the variables that we can understand, often do not like, and often we have enough knowledge to make educated guesses or produce (risk) assessments of the likelihood and consequence of their occurrence. Essentially we have two components that are part of this uncertainty. The first is due to the fact we do not know whether these events will happen or not. The second is due to the conceptual or mental models we use to estimate their effects. We can indeed strive to make such guesses, estimates and risk assessments as accurate as possible using a variety of models, yet they will always include a high level of intrinsic uncertainty which goes beyond the fact that any model we use is intrinsically limited. Often, the uncertainty will be larger than the values we can forecast, and the further we try to project into the future, the larger the uncertainty will be. In addition, the uncertainty in the forecast in a situation of high volatility is extremely high and the statistical noise may become higher than the signal. We are living in a situation of extreme volatility, because of the rapidly changing environmental conditions, due to climate change and environmental deterioration, and because of the many side effects such as financial instability, mass migration and pandemics such as COVID-19. Thus, estimating the level of uncertainty, that is very challenging at the best of times and many times may even be impossible, is exceedingly problematic in the current situation.

A second important source of uncertainty derives from the intrinsic inability of our simplified models to match the true complexity of relationships in the ecosystems and from the fact that any model that evaluates the economic value of biodiversity suffers from the well-known 'garbage in garbage out' issues. The fact that we still do not know the dynamics underlying the relationship between ecosystem functioning and biodiversity sufficiently well and the fact we do not have a sufficient amount of data on a sufficient number of cases contribute to the 'garbage in' part of the equation.

There are then the known unknowns. While an exhaustive discussion of all the known unknowns would be out of the scope of the paper, there is one type of known unknowns that is of particular relevance – those values that we know cannot be known in market terms. In particular, to supplement the discussion, we focus on two types of agricultural biodiversity values not derived from the market and not included at all in current ecosystem service valuations. Protected values and non-use values are known to be excluded from almost all valuations.

Protected values are those values based on alternate value systems that do not fit into the utilitarian market approach. Market values are utilitarian, which is one of the primary dangers of applying economic abstraction to living beings: they can be traded-off against each other. We see this in how production functions normally include agricultural biodiversity as one of many equivalent inputs to productivity. In the utilitarian approach, in essence we could decrease biodiversity if we increased another factor, say fertilizer, to achieve the same productive outputs. What happens then if you have fundamentally 'protected' values, those that cannot be traded off (Morris et al., 2020; Pearson et al., 2012)? These protected values are like moral obligations, religious beliefs, love or rules of law, or they may even derive from the application of the precautionary principle or internationally agreed frameworks of inalienable rights, such as the International Bill of Human Rights or the Sustainable Development Goals (SDGs). These values cannot be marginally increased or decreased for trading-off against other outputs. For example, protected values such as clean air and access to clean water for all, or protecting endangered species, may be seen as fundamental moral obligations linked to inalienable rights.

Non-use values are the second type of known value not included in ecosystem valuations generally, but fit within the utilitarian perspective. As part of the total economic value concept, these non-use values encompass the benefits people receive from agricultural biodiversity that are not purely economic. For example, non-use values cover aesthetic value, psychological values of well-being, sense of place and community or values of bequest and benevolence from knowing its existence (see Pearson et al. (2010) for a list of values not derived from the market). In addition, these values may extend beyond individuals who might benefit directly from agricultural biodiversity, for example, people in urban areas, who derive value from the knowledge of agricultural biodiversity in action.

The final group of limitations for ecosystem valuations is 'unknown unknowns'. These include our understanding of both human and ecosystem thresholds. As an example, extreme events such as pandemics may result in significant changes in behaviour and consumer choices, altering the amount and type of value that is placed on agricultural production and ecosystem services. Similarly, work has identified that current global systems may be approaching a threshold, where a return to current functioning is not linear but either asymptotic, exponential or not possible. Such thresholds have been identified both at the planetary scale and at more local agroecosystem scales. The best approach to deal with unknown factors in ecosystem valuation is to clarify the assumptions under which the valuation occurs. Attempting then to undertake a sensitivity analysis to determine how far these unknown changes might affect your valuation may offer some support to decision-making processes. Yet, in the case of systems affected by high volatility, sensitivity analysis requires pushing the boundaries of the envelope of the analysis to extreme cases, where the system may be unable to function: as such they may yield paradoxical results[3].

7 The investor perspective: the natural environment as a legally structured persona

What is next then for biodiversity evaluation, and more specifically, for agrobiodiversity evaluation?

The still incomplete understanding of the complex systemic issues we have outlined in this chapter makes any evaluation of the costs and benefits of biodiversity in general, and of agrobiodiversity in particular, highly approximate, imprecise and affected by uncertainty. It is likely that the true value of the

3 A typical example of a paradoxical result that sensitivity analysis may yield is that of greenhouse gas emissions from human activities decreasing to zero (a 'good' outcome) because everybody is dead and the human species is extinct.

benefits provided by biodiversity and agrobiodiversity is much greater than what we currently estimate.

Within this context, it is important to note that any value that is attributed to biodiversity and agrobiodiversity is negotiated; that is, it is the outcome of a social, political and economic negotiation between different parties such as policymakers, decision-makers, investors, farmers, local communities, non-governmental organisations, practitioners and experts (including economists) representing competing interests.

One option is to give the natural environment a voice and a legally structured persona.

While there is a strong ethical argument for doing so, this would be outside the scope of the present chapter. Thus, in this section we focus on how this could represent a historical shift in the way we value biodiversity and agrobiodiversity towards an approach which would bring us closer to understanding its real value, embedding also many of the elements that are currently missing, including the social considerations.

Giving the natural environment a voice and a legally structured persona would open up the possibility of the natural environment being represented in negotiations through legally appointed representatives. The general concept of this type of representation is well-embedded in legal frameworks across many cultures and economic systems.

Private enterprises and corporations are a perfect case in point. Corporations are abstract entities that do not exist as physical persons, yet they assume an artificial persona, which allows them to be enshrined into law, given the rights and legal obligations under provisions of the law. They are given the opportunity to defend and enforce their legal rights, to execute legally binding economic and financial transactions and to be represented in negotiations by legally appointed representatives.

In this case, this model can be developed further by giving the natural environment fundamental, that is, inalienable, rights, such as the right to exist, or the right to have its fundamental needs met, rights that should be enshrined into a Natural Environment Bill of Rights similar to the framework of the International Bill of Universal Rights[4].

These rights would include the right to be represented legally, to execute legally binding economic and financial transactions through representatives and act as investors with the same rights we give other investors, not least the right to have a voice in how its assets are valued, the right to withdraw

4 The international bill of universal rights (and duties) started with the approval of the Universal Declaration of Human Rights (UDHR) in Paris on 10 December 1948 with the UN General Assembly resolution 217 A, and was completed by the International Covenant on Civil and Political Rights, and the International Covenant on Economic, Social and Cultural Rights which entered into force in 1976.

representatives from unsatisfactory negotiations and the right to the best possible return on investment, to name but a critical few.

This would open up the possibility of the natural environment being fully and independently represented on company boards with the remit to decide on how maximisation of its assets would work, which would include agricultural biodiversity. Not only giving the natural environment a legal persona whose rights are enshrined into a Bill of Rights has two important consequences. First, the natural environment person would be able to represent its stakeholder interests in any negotiation about the value of biodiversity and agrobiodiversity, which, in the case of an economic endeavour would be shareholder interests. Second, it would have the right to be protected from all hostile acts[5].

Such a Bill of Rights would also enable us to bring into the equation the achievement of climate-change targets and SDGs, in particular how fast we achieve them, and what priority they have in the economic agenda.

It must indeed be understood that the natural environment is an investor in all human economic activities, and that some of the assets brought by this investor are essential, such as oxygen for breathing. Thus, these economic activities would not be possible without this investor, who becomes the holder of a 'golden share', which is effectively a controlling share.

Within the context of this change in approach, it is now necessary to develop a social conversation on such Bill of Rights, on representativity and not least on enforceability, that is, on what it means to enforce the rights of this legal persona.

8 Conclusions

It must be understood, first of all, that the value of agricultural biodiversity has so far only been connected to the value of the final goods or services, hence biodiversity has not been given a value per se. Secondly, system boundary matters: the tendency has been, so far, to only value what we can link in causal relationships, that is, what we can clearly understand and map, where we can clearly see cause and effects, and a connection between ecosystem function and ecosystem services and goods is provided. What we understand is only a small part of a very complex whole (in the sense of complex systems), and the true value is likely to be much greater than what we estimate with current approaches. Within this premises, we argue that the following considerations should apply.

First, valuing biodiversity, whether in general or in agriculture, in a pure economic sense is dangerous and should be avoided, as it is the case for all living and life-sustaining systems. Ethical considerations and, as a

5 The authors are indebted to the late Polly Higgins for this fundamental insight.

consequence, social and political decision-making and policymaking should assert the priority of a robust environment for all independently from economic evaluation. This in turn requires a massive shift in our collective mindset and prioritisation system, which has been strongly resisted so far. Yet, under the combined pressure of climate change, global warming, the financial crisis and its geopolitical consequences, inequality and its social consequences at a global level, and of rapidly progressing environmental degradation, it may be the right time to reconfigure our Weltanschaung and mindsets. The idea that an absolute rational economic value per se exists has been proven to be irrational. We argue that some choices and some decisions, in particular those that are linked to the survival of the species as a whole, should be part of a Bill of Rights, beyond the reach of any economic calculation of costs and benefits.

Secondly, a systemic perspective should always be applied, both in the sense that agricultural biodiversity is part of a complex system in a constant flux of change and it should only be investigated through the lens of complex system theory. The idea that we can capture the value of a small fragment of the system abstracted from the rest is unrealistic and very dangerous. It relies on the mistaken assumption that ecosystems are static and function rigidly in a specific well-defined way that will always be the same. There is so far in mainstream economics an intrinsic inability to understand and interpret complexity, and some of the dynamics of the system, for example, non-linear effects, feedback effects, emerging properties and relations of dependency. This makes pure economic valuation, which relies on understanding the system dynamics completely and beyond the boundaries of the economics subsystem, very unreliable. Second and higher-order effects and benefits, indirect effects and benefits, feedback effects and its costs, should always be included, and intangible and immaterial effects should be included in the evaluation. Uncertainty should always be recognised, made explicit in all stages of the calculations, communicated to the stakeholders and its importance should be clearly stressed.

Thirdly, even in those situations where attempting to produce an economic value is mandatory, conventional reasoning in terms of discount rate is dangerous and should not be applied. In other words, the future should be considered to be of the same value as the present. The asteroid thought experiment[6], often used to justify the choice of rather high discount rates, seems

6 In summary, the asteroid thought experiment is often used to argue that we need to maintain a higher than zero discount rate because an asteroid may strike at any time in the future, which is a metaphor for an unknown species extinction level event of planetary magnitude which would affect our ability to live on the planet. It goes without saying that it would affect the ability to perform any economic activity (dead people don't trade). If that is the case, the argument goes, then the present is worth more than the future. In a system hovering around a tipping point, the passing of which may well cause the system to enter a metastable state unfriendly to human life, adopting behaviours that may cause the system to tip over is equivalent to 'inviting the asteroid'. Economic short termism, which essentially consists in considering today (in particular, today's gain) of much higher value than the future, invites and promotes behaviours that may tip the system, already stressed by climate change and environmental degradation, over into the new unfriendly state, hence inviting the metaphorical asteroid.

to be nonsensical in the context of effects on a system that is hovering around tipping points, because it would essentially mean inviting the metaphorical asteroid to hit. As such, ensuring that biodiversity in general, and agricultural biodiversity specifically, is sustained in the future should be a policymaking priority independently of present value in economic terms. This could be seen as an application of the precautionary principle, because at a time of high volatility in a broad sense, where we may be approaching a systemic tipping point at planetary scale, we do not know what could be the future value of key ecosystem functions.

9 References

Bateman, I. J., Harwood, A. R., Mace, G. M., Watson, R. T., Abson, D. J., Andrews, B., Binner, A., Crowe, A., Day, B. H., Dugdale, S., Fezzi, C., Foden, J., Hadley, D., Haines-Young, R., Hulme, M., Kontoleon, A., Lovett, A. A., Munday, P., Pascual, U., Paterson, J., Perino, G., Sen, A., Siriwardena, G., van Soest, D. and Termansen, M. 2013. Bringing ecosystem services into economic decision-making: land use in the United Kingdom. *Science* 341(6141), 45–50. doi:10.1126/science.1234379.

Castro e Silva, M. and Teixeira, A. A. C. 2011. A bibliometric account of the evolution of EE in the last two decades: is ecological economics (becoming) a post-normal science? *Ecological Economics* 70(5), 849–62. doi:10.1016/j.ecolecon.2010.11.016.

Costanza, R. 1991a. Ecological economics: a research agenda. *Structural Change and Economic Dynamics* 2(2), 335–57. doi:10.1016/S0954-349X(05)80007-4.

Costanza, R. 1991b. *Ecological Economics: The Science and Management of Sustainability*. Columbia University Press.

Dean, G., Clarke, F. and Capalbo, F. 2016. Pacioli's double entry – part of an intellectual and social movement. *Accounting History Review* 26(1), 5–24. doi:10.1080/21552851.2015.1129083.

Di Falco, S. 2012. On the value of agricultural biodiversity. *Annual Review of Resource Economics* 4(1), 207–23. doi:10.1146/annurev-resource-110811-114543.

Ecosystems, T.-T.E. of Biodiversity 2018. Measuring What Matters in Agriculture and Food Systems: A Synthesis of the Results and Recommendations of TEEB for Agriculture and Food's Scientific and Economic Foundations Report. UN Environment Geneva.

FAO 1999. Agricultural biodiversity, multifunctional character of agriculture and land conference. FAO, Maastricht, the Netherlands.

Fischer, M. J. 2000. Luca pacioli on business profits. *Journal of Business Ethics* 25(4), 299–312. doi:10.1023/A:1006281415873.

Ford-Thompson, A., Bell, S., Capelli, S., Naylor, B., Raffaelli, D., Ritson, C., Stokowska, C., Timmis, J., Topi, C. and White, P. 2015. Ecosystem interactions ecosystem service interactions among lowland mixed-use agricultural, river-wetland and urban systems (No. NR0151: Ecosystem Interactions). DEFRA, London, UK.

Haines-Young, R. and Potschin, M. 2010. *Proposal for a Common International Classification of Ecosystem Goods and Services (CICES) for Integrated Environmental and Economic Accounting*. European Environment Agency.

Haines-Young, R. and Potschin, M. 2011. Common international classification of ecosystem services (CICES): 2011 update. Report to the European Environmental Agency, Nottingham.

Haines-Young, R. and Potschin, M. 2013a. CICES V4.3 – Revised report prepared following consultation on CICES version 4, August-December 2012. EEA Framework Contract No. EEA IEA/09/003.

Haines-Young, R. and Potschin, M. 2013b. Common International Classification of Ecosystem Services (CICES): Consultation on Version 4, August-December 2012.

Haines-Young, R. and Potschin, M. 2018. Common International Classification of Ecosystem Services (CICES) V5.1 and Guidance on the Application of the Revised Structure, 53.

Ikerd, J. E. 1993a. The need for a system approach to sustainable agriculture. *Agriculture, Ecosystems and Environment* 46(1–4), 147–60. doi:10.1016/0167-8809(93)90020-P.

Ikerd, J. E. 1993b. The need for a systems approach to sustainable agriculture. In:. *Agriculture, Ecosystems and Environment*. Elsevier, pp. 147–60. doi:10.1016/0167-8809(93)90020-P.

Khangura, S., Konnyu, K., Cushman, R., Grimshaw, J. and Moher, D. 2012. Evidence summaries: the evolution of a rapid review approach. *Systematic Reviews* 1, 10. doi:10.1186/2046-4053-1-10.

Kirschenmann, F. 1991. Fundamental fallacies of building agricultural sustainability. *Journal of Soil and Water Conservation* 46, 165–8.

Kumar, P. 2010. *The Economics of Ecosystems and Biodiversity: Ecological and Economic Foundations*. UNEP/Earthprint.

Maes, J., Teller, A., Erhard, M., Grizzetti, B., Barredo, J. I., Paracchini, M. L., Condé, S., Somma, F., Orgiazzi, A. and Jones, A. 2018. *Mapping and Assessment of Ecosystems and Their Services: An Analytical Framework for Ecosystem Condition*. Publications Office of the European Union, Luxembourg, pp. 1–78.

Millennium Ecosystem Assessment 2005a. *Ecosystems and Human Well-Being*. The World Resources Institute, Washington, DC.

Millennium Ecosystem Assessment 2005b. *Ecosystems and Human Well-Being: Biodiversity Synthesis*. The World Resources Institute, Washington, DC.

Millennium Ecosystem Assessment 2005c. *Ecosystems and Human Well-Being: Opportunities and Challenges for Business and Industry*. The World Resources Institute, Washington, DC.

Millennium Ecosystem Assessment 2005d. *Ecosystems and Human Well-Being: Synthesis*. The World Resources Institute, Washington, DC.

Millennium Ecosystem Assessment (Program) (Ed.) 2005. *Ecosystems and Human Well-Being: Synthesis*. Island Press, Washington, DC.

Morris, J., Ensor, J. E., Pfeifer, C., Marchant, R., Mulatu, D. W., Soka, G., Ouédraogo-Koné, S., Wakeyo, M. B. and Topi, C. 2020. Games as boundary objects: charting trade-offs in sustainable livestock transformation. *International Journal of Agricultural Sustainability* 0, 1–24. doi:10.1080/14735903.2020.1738769.

Omer, A., Pascual, U. and Russell, N. P. 2007. Biodiversity conservation and productivity in intensive agricultural systems. *Journal of Agricultural Economics* 58(2), 308–29. doi:10.1111/j.1477-9552.2007.00091.x.

Pearson, C. J. 2007. Regenerative, semiclosed systems: a priority for twenty-first-century. *Agriculture BioScience* 57(5), 409–18. doi:10.1641/B570506.

Pearson, L. J., Kashima, Y. and Pearson, C. J. 2012. Clarifying protected and utilitarian values of critical capital. *Ecological Economics* 73, 206–10. doi:10.1016/j.ecolecon.2011.10.027.

Pearson, L. J., Pearson, L. and Pearson, C. J. 2010. Sustainable urban agriculture: stocktake and opportunities. *International Journal of Agricultural Sustainability* 8(1-2), 7-19. doi:10.3763/ijas.2009.0468.

Potschin, M. B. and Haines-Young, R. H. 2011. Ecosystem services: exploring a geographical perspective. *Progress in Physical Geography: Earth and Environment* 35(5), 575-94. doi:10.1177/0309133311423172.

Power, A. G. 2010. Ecosystem services and agriculture: tradeoffs and synergies. *Philosophical Transactions of the Royal Society of London. Series B, Biological Sciences* 365(1554), 2959-71. doi:10.1098/rstb.2010.0143.

Qualset, C. O., McGuire, P. E. and Warburton, M. L. 1995. *"Agrobiodiversity" Key to Agricultural Productivity*.

Sangster, A. 2018. Pacioli's lens: god, humanism, euclid, and the rhetoric of double entry. *Accounting Review* 93(2), 299-314. doi:10.2308/accr-51850.

Schifino-Wittmann 2000. Agrobiodiversity: characterization, utilization and management. *Plant Pathology* 49(1), 167. doi:10.1046/j.1365-3059.2000.0423f.x.

Seppelt, R., Beckmann, M., Ceauşu, S., Cord, A. F., Gerstner, K., Gurevitch, J., Kambach, S., Klotz, S., Mendenhall, C., Phillips, H. R. P., Powell, K., Verburg, P. H., Verhagen, W., Winter, M. and Newbold, T. 2016. Harmonizing biodiversity conservation and productivity in the context of increasing demands on landscapes. *Bioscience* 66(10), 890-6. doi:10.1093/biosci/biw004.

Smale, M., Már, I. and Jarvis, D. I. (Eds). 2002. The economics of conserving agricultural biodiversity on-farm. Proceedings of the a Workshop. Godollo, Hungary.

Sutton, A., Clowes, M., Preston, L. and Booth, A. 2019. Meeting the review family: exploring review types and associated information retrieval requirements. *Health Information and Libraries Journal* 36(3), 202-22. doi:10.1111/hir.12276.

TEEB. 2008. *The Economics of Ecosystems Biodiversity: an Interim Report*. A Banson Production, Cambridge, UK. Ed. European Communities, Germany.

TEEB. 2009. *TEEB Climate Issues Update*. United Nations Environment Program, Nairobi.

TEEB. 2009. *The Economics of Ecosystems and Biodiversity for National and International Policy Makers–Summary: Responding to the Value of Nature 2009*. Welzel and Hardt, Wesseling, Germany.

TEEB. 2012. *The Economics of Ecosystems and Biodiversity in Business and Enterprise*. Earthscan Publications, UK and New York, NY.

TEEB. 2015. *TEEB for Agriculture & Food: An Interim Report*. The Economics of Ecosystems and Biodiversity and United Nations Environment.

TEEB. 2018. *TEEB for Agriculture & Food: Scientific and Economic Foundations*. UN Environment, Geneva, Switzerland.

TEEB - The Initiative [WWW Document] n.d. *TEEB*. Available at: http://www.teebweb.org /about/the-initiative/ (Accessed on 21 April 2020).

Wittmer, H. and Gundimeda, H. 2012. *The Economics of Ecosystems and Biodiversity in Local and Regional Policy and Management*. Routledge.

Chapter 4

Functional biodiversity for the provision of agroecosystem services

Paolo Bàrberi and Anna-Camilla Moonen, Institute of Life Sciences – Scuola Superiore Sant'Anna, Italy

1 Introduction

Agriculture is the main land-use type in many parts of the world; in Europe, it covers approximately 50% of the global surface (European Commission, 2015). As such, the impact of agricultural activities on the environment can be profound. In fact, agriculture is often referred to as one of the major contributors to environmental degradation (Matson et al., 1997), greenhouse gas emissions (Garnett et al., 2013) and biodiversity loss (Henle et al., 2008). However, agriculture itself can suffer from environmental degradation caused by point and non-point pollution sources like industry, traffic and urbanization (Agrawal et al., 2003). Many of the accusations done to agriculture actually refer to the detrimental effects of large-scale industrial agriculture on the environment and biodiversity, which are undeniable (IPES-Food, 2016). The other side of the coin is that agricultural systems based on sound sustainability principles (e.g. organic farming and agroecology) have large potential to maintain and even increase habitat quality and biodiversity conservation (Hole et al., 2005). A fair discussion on the relationship between agriculture and biodiversity should therefore consider both sides of the problem.

Much of the interest on the side effects of agriculture on biodiversity, at least in Europe, stems from data that showed a 30% decline in the population of farmland birds occurred in the 1980–2002 period versus a relatively steady population of woodland birds (European Environment Agency, 2004). Since

http://dx.doi.org/10.19103/AS.2020.0071.12

then, birds are considered an iconic indicator of the state of biodiversity in agricultural systems. In the EU, interest on the potential detrimental effects of agriculture on environment and biodiversity has triggered policy interventions aimed to subsidise agricultural practices conserving species and habitats in the agricultural mosaic. One major example of these policies are the so-called agri-environmental schemes developed as part of the EU's Common Agricultural Policy (CAP). However, these schemes have met the goal of fostering species conservation in agroecosystems only partly, for example, in 54% of the 62 cases examined by Kleijn and Sutherland (2003). Overall, the effect of agri-environmental schemes on species and habitat conservation seems to be highly context- and taxon-dependent and has not been substantially improved by subsequent EU policy reviews (Batáry et al., 2015).

More recently, the rise of organic farming and agroecology has shed light on a different perspective on agriculture-biodiversity relationships: judicious use of biodiversity components (often referred to as 'functional biodiversity') can provide several agroecosystem services, directly or indirectly related to agricultural production, while preserving the environment (Duru et al., 2015). There are large avenues for research to understand the mechanisms driving functional biodiversity across different contexts and scales, yet, this seems to be a very promising approach to meet global sustainability goals, including, *inter alia*, food security and sovereignty, climate mitigation and adaptation, and clean water supply (Moonen and Bàrberi, 2008).

The greatest challenge for the agriculture of the future will be to meet increasing food demands and dietary requirements of a rising global population without impinging on the environment. Unravelling the links between biodiversity and the provision of agroecosystem services (including yield and produce quality) could give a substantial contribution in this direction and help design truly sustainable farming and food systems. However, this would require a paradigm shift in the entire food production and supply chain. On the one hand, consumers may need to reconsider the elements of a healthy and sustainable diet and move from quantity to quality and from animal protein- to plant protein-based diets (Muller et al., 2017). On the other hand, producers may need to change the way they evaluate their products shifting from a mere yield viewpoint to sustainable farming systems that may produce less yield, but that produce more agroecosystem services such as soil fertility, biological pest regulation, clean water and attractive landscapes all over the world.

2 The rise of studies on agriculture: biodiversity relationships

Undoubtedly, there is a growing interest on research addressing the interplay between agriculture and biodiversity. To investigate this trend in detail, we conducted a scientific literature search using the Elsevier Scopus® database. We used three combinations of search terms: (i) agricultur* AND *biodiversity, (ii) agricultur* AND "*ecosystem service*" and (iii) agricultur* AND "functional *biodiversity". We did not set any time limits to the search, and we also searched the general terms "agricultur*" and "*biodiversity" for comparison purposes. The results are shown in Figs. 1 and 2 and in Table 1.

No single record for any of the three search combinations is present in the database until 1990. It should be pointed out that only part of the literature produced before the 1990s has been inserted in digital databases and therefore there may be a slight bias in the year when these keywords first appeared in the literature. Data in Fig. 1 show three clear trends:

- The rise of records for the agricultur* AND *biodiversity combination (the one mostly represented) follows shortly the establishment of the United Nation's Convention on Biological Diversity (CBD) at the Rio de Janeiro Earth Summit during 3–14 June 1992 (United Nations, 1992).
- The rise of records for the agricultur* AND "*ecosystem service*" combination follows shortly the end of the Millennium Ecosystem Assessment (MEA) in 2005 (Millennium Ecosystem Assessment, 2005).

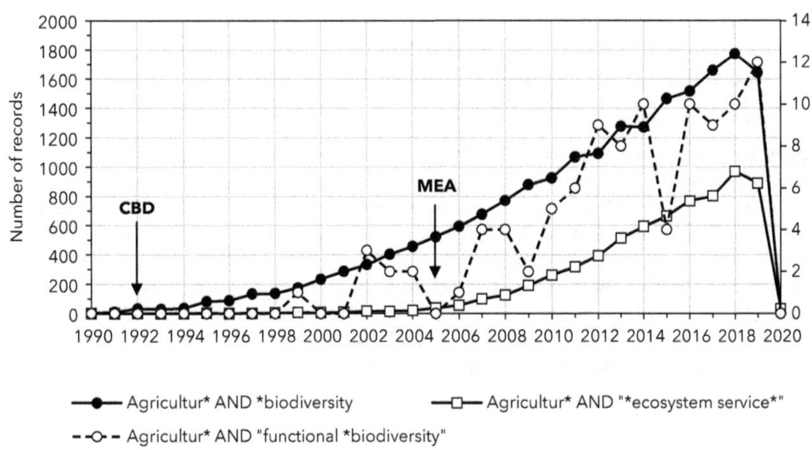

Figure 1 Number of records retrieved for each of the three term combinations used in the literature search (see details in text). Values for the Agricultur* AND "functional *biodiversity" combination are read on the right Y axis, while those for the other two combinations are read on the left Y axis. Arrows indicate the year when the two milestone events that boosted agricultural biodiversity studies took place (United Nation's Convention on Biological Diversity - CBD: established in 1992; Millennium Ecosystem Assessment - MEA: finished in 2005). Source: Scopus© database (accessed 20 November 2019), personal elaboration (search period: 1990–to date).

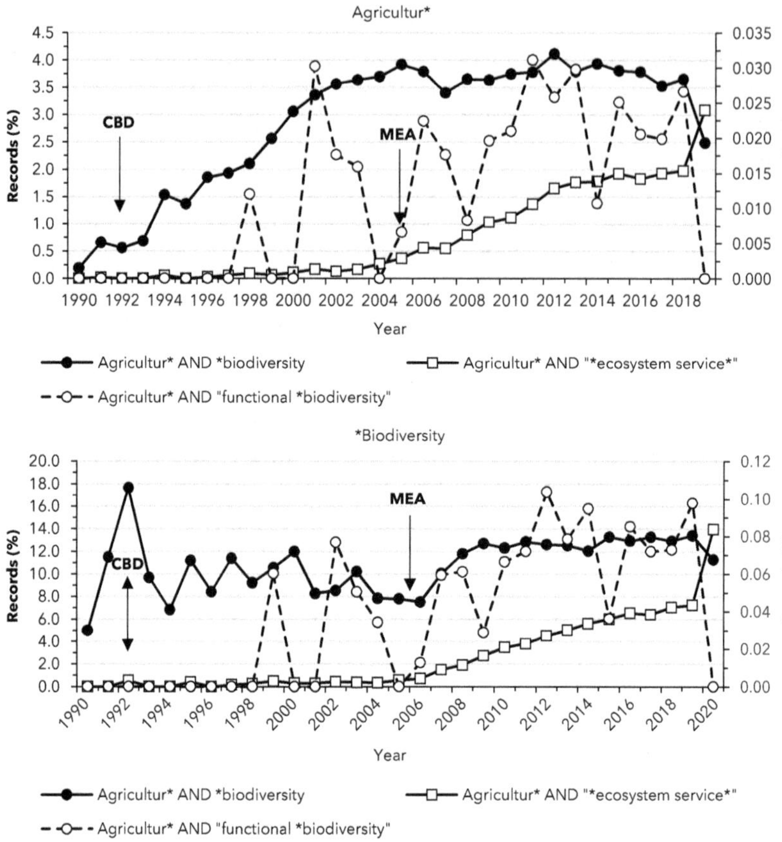

Figure 2 Percentage of records retrieved for each of the three term combinations used in the literature search (see details in text) as relative to the total number of records retrieved for the Agricultur* (top) and the *Biodiversity (bottom) terms. Values for the Agricultur* AND "functional *biodiversity" combination are read on the right Y axis, while those for the other two combinations are read on the left Y axis. Arrows indicate the year when the two milestone events that boosted agricultural biodiversity studies took place (United Nation's Convention on Biological Diversity - CBD: established in 1992; Millennium Ecosystem Assessment - MEA: finished in 2005). Source: Scopus© database (accessed 20 November 2019), personal elaboration (search period: 1990-to date).

- The records dealing with the combination of agricultur* AND "functional *biodiversity", although showing an increasing trend, are so far a minority, indicating that the term 'functional biodiversity' is yet to become of widespread use in agricultural research.

To purge these trends from the effect of the increasing number of overall scientific publications, the number of records retrieved for each search-term

Table 1 Number of records retrieved for each of the three term combinations used in the literature search (see details in text), broken down by subject area, publication source and country

Category	Term combination		
Subject area	*Agricultur* AND *biodiversity*	*Agricultur* AND "*ecosystem service*"*	*Agricultur* AND "functional *biodiversity"*
Agricultural and Biological Sciences	11 192 (#1)	3735 (#2)	75 (#1)
Environmental Science	10 152 (#2)	4446 (#1)	56 (#2)
Social Sciences	2507 (#3)	1211 (#3)	14 (#3)
Biochemistry, Genetics and Molecular Biology	2,137 (#4)	407 (#5)	4 (#4)
Earth and Planetary Sciences	1329 (#5)	547 (#4)	3 (#5)
Immunology and Microbiology	957 (#6)	116 (#12)	2 (#8)
Multidisciplinary	882 (#7)	203 (#9)	0 (-)
Engineering	826 (#8)	342 (#7)	3 (#5)
Medicine	728 (#9)	119 (#11)	1 (#10)
Energy	638 (#10)	373 (#6)	3 (#5)
Economics, Econometrics and Finance	480 (#11)	333 (#8)	2 (#8)
Decision Sciences	243 (#16)	146 (#10)	1 (#10)
Nursing	81 (#24)	5 (#21)	1 (#10)
Others	2194	395	0
TOTAL	19 607	6804	102
Publication source			
Agriculture Ecosystems and Environment	625 (#1)	280 (#1)	10 (#1)
PLoS ONE	441 (#2)	134 (#3)	0 (-)
Biological Conservation	372 (#3)	81 (#13)	2 (#6)
Biodiversity and Conservation	297 (#4)	39 (#25)	0 (-)
Journal of Applied Ecology	283 (#5)	134 (#3)	6 (#2)
Science of the Total Environment	245 (#6)	181 (#2)	3 (#4)
Science	223 (#7)	23 (#50)	0 (-)
Ecological Indicators	209 (#8)	134 (#3)	1 (#14)
Forest Ecology and Management	190 (#9)	35 (#30)	0 (-)
Conservation Biology	172 (#10)	25 (#47)	0 (-)
Land Use Policy	172 (#10)	123 (#7)	1 (#14)
Sustainability (Switzerland)	118 (#20)	134 (#3)	1 (#14)
Ecosystem Services	29 (#99)	114 (#8)	0 (-)
Shengtai Xuebao Acta Ecologica Sinica	134 (#19)	109 (#9)	1 (#14)
Journal of Environmental Management	156 (#14)	107 (#10)	0 (-)
Acta Horticulturae	135 (#18)	20 (#58)	6 (#2)

(*Continued*)

Table 1 (*Continued*)

Category	Term combination		
Publication source	*Agricultur* AND *biodiversity*	*Agricultur* AND "*ecosystem service*"*	*Agricultur* AND "functional *biodiversity"*
Landscape Ecology	138 (#17)	54 (#19)	3 (#4)
Applied Soil Ecology	71 (#32)	28 (#38)	2 (#6)
Ecology and Society	66 (#36)	61 (#15)	2 (#6)
Biomass and Bioenergy	44 (#64)	22 (#52)	2 (#6)
Weed Research	32 (#91)	9 (#97)	2 (#6)
European Journal of Agronomy	<20 (#>159)	<8 (#>102)	2 (#6)
Journal für Kulturpflanzen	<20 (#>159)	<8 (#>102)	2 (#6)
Netherlands Journal of Agricultural Sciences	<20 (#>159)	<8 (#>102)	2 (#6)
Country			
United States	4130 (#1)	2090 (#1)	15 (#4)
United Kingdom	2411 (#2)	994 (#2)	16 (#3)
Germany	1953 (#3)	837 (#4)	15 (#4)
France	1473 (#4)	554 (#5)	15 (#4)
Australia	1422 (#5)	510 (#6)	8 (#8)
China	1326 (#6)	846 (#3)	4 (#14)
Italy	1149 (#7)	427 (#7)	17 (#1)
Brazil	1031 (#8)	308 (#12)	5 (#11)
India	1001 (#9)	169 (#16)	1 (#25)
Canada	933 (#10)	335 (#10)	3 (#16)
The Netherlands	930 (#11)	399 (#8)	17 (#1)
Spain	855 (#12)	378 (#9)	8 (#8)
Switzerland	772 (#13)	285 (#13)	14 (#7)
Belgium	377 (#18)	153 (#20)	6 (#10)

The ten most common cases in each category for each term combination are shown (numbers in parentheses are ranks).
Source: Scopus© database (accessed 20 November 2019), personal elaboration (search period: 1990-to date). The sum of records by country does not correspond to the total number of records because some studies have been carried out in more than one country.

combination was standardised on the total number of records retrieved for the agricultur* or *biodiversity general categories (Fig. 2). The analysis reveals that:

- The percentage of records for the agricultur* AND *biodiversity combination increased until 2002 (as related to the agricultur* category) or 2009 (as related to the *biodiversity category, with greater year-to-year fluctuations), and then stabilised at approximately 3.5–4% and 12–14% of the total records, respectively.

- Records of the agricultur* AND "*ecosystem service*" combination showed the same increasing trend observed for the unstandardised data, and in 2018 represented 2% and 7% of the total records in the agricultur* and *biodiversity categories, respectively.
- Records of the agricultur* AND "functional *biodiversity" combination, although showing roughly the same increasing trend of the unstandardised data, still represent only <0.03% and <0.10% of the total records in the agricultur* and *biodiversity categories, respectively.

Table 1 shows the partitioning of records by subject area, publication source and country. The major outcomes of this analysis are:

- Agriculture-biodiversity interactions are clearly a cross-cutting research theme, since records are found in >13 of the Scopus® subject areas, although – unsurprisingly – most records belong to the 'Agricultural and Biological Sciences' and 'Environmental Science' areas for all three search-term combinations.
- This research theme is vastly popular among authors and editors, as shown by the large number of journals willing to publish related articles. Among them, there seems to be some specialization based on focus: for example, more applied or interdisciplinary journals seem to yield more papers focussing on ecosystem services. This is reflected by the relatively different journal ranking by number of records for the three search-term combinations except *Agriculture, Ecosystems & Environment*, which is the most popular journal in all cases.
- Although based on a limited total record number, the term 'functional biodiversity' seems to be already popular in countries like Italy and the Netherlands (where it ranks first), while the United States and the United Kingdom are the countries where the highest number of studies pertaining to the other two search-term combinations were carried out.

As previously pointed out, our literature search shows clearly that the dynamics of agriculture-biodiversity research has been strongly influenced by two distinct events: the United Nations' Convention on Biological Diversity (established in 1992) and the Millennium Ecosystem Assessment (ended in 2005). We will then illustrate this dynamics by partitioning it in three periods: pre-CBD, post-CBD (pre-MEA) and post-MEA. The criteria used to synthesise research trends and major outcomes in each period are detailed in the next three sub-sections.

2.1 The early days (1990-1992)

No paper on agriculture and biodiversity appeared until 1990. In the three years that led to the Earth Summit in Rio de Janeiro (1990–1992) only 45 papers were published, most of them in 1992.

A strong impetus to this field of research was given by raising awareness of the environmental damage and biodiversity loss caused by industrial agriculture and the associated management practices and business model at any scale from field to farm and landscape (Orians and Lack, 1992). In the same line of thought, other studies focussed on the negative effects of national governmental policies favouring industrial agriculture over biodiversity-friendly one, for example, in India (Gadgil, 1992), and on the spill-over effect of industrial agriculture practised in developed countries on biological diversity in developing countries (McNeely and Norgaard, 1992). The lock-in against sustainable agricultural management due to the dominant reductionist and technology-centred knowledge system was also pointed out (Norgaard, 1992).

The detrimental effects on biodiversity of large-scale agricultural land-use change due to development programmes promoting deforestation (de Miranda and Mattos, 1992; Johns, 1992) or biofuel production (Cook et al., 1991) were also highlighted. Genetic erosion of agrobiodiversity, that is, the loss of either neglected cultivated species (e.g. mountain native crops in Peru; Zimmerer, 1992) or varieties (e.g. potato germplasm in the Andes; Brush et al., 1992) triggered further attention towards the need of conserving biodiversity to grant sustainable agricultural development.

In those days, the importance of biodiversity conservation to provide agroecosystem services was still largely undervalued, although the first works highlighting the role of soil biota (e.g. microarthropods) biodiversity for the improvement of soil quality and nutrient cycling appeared (Crossley et al., 1992; Hendrix et al., 1992). Similarly, the importance of semi-natural habitats like field margins for the conservation of diversity in natural enemies complexes and in overall arthropod populations (Dennis and Fry, 1992), and of assuring social equity in the business of classical biological control agents (Altieri, 1991) put biological pest control in the spotlight as another important biodiversity-related agroecosystem service.

Although at that time agriculture was not considered among the most important domains of biodiversity research, these papers contributed to support the idea that an international convention for the conservation of biodiversity was urgently needed (Swanson, 1992). Biotechnology was advocated as a promising field to help biodiversity conservation (Mannion, 1992), probably more than it is perceived nowadays.

Unravelling the basic mechanisms regulating biological diversity was considered key to avoid potential conflicts between agriculture and biodiversity conservation (Huston, 1992). This is a goal of utmost importance that can only be achieved through cooperation between agriculturalists and conservationists, as it was already clear in the early 1990s (Gall and Staton, 1992).

2.2 First milestone: the United Nations' Convention on Biological Diversity (1993–2005)

The CBD is a standing convention launched in June 1992 after four years of work and negotiations promoted by the United Nation's Environmental Programme (UNEP), inspired by the growing universal commitment to sustainable development. The CBD has been ratified by 168 countries and entered into force on 29 December 1993. It represented a crucial step forward for global action on the conservation of biological diversity, the sustainable use of its components, and the fair and equitable sharing of benefits arising from the use of genetic resources. Details on the history, structure, work programmes, cross-cutting initiatives and achievements of the CBD can be found on the official website (www.cbd.int).

A total of 2917 documents reporting explicitly the terms Agricultur* AND *biodiversity were published in the 13 years comprised between the releases of the CBD and of the MEA. Among these, 837 reported the term Agricultur* as keyword. To capture the essence of research done in this period, we have focussed on the most cited papers, that is, those with >500 citations each (n = 60).

Many of the most influential papers highlighted the damage done by industrial agriculture on ecosystems and biodiversity, advocating for a drastic change in agricultural management systems worldwide. Expansion of agricultural land use and management intensification through increased fertiliser, pesticide, irrigation and energy use have altered biotic interactions and resource availability with serious environmental effects at different spatial scales (Matson et al., 1997; Tilman, 1999). In this respect, a strong emphasis was placed on the detrimental effects of high-input agriculture on water ecosystems, including eutrophication, point and nonpoint pollution of surface and running waters by inorganic and organic compounds (Carpenter et al., 1998; Malmqvist and Rundle, 2002), surface runoff, subsurface flow and sediment loss (Stoate et al., 2001), and estuarine ecosystem damage (Kennish, 2002).

In this period, several emblematic studies pointed out the decline in bird populations as a key indicator of biodiversity loss in agroecosystems. This is a direct and indirect consequence of agricultural intensification, due to high-external input use (Krebs et al., 1999; Chamberlain et al., 2000), habitat loss and fragmentation, paucity of feed resources (e.g. wild plant seeds and seedlings), increased livestock density and grazing intensity (Vickery et al., 2001) or a combination of these factors (Donald et al., 2001). Farmland birds have also been proposed as surrogate indicators for trends in other biodiversity elements in agroecosystems (Gregory et al., 2005). As already pointed out, these studies were instrumental in developing dedicated European agri-environmental schemes like set-aside, vegetated strips/margins and targeted stubble management (Marshall and Moonen, 2002; Science for Environment Policy, 2017).

Besides agricultural intensification, land-use change towards the expansion of large-scale agriculture was indicated as another important cause of loss of biodiversity and associated services (Foley et al., 2005). These changes often increase (agro)ecosystem vulnerability due to declining soil fertility and water availability, and the risk of forest fires, as it has occurred in Mediterranean and mountain areas (Schröter et al., 2005). Another consequence of land-use change driven by industrial agriculture has been the progressive abandonment of marginal agricultural land, with severe impacts on biodiversity, environmental degradation and the socio-economic viability of territories (MacDonald et al., 2000). In tropical regions, the encroachment of biodiversity hotspots in natural park areas was proposed as an effective way to protect biodiversity from intensive agriculture (Bruner et al., 2001; Klink and Machado, 2005). This proposal triggered the subsequent debate on the 'land sparing' versus 'land sharing' approach for concomitant food production and biodiversity conservation (Green et al., 2005; Perfecto and Vandermeer, 2010; Phalan et al., 2011; Tscharntke et al., 2012; Fischer et al., 2014). A less-known consequence of land-use change like deforestation, damming, and the substitution of forests with cropland and rangeland is the global increase in morbidity and mortality from emergent human parasitic diseases like malaria, Leishmaniosis, giardiasis, schistosomiasis, lymphatic filariasis, and dracunculiasis (guinea worm). This is due to the creation of new habitats that are more favourable for parasites and their host vectors (Patz et al., 2000; Sattenspiel, 2000) or to detrimental effects on water quality, like in the case of the cholera vibrio (Sattenspiel, 2000).

Part of the detrimental consequences of agricultural intensification and land-use change are related to habitat degradation, of which one of the most important facets is habitat fragmentation. For example, limited habitat connectivity reduces the number of insect species in agricultural landscapes, including substantial reduction of natural enemies' populations. Compared to non-isolated populations, the parasitism rate of isolated herbivore pests was 40–81% lower (Kruess and Tscharntke, 1994). Landscape heterogeneity was suggested as the key factor in restoring and sustaining biodiversity in temperate agricultural systems (Benton et al., 2003). However, in landscapes with higher heterogeneity biodiversity gains due to more environmentally benign management systems like organic farming are less likely to become evident, when compared to the conversion to biodiversity-friendly farming practices applied to more homogeneous agricultural landscapes (Bengtsson et al., 2005). Seen from another perspective, by enhancing local agroecosystem diversity, structurally complex landscapes may buffer the negative effect of highly intensive management. At landscape scale, biodiversity patterns and ecosystem services are largely driven by organisms with high-dispersal ability, which can recolonise habitats and access larger resource pools (Tscharntke et al., 2005). Therefore, there is a spill-over of biodiversity from semi-natural habitats to intensively managed cropped fields in landscapes with a higher proportion of semi-natural habitats. This process reduces the proportional positive effect that, for example, organic farming has on biodiversity conservation in heterogeneous landscapes as compared to homogenous landscapes that are dominated by arable land (Rundlöf and Smith, 2006).

Alongside the CBD, the speed up in the advancement of novel techniques for genetic, genomic and molecular analysis boosted the interest on the exploration and conservation of genetic and phenotypic diversity to meet agricultural sustainability goals. These techniques also represent quicker ways to fix traits conferring crops for better adaptation to abiotic stresses exacerbated by climate change (Tanksley and McCouch, 1997). However, these advancements also pose a risk to overlook the complexity underneath the phenotypic expression of some traits. For example, despite the complexity of salt tolerance inheritance in plants, many researchers claimed that it would be enough to transfer only one or few genes to obtain a salt-tolerant crop, increasing the risk of failure when lab-made solutions are transferred to the real world (Flowers, 2004).

The post-CBD and pre-MEA periods saw the publication of three landmark papers that highlighted the importance of biodiversity for sustainable production, and paved the road for subsequent functional biodiversity studies. In his 1999 iconic paper, Altieri pointed out the manifold ecological services played by biodiversity in agroecosystems, for example, food production, nutrient (re)cycling, climate and water regulation, pest/disease/weed suppression, providing examples of how to meet these goals through management (Altieri, 1999). Zhu et al. (2000) provided the first clear evidence of the effect of genetic heterogeneity (diversity) on large-scale disease suppression, using rice blast (*Magnaporthe grisea* (T.T. Hebert) M.E. Barras) as a case study. They showed that mixtures between disease-susceptible and disease-resistant rice varieties had 89% greater yield and 94% disease reduction than the same varieties grown in monoculture, requiring no fungicide sprays. The third landmark paper refers to a long-term grassland experiment where biomass productivity of swards, including a progressively higher number of species was measured across 7 years (Tilman et al., 2001). Plant diversity and niche complementarity among plant functional groups increased plant biomass substantially, which was 2.7-fold higher in 16-species plots than in monoculture plots.

In the pre-MEA period, research on biodiversity-driven agroecosystem services was still relatively uncommon although some papers, which then became influential, addressed the importance of biodiversity and low-intensity agricultural management for halting a decline in pollinator populations (Allen-Wardell et al., 1998; Kremen et al., 2002) and for maintaining or increasing soil fertility. In this latter case, the major foci were soil microbial diversity (Kennedy and Smith, 1995), microbial synergisms in the rhizosphere (Barea et al., 2005), arbuscular mycorrhizal fungi (Jeffries et al., 2003), and the importance of organic farming to preserve soil biodiversity and associated ecosystem services (Mäder et al., 2002). Less known yet ubiquitous organisms like microbial endophytes were suggested as an important source of future products, spanning from human and veterinary medicine (e.g. novel antibiotics, antimycotics, immunosuppressants, and anticancer compounds) to agriculture and industry (Strobel and Daisy, 2003).

The rising interest on agrobiodiversity worldwide drove the Conference of the Parties of the CBD to adopt a dedicated work programme on agrobiodiversity in 2000, at its fifth meeting (www.cbd.int/agro). This programme includes four elements (assessment, adaptive management, capacity building and mainstreaming) and three cross-cutting initiatives (on pollinators, soil biodiversity and biodiversity for food and nutrition). These initiatives reflected the above-mentioned major foci highlighted by research. The CBD agricultural biodiversity work programme is to be implemented upon an ecosystem approach, with due recognition of the cultural, socio-economic and environmental components. One year later, the Organisation for Economic Co-operation and Development (OECD) organised an Expert Meeting on Agri-Biodiversity Indicators, whose background paper illustrated the three key components of agrobiodiversity (genetic, species and habitat/ecosystem diversity), reprising and expanding the CBD definition (Parris, 2001). All these activities and outcomes fuelled the Millennium Ecosystem Assessment and contributed to the explosion of research on biodiversity and agroecosystem services, which is still ongoing.

2.3 Second milestone: the Millennium Ecosystem Assessment (2006–2019)

The Millennium Ecosystem Assessment was a huge research exercise intended to assess the consequences of ecosystem change for human well-being. From 2001 to 2005, the MEA engaged more than 1360 experts worldwide, whose efforts produced a series of reports with the state-of-the-art scientific appraisal of the state and trends in the world's ecosystems and associated services. On top of this, the MEA also provided the scientific basis for action aimed at conservation and sustainable use of ecosystems. One clear landmark brought forward by the MEA has been the classification of ecosystem services in four categories, that is, supporting, provisioning, regulating and cultural (Millennium Ecosystem Assessment, 2005), which have been used as common reference since then. Further details on the MEA and the whole set of reports can be found on the official website (www.millenniumassessment.org).

The research focus on agriculture and (agro)ecosystem services and the use of this latter term emerged strongly after the release of the MEA and shows a continuously increasing trend, totalling up to 6673 records in the period from 1 January 2006 to 20 November 2019. We considered the papers published in this period that were already cited >200 times (Agricultur* AND "*ecosystem service*" search-term combination, $n = 112$) as the most influential. The most recent of them were published in 2016.

This period has seen tremendous scientific advancements on this subject. For example, research provided better insights on the contribution of agricultural management practices on enduring biodiversity decline (Geiger et al., 2010). Considerable work has been done on strengthening the scientific base of biodiversity-ecosystem services interactions. While it is generally recognised that more diversified systems, for example, organic agriculture, are able to provide more ecosystem services (Kremen and Miles, 2012; Reganold and Wachter, 2016), it also became clear that the magnitude of this effect is strongly context- and service-dependent. For example, higher gains in biodiversity-related services are expected in low-fertility or degraded land than on fertile land, as, for example, shown for C and N cycling as mediated by plant functional composition (Fornara and Tilman, 2008). This can be explained in more general terms by the asymptotic relationship between the increase in (functional) biodiversity and the response in ecosystem service provisioning, as shown by Bengtsson (1998) and Tilman et al. (2001). If the system has low initial levels of biodiversity, a higher increase in the magnitude of the service provision through biodiversity-friendly management can be expected. It should be stressed that human-induced land-use changes can strongly affect the level of both planned and inadvertent food supply to animals (including pests), triggering cascading effects on the service to disservice provisioning ratio (Oro et al., 2013).

One key question for agrobiodiversity research is how much biodiversity is needed to obtain a significant amount of service(s). In general, the abundance of generalist species associated with a given service seems more important than species diversity to drive the extent of service provisioning. This has been demonstrated for pollinators but is thought to be a general ecological pattern applicable also to other services (Winfree et al., 2015). Quantification of ecosystem services in economic terms is important to convince policymakers to invest in targeted biodiversity conservation measures. For example, the overall total economic value of pollination has been estimated as 153 billion

€, corresponding to 9.5% of the total world food production value in 2005 (Gallai et al., 2009). Demonstrating that diversified cropping and farming systems provide ecosystem services while maintaining or increasing crop yield and profitability would be another trigger for the development of policies targeting agrobiodiversity. There is increasing evidence that this goal is feasible. Using data from a long-term experiment in the United States, Davis et al. (2012) showed that a 4-year maize-soybean-small grain crop-lucerne rotation maintained or increased grain yields and profit over a conventional 2-year maize-soybean rotation, while reducing the use of external inputs substantially. As a consequence, freshwater toxicity was reduced by two orders of magnitude.

So far, the three most popular agroecosystem services in this domain of science have been biological control of crop pests, pollination and soil quality. Regarding biological control, research done in the last 13 years highlighted the crucial role of landscape diversity to increase the potential provision of this service. Enhancement of natural enemies' populations can be successfully attained by a combination of fine-tuned farming practices and environmental manipulation aimed at providing habitat and resources to specific natural enemies of target pests: this is termed 'conservation biological control' (Meyling and Eilenberg, 2007). A complementary approach is to use habitat agrobiodiversity as to either repel ('push') or attract ('pull') crop pests, with the objective of preventing or limiting their contact with the crop, through pest avoidance or elimination (Cook et al., 2007). Habitat modification influences food web relationships, thereby modulating provision of the biological control and pollination services. In tropical environments, generalist parasitoids may change behaviour by becoming more specialised in response to habitat disturbance (Tylianakis et al., 2007). In the United States, reduction in landscape diversity caused by the expansion of biofuel crops reduced the number of aphid natural enemies in soybean crops, with an estimated 24% reduction of the biocontrol service; this corresponds to an economic loss of 58 million USD year^{-1} due to grain yield decline and higher pesticide use (Landis et al., 2008). Diversity in the agricultural landscape matrix can also help to fight back invading pests (Gardiner et al., 2009). A meta-analysis revealed a significant general effect of natural enemy richness on top-down control of herbivores across a variety of mesocosm and open-field studies in tropical and temperate agroecosystems, although such effect was not so evident in non-agricultural habitats (Letourneau et al., 2009). A subsequent meta-analysis confirmed the positive response of natural enemies to landscape complexity, but highlighted a difference in the response of generalist versus specialist enemies, with the former showing consistent positive responses across many spatial scales, and the latter having a stronger positive response to landscape complexity at smaller scales (Chaplin-Kramer et al., 2011).

The contribution of generalist versus specialist species to service (potential) provision has been an important focus of research also in pollination studies. Holzschuh et al. (2007) observed that generalist bee diversity in wheat fields across different regions of Germany was mainly influenced by farming system, being higher under organic management, with higher gains in homogeneous landscapes. However, conservation of more specialised bee species in semi-natural habitats would require more targeted measures, going beyond the general support of organic farming. In watermelon, native bees alone have been estimated to provide enough pollination in > 90% of the 23 farms studied, thereby potentially buffering the possible decline in honey bee colonies (Winfree et al., 2007). Several indicators of landscape heterogeneity have been investigated, aiming to draw general patterns linking habitat diversity with pollinator abundance and activity. One important indicator is the distance from natural or semi-natural habitats. Pollinator species richness and native visitation rate declined exponentially with distance from such habitats. The strongest decline was observed for visitation rate (50% less at 0.6 km distance versus 1.5 km for species richness), especially in tropical environments and for social bees (Ricketts et al., 2008). However, there was not a clear relationship between these trends and fruit or seed set, suggesting that other factors are important to turn a potential service into an actual service. These results are partly in contrast with those of a subsequent study, where distance from natural habitats affected only wild pollinator richness and visitation rate (34% and 27% decline at 1 km distance, respectively), with negative effects on fruit set (16% reduction), while honey bees (>25% of total crop visits in 21 out of 29 studies) were unaffected by isolation (Garibaldi et al., 2011). Research done on 39 different cropping systems indicated that landscape configuration (patch shape and connectivity, and habitat aggregation) was not as important as landscape composition (habitat quality) for wild bee populations (Kennedy et al., 2013). It was also noticed that management measures focussed on the provision of the pollination service might not favour rare wild pollinator species, which would need specific measures to ensure conservation (Kleijn et al., 2015). These findings show that, although landscape diversity is clearly important for pollination, there is still uncertainty on which are the most important drivers and where, and on how they interact to modulate the service expression.

It is widely recognised that habitat loss and fragmentation are the major threats to bee diversity, with stronger effects than invasive/alien species, emerging pathogens, pesticides and climate change (Brown and Paxton, 2009; Potts et al., 2010). Neonicotinoid insecticides are suspected of being an important cause of the bee colony collapse syndrome. Although imidacloprid residues in nectar under field conditions are unlikely to have lethal effects, they can reduce honey bees performance between 6% and 20% (Cresswell, 2011), thereby decreasing the magnitude of the pollination service. Severe reduction of pollinator abundance and activity is also expected by chronic exposure to pesticide combinations, a subject so far poorly studied. Field-level concentrations of neonicotinoid and pyrethroid insecticides impaired natural foraging behaviour (e.g. pollen collection efficiency) and increased mortality of bumblebee workers, causing strong reductions in colonies (Gill et al., 2012). It is becoming evident that conservation of pollinators and of the pollination service requires both targeted habitat management and low-input, diversified cropping system management.

Regarding soil quality, it is now crystal clear that soil biodiversity is of utmost importance for the provision of this service (Kibblewhite et al., 2008). Overall, soil biodiversity gives stability to abiotic stress and disturbance, although the mechanisms have not yet been fully unravelled. Biotic stress to crops (e.g. attack from soilborne pathogens) can be alleviated by soil microbial diversity, in combination with sound management (Brussaard et al., 2007). In addition, arbuscular-mycorrhizal fungi play a key role as 'agroecological engineers' for the provision of services such as nutrient (mainly N and P) and water-use efficiency in many agroecosystems (Gianinazzi et al., 2010). Soil fauna, for example earthworms, are also very important for a plethora of services, obtained through their positive effect on soil structure (Blouin et al., 2013). Lastly, it should not be forgotten that biodiversity-friendly management has huge potential to improve soil C cycling, with clear positive effects not only on crop growth, (Powlson et al., 2011) but also on climate change mitigation through sequestration (Lal et al., 2007).

Agrobiodiversity research is moving forward at a fast pace. After the MEA, the scientific community has become progressively aware of the importance of interdisciplinary research to tackle such a complex subject. One novel methodological approach that has been used in this domain is the analysis of food webs, that is, networks of feeding interactions, to investigate the structure of ecological communities, their stability and function, and the services they can provide. For example, Tylianakis et al. (2007) used a food-web approach to analyse 48 quantitative food webs based on the interactions among cavity-nesting bees, wasps and their parasitoids in five different tropical habitats. Improved indicator systems to highlight the links between agricultural practices, biodiversity and ecosystem services have been proposed (Dale and Polasky, 2007). Likewise, research has elucidated the role of less-known taxa (e.g. bats) as providers of ecosystem services and hence as bioindicators (Kunz et al., 2011).

Awareness is also raising on the importance of attaining multiple ecosystem services from agroecosystems instead of focussing on individual ones. These 'stacked' ecosystem services (Fiedler et al., 2008) or ecosystem service 'bundles' (Raudsepp-Hearne et al., 2010) would favour the transition from input-based agriculture to service-based cropping and farming systems, with minimum reliance on fossil fuel-based non-renewable inputs. Development and context-based fine-tuning of such practices form the backbone of emerging paradigms for designing sustainable food and farming systems, like agroecology (Wezel et al., 2009), 'ecological intensification' (Bommarco et al., 2013; Tittonell, 2014) and 'biodiversity-based agriculture' (Kremen et al., 2012). Legume-based cropping systems (Jensen et al., 2012), intercropping (Brooker et al., 2015) and agroforestry (Jose, 2009; Tscharntke et al., 2011) are excellent examples of systems and practices that are able to provide multiple ecosystem services. Understanding the across-scale dependency of multiple ecosystem service provision, as driven by the interactions among taxa, habitat elements and agricultural management, would allow designing more targeted agri-environmental schemes and practices and maximizing their effectiveness (Holzschuh et al., 2007; Gabriel et al., 2010). These findings, which may also take advantage of improved land-use metrics (Verburg et al., 2009), would be instrumental to inspire novel and better policies. These would need to encompass measures encouraging collaboration among farmers, land managers and other stakeholders at the landscape level, promoting an integrated landscape approach to sustainable management (Chazdon et al., 2009) based on shared views and commonly defined priorities (Sayer et al., 2013).

Aiming at multiple (agro)ecosystem service provision seems a logical approach, also policy-wise. However, this goal is not easy to meet, because there are often trade-offs among services (Kareiva et al., 2007; Gordon et al., 2010). These are partly due to ecological drivers like niche overlapping or different habitat requirements for species associated to different services, and partly to socio-economic drivers leading to contrasting human interests and priorities.

The high complexity of among-service interactions requires novel and improved methodologies to study trade-offs and highlight their ecological and socio-economic consequences (Farber et al., 2006). Examples of these methodologies are the use of model systems like farmland birds (Butler et al., 2007), the development of novel indicator/decision-support systems (Dale and Polasky, 2007; Goldstein et al., 2012) and spatial analysis methods (Anderson et al., 2009). Novelty is required on the theoretical side as well, moving away from science containment (biodiversity-only or agriculture-only focus) and developing integrated approaches that aim to find the best compromise between biodiversity conservation and agricultural production, for example, the 'eco-agriculture' approach (Scherr and McNeely, 2008). In this line of action, one interesting approach is re-considering the importance of native plants as providers of ecosystem services. For example, native plants are often better than introduced plants at supporting arthropod-mediated ecosystem services (Isaacs et al., 2009), and their maintenance also plays for local adaptation, and native biodiversity (species and habitat) conservation.

Biodiversity conservation *per se* may not be easy to reach where there is a strong focus on ecosystem service provision. However, highlighting the positive and negative correlations between biodiversity conservation and ecosystem services can substantially reduce trade-offs. A spatial analysis showed that targeting ecosystem services directly facilitates the provision of multiple ecosystem services, but would result in 44% biodiversity loss compared to targeting biodiversity conservation only. In contrast, targeting biodiversity conservation plus the positively associated services only would reduce biodiversity losses to 7% (Chan et al., 2006).

In agroecosystems, trade-offs often take the form of a balance between services and disservices, the latter being the harmful effects of some biodiversity components on crops, trees and livestock (e.g. pests, weeds and diseases). This balance depends on agroecosystem management at different scales (field, farm, landscape) and on the diversity, composition and functioning of (semi)natural areas present in the agricultural landscape (Zhang et al., 2007). Across-scale interactions and economic externalities are important drivers of this balance. Agricultural management can also create disservices beyond the agroecosystem boundary, for example, wildlife habitat loss, nutrient runoff, waterway sedimentation, greenhouse gas emissions, and pesticide-related pollution and poisoning. Better knowledge of the relationships between biodiversity, agriculture and ecosystem services would help design more environmentally friendly agricultural systems and achieve win-win situations (Power, 2010).

Scientists have also started to acknowledge the relevance of large-scale global policies on (agro)ecosystem service provision. A land-use scenario analysis by 2051 for the United States showed that large cropland extension consequent to increased crop demand would increase carbon storage and timber and food production, but would result in substantial habitat loss (>10%) for 25% of the modelled species (Lawler et al., 2014). The authors claimed that aggressive global land-use policies are needed to shift land-use change trends and, consequently, the trajectory of ecosystem service provision.

2.4 Functional biodiversity creeps in

The thrust of studies on the link between biodiversity and (agro)ecosystem services, consequent to the release of the Millennium Ecosystem Assessment, has also facilitated the emergence of studies on functional biodiversity in agroecosystems. Although the term 'functional biodiversity' is still seldom used in papers and at times with different meanings, there is an ongoing trend associating it with the provision of agroecosystem services, more clearly than it occurred in the past.

The analysis of the 102 papers explicitly referring to functional biodiversity in an agricultural context shows a variety of foci. Part of them focussed on organisms associated to the provision of services ('functional groups'), for example, bacteria (Topp, 2003; Zhao et al., 2010; Cycoń et al., 2013), nematodes (Mulder et al., 2003) or birds (French and Picozzi, 2002; Nereu et al., 2018), without necessarily addressing the level of diversity within a functional group. Other papers focussed instead on the provision of target services, for example, soil quality (Babujia et al., 2016; Schorpp and Schrader, 2016; Furtak et al., 2017), bioremediation (Sloots and van der Vlies, 2007; Serra et al., 2019) or weed suppression (Rinaudo et al., 2010; Petit et al., 2015), sometimes trying to quantify the expression of services in terms of crop yield or income (Sasaki et al., 2011; Bowman and Zilberman, 2013). A third group of papers focussed on practices that are able to provide multiple agroecosystem services, for example, intercropping (Baumann et al., 2004; Van Mele and Van Chien, 2004) or agroforestry (Tscharntke et al., 2011; Pardon et al., 2019). Finally, a fourth group of papers focussed more on the biodiversity level (as defined by the CBD) and its relationship with service provision, that is, the genetic level (Delmont et al., 2012; Costanzo and Bàrberi, 2016; Lazzaro et al., 2018; Carranza-Gallego et al., 2019), the species level (Verheyen et al., 2014; Adams et al., 2017; Njeru et al., 2017) or the habitat level (Anderson et al., 2013; Chamorro et al., 2016; Sutter et al., 2018; Martin et al., 2019). Several papers show a certain degree of overlapping among the above-mentioned foci: this is not surprising, considering the inherent complexity of biodiversity studies.

Despite this variety, Altieri's breakthrough paper (Altieri, 1999) contributed to strongly characterise functional biodiversity as a set of practices at field, farm and landscape scale mainly aimed to improve biological control of insect pests through the provision of habitat quality for indigenous natural enemies (i.e. 'conservation biological control'). In fact, 46% of the 102 papers explicitly addressing functional biodiversity deals with biological control of crop pests and (to a lower extent) diseases. The management of semi-natural habitat at a landscape scale is deemed fundamental for the optimisation of conservation biological control, considering the fact that most elements in the pest-beneficial complex are mobile species. For many researchers, this identity is so strong that a very active working group on Landscape Management for Functional Biodiversity was founded in 2003 under the umbrella of the International Organisation for Biological and Integrated Control (IOBC; www.iobc-wprs.org/expert_groups/19_wg_landscape_management.html). This working group mainly deals with conservation of biological control, although at times other ecosystem services (e.g. pollination and weed reduction) are also addressed.

An interesting, more recent trend in functional biodiversity research is the increase in interdisciplinary studies. Biodiversity scientists are becoming aware of the importance of studying organism interactions across different trophic levels to highlight opportunities to obtain multiple agroecosystem services or a better balance between services and disservices. Good examples of such research are the papers published in *Ecology and Society*'s 2012 Special Issue on diversified farming systems (Kremen et al., 2012), Weisser et al.'s (2017) synthesis of the Jena Experiment, a long-term German study aimed to investigate the effects of plant diversity on soil-nutrient cycling and trophic interactions, and Baeten et al.'s (2019) study on the relationship between species composition and multiple ecosystem services in European forests.

The rising interest in the functional aspects of biodiversity has also stimulated the discussion on how to implement policies to support them. Most of this discussion is taking place in the EU, thanks to a general policy framework supporting the 'greening' of agricultural production (Delbaere et al., 2014). However, much is still to be done to favour the acceptance of diversified cropping and farming systems from farmers and other stakeholders, which demands more and clearer information on the advantages of such systems from a productive, economic and environmental perspective (Bianchi et al., 2013). Here, research is instrumental to better shape policy instruments and management measures aimed to improve the delivery of biodiversity-based agroecosystem services in target contexts (Bowman and Zilberman, 2013; Nilsson et al., 2019). The role of functional biodiversity in the larger policy discussion on land sharing versus land sparing (Tscharntke et al., 2012) and its place in the mainstream commodity-centred neoliberal economic regime (Lockie, 2009) are also being debated.

Since functional biodiversity is an emerging concept yet still somewhat blurred, in the next section we will (i) focus on its various definitions and their limitations, (ii) propose a more focussed and comprehensive definition and its application to an agroecosystem context and (iii) delineate the 'trait-based' approach, its trend and its potential to build bridges across disciplines engaged in biodiversity research.

3 Functional biodiversity and the delivery of agroecosystem services

3.1 Defining functional biodiversity: a multifaceted world

Analysis of the scientific literature reveals the existence of several definitions of functional biodiversity, varying upon the disciplinary background of authors and their main interests. In general, ecologists and agronomists have somewhat diverging views, while - expectedly - agroecologists have an intermediate view. To illustrate this, we present some examples in the following paragraphs. Nevertheless, it is interesting to note that, recently, previously different views tend to converge following the emergence of a trait-based approach to functional biodiversity research: this issue will be treated in Section 3.3.

An example of ecologically centred definition can be found in the work of Pearce and Moran (1994), who refer to 'functional diversity' as the relative abundance of organisms expressing different ecological functions. No mention was made either to specific functions or to their relevance to (agro) ecosystem services, possibly because the interest on this issue was still to come.

Gurr et al. (2003) used the term 'multi-function agricultural biodiversity' to indicate the positive cascading effect between plant diversity and biological pest control across different spatial scales, from the field to the landscape. Although their definition acknowledges the importance of biodiversity to modulate the effect of functional interactions between organisms across spatial scales and trophic levels, it refers only to a specific service, reiterating the classical above-mentioned narrow view identifying functional biodiversity with conservation biological control (see also Barbosa, 1998).

Peeters et al. (cited in Clergue et al., 2005) identified functional biodiversity (also called 'para-agrobiodiversity') with the diversity of *associated* organisms useful for production, for example, soil microorganisms, beneficial arthropods and spontaneous grassland flora. Although they addressed the link between biodiversity and service provision, their definition includes neither *planned* biodiversity (i.e. the elements deliberately introduced to provide agroecosystem services) nor the link between biodiversity and potential disservices.

It is worth noting that the previous definitions did not explicitly address the three levels of agrobiodiversity (genetic, species and habitat) as classified by the CBD and later reprised by the OECD. In turn, the OECD classification, although not specifically addressing the link between (agro)biodiversity and functionality, if seen from the lenses of planned and associated biodiversity shows two important gaps, because it does not include associated agrobiodiversity at gene level and planned agrobiodiversity at species level (Bàrberi, 2015). Associated genetic agrobiodiversity refers, for example, to the genetic variation within populations of crop pests and their natural enemies, pathogens or weeds, which is important to fully understand the extent and functional consequences of interactions among organisms (Crutsinger et al., 2008). Planned agrobiodiversity at the species level includes a plethora of elements which are crucial in sustainable agriculture, that is, the diversity of crops in a rotation, of cover crops and of grassland species, all elements capable of providing multiple services like productivity, soil fertility, pest/disease/weed suppression and pollination (Tilman et al., 2002).

The heterogeneity in the theoretical foundations of functional biodiversity shows that there is room to improve its definition as to facilitate pick-up of the concept and of related measures in both research and agricultural practice.

3.2 Re-defining functional biodiversity: a stronger focus on agroecosystem services

The definitions of functional biodiversity available approximately 15 years ago, that is, when the Millennium Ecosystem Assessment was released, were not fully appropriate to address the relationship between biodiversity and ecosystem services in managed ecosystems with clear productive goals like agroecosystems. In addition, in the scientific literature there was (and partly there still is) some confusion between the concepts of functional biodiversity and biofunctionality, that is, the simple presence of species potentially capable of delivering services ('functional groups'), without any clear reference to the importance of diversity within a functional group (Moonen and Bàrberi, 2008). For example, many of the papers published in the very interesting Special Issue of *Agriculture, Ecosystems & Environment* (vol. 74) on 'Invertebrate Biodiversity. Bioindicators of Sustainable Rural Landscapes' (1999) were actually referring to biofunctionality and bioindicators, and not to functional biodiversity. As such, we felt the need to shed some light on this topic and coined what we consider a fine-tuned and more precise definition of functional biodiversity, that is 'that part of the total biodiversity composed of clusters of elements (at the gene, species or habitat level) providing the same (agro)ecosystem service, that is driven by within-cluster diversity' (Moonen and Bàrberi, 2008). Our definition stresses the importance of diversity within a functional group to enhance the delivery of the target agroecosystem service, an aspect that was missing in previous definitions. Although we focussed on agroecosystems, this is a general definition that may be applied to any type of ecosystem.

One issue that is not yet well-rooted in the scientific debate around functional biodiversity is that a positive or negative 'function' cannot unequivocally label a given agrobiodiversity element, whose role can dramatically change depending on the context. For example, wild blackberry (*Rubus fruticosus* L.) can be a crop, an invasive weed, a windbreak, a semi-natural element supporting pests or beneficial arthropods, and a pleasant or unpleasant landscape element depending on context and human perception (Moonen and Bàrberi, 2008). The emerging discussion on trade-offs between provisioning services and other ecosystem services and disservices, and their dependence on spatial and temporal scale and agricultural management practices (Power, 2010) should incorporate this issue more explicitly. This should facilitate the identification of priority services with the engagement of local actors and stakeholders, and thus streamline action towards obtaining multiple agroecosystem services and win-win situations (Swinton et al., 2007).

Once we realise that (dis)services linked to agrobiodiversity components are strictly context-dependent, it follows that a clear definition of the context, its goals and priorities should be a key preliminary step (i) in any functional agrobiodiversity study or project. The next logical steps in this process should be (ii) the identification of priority services in the given context, (iii) the assemblage of the functional group(s), that is, the array of elements (at gene, species or habitat level) available in the context that are potentially able to provide the target service(s), and (iv) the creation of a short list of indicators that are able to highlight the actual provision of service(s) by the chosen elements once implemented, with suggestion of the most appropriate spatial and temporal scale. We believe that this four-step approach could serve as a useful guide for designing and implementing studies/projects aimed to foster agroecosystem services anywhere. An example of application of this approach is given in Table 2. It is worth noting that the same approach can be used either in studies completely managed by scientists or in other seeking interaction with local actors/stakeholders through participatory action-research.

3.3 The emergence of a functional traits approach

Once the priority services have been chosen (e.g. in step (ii) of our four-step approach), one key issue to be addressed is: what are the attributes that make an agroecosystem element potentially capable of providing the desired service hence worth being included in the functional group (step (iii) in our approach)?

An interesting way to tackle this problem is to take inspiration from an emerging field of plant ecology: the study of vegetation through the analysis of functional traits possessed by plant species. These traits are morphological or physiological features possessed by individual plants (Violle et al., 2007) that help explain why the corresponding species is present or abundant in a given (agro)ecosystem and how different species assemble in communities ('response traits'). Examples of these traits are seed mass, specific leaf area (SLA) and root length density. At the same time, these or other traits can highlight the potential of plant species to provide (agro)ecosystem services ('effect traits' *sensu* Garnier and Navas, 2012), but also to create disservices. An excellent compendium of the theory and application of functional trait analysis in plant ecology can be found in Garnier et al. (2015).

Although the definition of 'functional trait' is rather strict in plant ecology (and is mainly intended as a 'response trait' *sensu* Garnier and Navas, 2012), we could extend its meaning and – in an agroecosystem context – refer it to any attribute or property that makes a plant species potentially able to provide an agroecosystem service or create a disservice. An attempt made in this direction is the functional trait database developed by Bàrberi et al. (2018), which includes 17 'traits' (attributes) for 240 plant species occurring in the arable weed flora of organically farmed fields across 10 European countries. The 17 traits are 'hybrid' in the sense that some are 'real' traits (*sensu* Violle et al., 2007)

Table 2 Example of application of our four-step approach to the implementation of a functional agrobiodiversity study, including a participatory aspect[1]

Step	Description	Example	Participatory aspect
(i)	Definition of the context and of site-specific objectives	Improvement of soil health; diversification of cropping/farming systems; input reduction (fertilisers, pesticides); moving towards organic crop production	Setting up local actors/stakeholders team: participatory analysis of the agroecosystem state and co-definition of site-specific objectives (e.g. through focus groups)
(ii)	Definition of the priority agroecosystem service(s)	Improvement of soil fertility, pollution reduction, sustainable pest/disease management, water supply/quality	Co-definition of target services for each study site (e.g. through focus groups or semi-open interviews)
(iii)	Definition of the agroecosystem functional group(s) and of key traits/attributes associated to the delivery of priority agroecosystem service(s)	Local legume plants (e.g. as cover crops/intercrops), mulches, natural enemies of the target pests, wild plant species and habitat structures (headlands, hedgerows, woodland, multi-layer agroforestry etc.) attracting natural enemies or limiting disease spread, crop cultural practices known to affect soil fertility and pest/disease containment (genotype choice, planting, fertilisation, pest/disease/weed management etc.)	Co-definition of local resources and practices (plants, habitat elements, management techniques etc.) potentially able to provide the target services, taking into account farmers' preferences; setting-up on-farm trials
(iv)	Definition of space and time boundaries, of the best set of soil health indicators, and of measurement details	Field, farm and landscape scale, fallow and cropping phase, visual estimate and on-field measurement of soil and plant status (e.g. physical, chemical and biological soil health estimated by the spade test, SPAD-based plant chlorophyll content, visual estimate of pest/disease attack), soil and plant sampling and subsequent laboratory analyses (e.g. soil structure stability, organic carbon and NPK content, earthworm abundance), presence and abundance of natural enemies (sample), vegetation/mulch cover (sample), crop growth and marketable yield (sample), type and abundance of surrounding wild vegetation/habitats (sample), number, type and rates of pesticides applied etc.	Co-definition of soil health and crop performance indicators; co-evaluation of on-farm trials (e.g. through focus groups and targeted evaluation sheets)

[1]Modified after Bàrberi (2018).

expressing the potential of plant species to create disservice (e.g. through resource competition), while others are attributes that indicate species capable of providing ecosystem services, for example, attractiveness to pollinators (link to pollination service) or establishing symbiosis with arbuscular mycorrhizal fungi (link to soil health service). Two major difficulties in populating and using such a database are the lack of information on some traits (despite the existence of some valuable plant trait databases available online) and the high heterogeneity in trait value within a species (see e.g. Iannetta et al., 2007 for *Capsella bursa-pastoris*). More basic research is therefore needed to improve the completeness of plant trait databases, especially regarding species that are typical of agroecosystems, and thus their usefulness.

In principle, this approach to functional trait analysis can be applied to any agroecosystem element potentially linked to services or disservices (the 'functional group' *sensu* Moonen and Bàrberi, 2008). However, so far it has been mainly applied to plants, partly because they are at a key crossroad along the highway of trophic interactions, and to the species level of agrobiodiversity, and partly because the functional traits of most arthropods and other invertebrate groups are largely unknown. In order to create functional trait databases for the most relevant invertebrates inhabiting our agroecosystems, a huge research effort should be made to study and describe their key traits for ecosystem functioning. This is further complicated by the shrinking number of taxonomists across biological sciences (Walter and Winterton, 2007).

Studies explicitly based on a functional trait approach to agroecosystem services are still scarce and very recent, the majority of which having been published in or after 2014. Probably the first example of such studies is van Bueren et al.'s (2002) call for developing an ideotype for organic crops. Modern cultivars often lack traits that would be desirable under organic management, for example, adaptation to organic soil-fertility management, better resource-use efficiency through improved root system and the ability to interact with beneficial soil microorganisms, the ability to compete with weeds, reasonable yield and high-yield stability, and good product quality. Also, indirect defence mechanisms to attract beneficial insects when attacked by a crop pest would be an important functional trait (Tamiru et al., 2011). All these traits would obviously facilitate the provision of production-related agroecosystem services in a sustainable way.

In agreement with the prevalence of conservation biological control as subject of functional agrobiodiversity research, some studies have focussed on plant traits related to floral pollen or nectar provision to support beneficial arthropods like pollinators or natural enemies of crop pests (Carrié et al., 2012; van Rijn and Wäckers, 2016). From a different perspective, Winqvist et al. (2014) showed that ground beetles' (Coleoptera: Carabidae) traits were affected by agricultural intensification and landscape patterns, being phytophagous and omnivorous beetles less species rich in high-yielding farms and (the latter type) more abundant in landscapes with higher proportion of arable land. Indeed, landscape heterogeneity - both in composition and configuration - is an important ecological filter of species traits (Duflot et al., 2014), just like agricultural management. In general, more diversified cropping systems are expected to increase functional trait diversity in agroecosystems, as seen in short-rotation coppice plantations (Verheyen et al., 2014), however, not necessarily more simplified cropping systems are shifting wild plant traits towards creation of disservice. For example, a simulation study in genetically modified glyphosate-resistant maize showed that this system did not select for more competitive weed traits, although individual management practices (e.g. crop sowing time) may modulate this effect (Colbach et al., 2017).

In functional agrobiodiversity studies, another crucial question is whether service provision can be obtained more easily by inclusion of an element (e.g. plant species) possessing a single, very effective functional trait or by inclusion of more elements possessing complementary traits. In other words, how important is functional trait diversity in providing the desired agroecosystem service(s)? In practice, there are various ways by which functional agrobiodiversity can be usefully included in agroecosystems. Linking, for example, crop traits to agroecosystem services could facilitate the identification of concrete options for farmers and policymakers. With this intention, Costanzo and Bàrberi (2014) characterised agrobiodiversity into functional identity, functional composition and functional diversity. Their work focussed on wheat, but it can be generalised to any cropping system and to all agrobiodiversity levels. They defined *functional identity* as the presence of a set of homogeneous phenotypic traits that are related to the expression of given agroecosystem services (e.g. a fast growing crop or cover crop that improves weed suppression). *Functional composition* is the complementary effect of different traits expressed by co-occurring elements (e.g. higher total yield or nutrient use efficiency provided by cereal-legume intercropping). *Functional diversity* s.s. is the direct effect of heterogeneity (genetic diversity) in the crop stand (e.g. the positive effect of wheat cultivar mixtures on disease suppression).

Evidence suggests that service provision can be optimised by one or another of these three mechanisms, depending on context. For example, in organic wheat functional identity was more important to drive weed reduction and grain yield, however, wheat-subterranean clover living mulch (i.e. functional composition) reduced dicotyledonous weed abundance and weed biomass without reducing wheat yield except when wheat was poorly established (Costanzo and Bàrberi, 2016). This suggests that the different functional biodiversity strategies are not mutually exclusive and that there is room to combine them to optimise service provision. Biofunctionality turned out to be more important than functional diversity in driving weed suppression in mycorrhized sunflower (Rinaudo et al., 2010), while functional identity mainly drove mycorrhizal symbiosis and productivity in organic tomato (Njeru et al., 2017). However, when the goal is to optimise provision of multiple agroecosystem services, functional diversity may be a better option, as in wheat cultivar mixtures with medium to high number of components (Lazzaro et al., 2018). If well-characterised, enhanced complementarity and synergy among wheat traits would maximise exploitation of genetic agrobiodiversity to obtain target services, for example, improved weed suppression (Lazzaro et al., 2019). Despite functional identity can give best results in optimum situations, unpredictability as given by changing climatic conditions suggest that functional trait diversity may be preferable to increase the stability of service provision across time, in accordance with the insurance hypothesis linking biodiversity and service provision in case of species (or trait) redundancy (Bengtsson, 1998; Yachi and Loreau, 1999). Such an effect has been observed, for example, in the case of high-diversity cover crop mixtures to improve weed suppression and nutrient supply to the following cash crop (Ranaldo et al., 2020).

To date, the vast majority of the scientific literature on functional biodiversity in agroecosystems is polarised: studies either focus on how to manage agriculture to protect or enhance general biodiversity or target species/habitat (agriculture for biodiversity; A for B), or on how to use biodiversity to provide ecosystem services directly or indirectly related to agricultural production (biodiversity for agriculture; B for A) (Bàrberi et al., 2010). Although in principle there is nothing wrong in this dual approach, the risk of such polarisation is to exacerbate competition between conflicting interests and to favour acrimony in the scientific debate, drifting people away from what should be the uppermost goal for sustainable agriculture: providing enough food of good quality to everyone by replacing external inputs with biodiversity-driven agroecosystem services while protecting natural resources and general biodiversity. This is the very nature of the agroecological approach to agricultural production (see e.g. Wezel et al., 2009), which highlights the importance of managing agroecosystems as to provide goods (B for A) while protecting the environment (A for B). In other words, agroecosystem management should aim at providing multiple agroecosystem services, including protection of general biodiversity and other cultural services, that is, it should aim at 'multifunctional' agriculture (Renting et al., 2009). Luckily, there are signs that research is being increasingly geared towards obtaining multiple agroecosystem services from functional biodiversity, by favouring synergies and minimising trade-offs between services (Sutter et al., 2018). The rising interest on diversified farming systems and agroecology (Kremen et al., 2012), coupled with the increasing evidence that diversified systems like agroforestry can more easily provide multiple agroecosystem services (Kay et al., 2018), should strengthen this trend further. The development of indicator systems taking into account taxonomic and functional aspects of biodiversity at the same time (Trivellone et al., 2014) or explicitly addressing trait-based multiple services (Sharafatmandrad and Mashizi, 2020) may also facilitate this transition.

It is obvious that preservation of functional biodiversity (and of biofunctionality) largely depends on cropping and farming system management. In this respect, despite variation due to local environmental conditions, management intensity seems more important than farming system type (e.g. organic vs. conventional), as it has been observed for earthworms and fungi (Spurgeon et al., 2013) and the entomofauna (Costa et al., 2012; Gkisakis et al., 2016). Among management practices, pesticides are a major disturbing factor for functional biodiversity (Cycoń et al., 2013), therefore sustainable farming systems of the future should aim at a considerable reduction of pesticide use (or their elimination, where possible), unless target species-specific pesticides with lower environmental impact can be developed. However, also in that case, direct control measures should always be secondary to the application of preventive measures.

4 How can research contribute to the advancement of functional agrobiodiversity?

In the previous sections we have shown that functional agrobiodiversity has great potential to improve agroecosystem sustainability, especially when the goal is to provide multiple agroecosystem services through appropriate combination of all levels of agrobiodiversity, namely, genetic, species and habitat. However, despite increasing scientific evidence, knowledge and application of functional agrobiodiversity principles are still scarce. Why is that and how can research best contribute to place functional biodiversity at the centre of discussion on sustainable farming and food systems of the future?

Clearly, the inherent complexity of (agro)biodiversity as a science does not facilitate pick up of concepts and related practices from farmers, policymakers and the society at large. Correct science communication, especially on issues of global importance like biodiversity, is very difficult, yet urgently needed, to fill in the gap between science and action (Requier et al., 2020). Scientists' engagement in communicating research to the wider public would facilitate the convergence between policy, science and public opinion objectives. Unfortunately, the current system of research evaluation does not give value to this role, thus the majority of scientists refrain from communicating their results beyond the closed circle of their peers (Pham, 2016). Governance change should then quickly occur to revert this trend. Research-funding agencies can make a significant change by stimulating scientists to communicate research results to the wider public. The 'inspiration sheets' and 'practice abstracts' that the European Commission wants researchers to produce in Horizon 2020-funded research projects (see e.g. https://iwmpraise.eu/inspiration-sh

eet-combining-roller-crimpers-and-flaming-for-the-termination-of-cover-crops) are some good examples.

In these times there are some conjunctures that may play for a quick rise of interest on functional agrobiodiversity. The first is the emergence of agroecology as a new paradigm for sustainable agricultural and food systems that may solve the large-scale problems caused by mainstream industrial agriculture (IPES-Food, 2016). Agroecology is currently being promoted by some key international players like the FAO, and seems to be gaining momentum in the debate on the shaping of future agri-environmental policies, that is, in the European Union (France Stratégie, 2019). A second important conjuncture is the increasing evidence of human-driven climate change (IPCC, 2014). There is already clear scientific evidence that diversified agroecological systems can contribute to climate change mitigation (Altieri and Nicholls, 2017; Altieri et al., 2015; Kremen and Miles, 2012). Therefore, concentrating research efforts on the climate mitigation service offered by functional biodiversity should shed light on the overall concept, and should stress the fact that joint action directed towards improved functionality of agricultural systems can at the same time address two gigantically hot problems of present times, that is, biodiversity conservation and climate change mitigation. Regarding biodiversity conservation, further evidence, extended to other geographical zones, is needed to pinpoint the hypothesis (often working in tropical environments) that the land-sharing model, where the agricultural matrix is interspersed among natural areas, is likely a win-win situation for both agricultural production and biodiversity conservation, especially in a context of small-scale agroecological production (Perfecto and Vandermeer, 2010).

To make functional biodiversity further operational, efforts should be increased to highlight the conditions under which synergies and conflicts between services may occur. In this regard, optimisation of services would also benefit from increased research effort on unravelling the finer mechanisms behind service provision, like, for example, understanding and reducing intraguild predation among natural enemies of crop pests (Davey et al., 2013). In general, to boost the uptake of trait-based approaches, far more attention should be paid to the fundamental role of taxonomists in general and entomologist in particular who are identifying and studying the biology and ecology of key species regulating agroecosystem processes such as soil fertility, pest and disease control and crop pollination. Already more than 30 years ago, an article in Science (Holden, 1989) highlighted the discrepancy between the huge number of insect species (10 million) and the very low number of taxonomists able to identify these species (only 1000 worldwide to identify tropical species). At the same time, insect pests cause severe yield losses and in recent years, the interest in managing beneficial insects has increased. This augmented need for taxonomists and entomologists is not at all reflected in investments in this category of researchers. Instead, entomologists have major difficulty in finding funding opportunities and receiving academic recognition because they do not produce tools and solutions that can be readily used by society nor are they able to keep up with the fast pace of publishing requested by 'society' because their work needs time and patience. It would therefore be extremely important for the future of agroecology, that relies on 'nature-based solutions' and therefore on functional biodiversity, to establish a parallel way of doing science, slow-science (Frith, 2020), where there would be money and recognition for this important group of researchers. When the day arrives that we cannot identify our species anymore, the entire basis for functional agrobiodiversity-based solutions for sustainable agriculture will become vain.

To convince farmers and policymakers that functional biodiversity can act at any spatial scale and in any farm type, research should focus on biodiversity-based practices that can work and be economical on large-scale farms, where simplification of cropping systems is commonly practiced to minimise costs per surface unit. In times of economic and climate uncertainty there is room for change, also considering that farmers shifting to diversified ecological farming feel better for contributing to environmental protection (Bondy and Cole, 2019).

Finally, scientists should take more courage, get out of their comfort zone and explore the vast horizon of inter- and transdisciplinary research, which is a must for addressing a complex subject like (functional) biodiversity (Evans, 2019; Rai and Singh, 2020; Ndebele-Murisa et al., 2020). Among transdisciplinary research approaches, citizen science may also be explored (Spence et al., 2019; Larson et al., 2020). The European Community's research policy is very relevant in this regard because it provides strong stimuli in terms of financial support to research projects aimed at exploring these cutting-edge themes and research approaches. A very recent example is the Horizon 2020 project FRAMEwork (due to start in late 2020) that will follow a participatory and citizen science approach to explore ways to improve functional agrobiodiversity management in European cropping systems through identification of landscape-scale management strategies developed in local farmer clusters. In addition, knowledge about the interconnectedness of sustainable food production, environmental protection and human health (Willett et al., 2019) should be spread within and beyond the scientific community. Managing agricultural landscapes to provide enough supporting and regulating ecosystem services and to minimise disservices will require policy-relevant, interdisciplinary and participatory research (Zhang et al., 2007; DeClerck et al., 2011).

In conclusion, functional agrobiodiversity can provide scientific evidence as well as practical solutions to help shape future sustainable agroecosystems, but needs time and investment in complex inter- and transdisciplinary studies. One major difficulty, and a potential constraint for communication, would be that - unavoidably - functional agrobiodiversity solutions could never be standardised, because they need to be optimised in any given context.

Biodiversity scientists (including agroecologists) should work hard to convince farmers, policymakers and society as a whole that sustainable cropping systems can only be developed and implemented through a holistic agroecological approach that does not solely aim at increased crop production but reshapes the entire paradigm of food production and consumption. This approach can be considered as the alternative to the so-far more common approach that aims at boosting agricultural production to then look for a 'silver bullet' solution for the problems it created. The key challenge to reach this goal is to convince actors and stakeholders that it is important to reconcile the pace of economic systems with that of ecological systems (Bàrberi, 2013) in order to obtain multiple agroecosystem services that will eventually benefit both humans and the environment.

On a practical side, a key issue is that farmers need to be able to manage more complex systems based on functional agrobiodiversity. Any biodiversity element (be it at genetic, species of habitat level) added to a system has its specific requirements and farmers need to pay attention to each element added to the system. It is therefore of utmost importance for the success of agrobiodiversity-based solutions to reduce the amount of complexity in the systems to the level that maximise the magnitude of service provisioning. Careful selection of functional traits and their complementarity is needed to keep the system manageable while adding enough diversity to provide the desired services. This may create a conflict with the idea of adding redundant species in agroecosystems in order to provide resilience in case of major disturbances (e.g. climate change). It will be the difficult task of agroecologists to find the balance between these two needs: redundancy for resilience and minimum levels of complexity for farmers. However, the design of optimal patterns of spatial and temporal diversification may provide solutions for both, for example, through the inclusion/maintenance of diverse semi-natural habitats surrounding cropped fields with a limited number of crops in a diversified crop sequence. Agroforestry systems are other interesting solutions to meet both objectives. The future of sustainable agriculture will very much rely on co-designing such balanced solutions with farmers and other actors, and on thoroughly disseminating them.

5 Where to look for further information

5.1 Scientific and professional organisations

- Agricology – Practical Sustainable Farming without Labels: www.agricology.co.uk.
- AgriCultures Network – Mainstreaming Agroecology and Family Farming: https://magazines.agriculturesnetwork.org.

- Agroecology Europe – a European Association for Agroecology:
  www.agroecology-europe.org.
- Biodiversity, Agriculture, Soil & Environment (BASE UK):
  https://base-uk.co.uk/about.
- British Ecological Society – Agricultural Ecology Group:
  www.britishecologicalsociety.org/membership-community/special-interest-groups/agricultural-ecology.
- CIRAD – Tropical and Mediterranean Cropping System Functioning and Management:
  https://umr-system.cirad.fr/en/the-unit.
- EKLIPSE – Knowledge & Learning Mechanism on Biodiversity and Ecosystem Services:
  www.eklipse-mechanism.eu/expert_working_groups.
- European Environment Agency (EEA) – Exploring Europe's Changing Landscape:
  www.eea.europa.eu/highlights/explore-europe2019s-changing-landscape.
- FAO Agroecology Knowledge Hub:
  www.fao.org/agroecology.
- Global Action Plan for Agricultural Diversification (GAPAD):
  http://gapad.org.
- Intergovernmental Science-Policy Platform on Biodiversity and Ecosystem Services (IPBES):
  https://ipbes.net.
- International Treaty on Plant Genetic Resources for Food and Agriculture:
  www.fao.org/plant-treaty/en.
- IOBC Expert Group on Landscape Management for Functional Biodiversity:
  www.iobc-wprs.org/expert_groups/19_wg_landscape_management.html.
- Platform on Agrobiodiversity Research (PAR):
  http://agrobiodiversityplatform.org.
- United Nation's Convention on Biological Diversity (CBD):
  www.cbd.int.

5.2 Research projects

- AGFORWARD – Agroforestry that will Advance Rural Development:
  www.agforward.eu/index.php/en/home-redirect.html.
- DIVERSIFOOD – High Quality Food Systems:
  www.diversifood.eu.
- EcoStack – Sustainable Production through Stacking of Ecosystem Services:
  www.ecostack-h2020.eu.

- LANDMARK - LAND Management, Assessment, Research and Knowledge Base:
  http://landmark2020.eu.
- Landscape for People, Food and Nature Initiative:
  http://peoplefoodandnature.org.
- LIBERATION - Linking Farmland Biodiversity to Ecosystem Services for Effective Ecological Intensification:
  www.fp7liberation.eu.
- Millennium Ecosystem Assessment:
  www.millenniumassessment.org/en/index.html.
- QUESSA - Quantification of Ecological Services for Sustainable Agriculture:
  www.quessa.eu.
- SOLMACC - Organic Farmers Countering Climate Change:
  http://solmacc.eu.

5.3 Websites

- Agrobiodiversity.net - European Agrobiodiversity Network:
  www.agrobiodiversity.net.
- Agroforestry - Trees of Life, Trees for Life:
  www.agroforestry.org.
- Ecological Flora of the British Isles - database:
  http://ecoflora.org.uk.
- e-FLORA-sys - Grassland Plant Diversity - database:
  http://eflorasys.univ-lorraine.fr/index.php/en/pages/display/home.
- Global Biodiversity Information Facility (GBIF) - database:
  www.gbif.org.
- LEDA Plant Trait Database, University of Oldenburg, Germany:
  https://uol.de/en/landeco/research/leda.
- Online Vegetation and Plant Distribution Maps, University of Berkeley, USA:
  https://guides.lib.berkeley.edu/VegMaps.
- Oppla - EU Repository of Nature-Based Solutions:
  https://oppla.eu.
- ReFarm - Resilience Framework for Agriculture and Risk Management Database:
  http://agrobiodiversityplatform.org/refarm.
- TRY - Global Plant Traits Database:
  www.try-db.org/TryWeb/Home.php.
- United Nation's Sustainable Development Goals for 2030:
  www.un.org/sustainabledevelopment/sustainable-development-goals.

- World Overview of Conservation Approaches and Technologies (WOCAT): www.wocat.net/en.

5.4 Research centres

- Bioversity International (CGIAR): www.bioversityinternational.org.
- Georg-August Universität Göttingen – Agroecology Group, Germany: www.uni-goettingen.de/en/74726.html.
- German Centre for Integrative Biodiversity Research (iDiv): www.idiv.de/en/about_idiv/structure.html.
- Institut National de la Recherche pour l'Agronomie et l'Environnement (INRAE) – Biodiversity, France: www.inrae.fr/en/biodiversity.
- Wageningen University and Research Centre – Biodiversity, The Netherlands: www.wur.nl/en/Research-Results/Themes/Biodiversity.htm.

6 References

Adams III, P. R., D. B. Orr, C. Arellano and Y. J. Cardoza (2017). Soil and foliar arthropod abundance and diversity in five cropping systems in the coastal plains of North Carolina. *Environmental Entomology* 46: 771–83.

Agrawal, M., B. Singh, M. Rajput, F. Marshall and J. N. B. Bell (2003). Effect of air pollution on peri-urban agriculture: a case study. *Environmental Pollution* 126(3): 323–9.

Allen-Wardell, G., P. Bernhardt, R. Bitner, A. Burquez, S. Buchmann, et al. (1998). The potential consequences of pollinator declines on the conservation of biodiversity and stability of food crop yields. *Conservation Biology* 12(1): 8–17.

Altieri, M. A. (1991). Classical biological control and social equity. *Bulletin of Entomological Research* 81(4): 365–9.

Altieri, M. A. (1999). The ecological role of biodiversity in agroecosystems. *Agriculture, Ecosystems and Environment* 74(1–3): 19–31.

Altieri, M. A. and C. I. Nicholls (2017). The adaptation and mitigation potential of traditional agriculture in a changing climate. *Climatic Change* 140, 33–45.

Altieri, M. A., C. I. Nicholls, A. Henao and M. A. Lana (2015). Agroecology and the design of climate change-resilient farming systems. *Agronomy for Sustainable Development* 35: 869–90.

Anderson, B. J., P. R. Armsworth, F. Eigenbrod, C. D. Thomas, S. Gillings, et al. (2009). Spatial covariance between biodiversity and other ecosystem service priorities. *Journal of Applied Ecology* 46(4): 888–96.

Anderson, A., T. Carnus, A. J. Helden, H. Sheridan and G. Purvis (2013). The influence of conservation field margins in intensively managed grazing land on communities of five arthropod trophic groups. *Insect Conservation and Diversity* 6: 201–11.

Babujia, L. C., A. P. Silva, A. S. Nakatani, M. E. Cantão, A. T. R. Vasconcelos, et al. (2016). Impact of long-term cropping of glyphosate-resistant transgenic soybean [*Glycine max* (L.) Merr.] on soil microbiome. *Transgenic Research* 25: 425–40.

Baeten, L., H. Bruelheide, F. van der Plas, S. Kambach, S. Ratcliffe, et al. (2019). Identifying the tree species compositions that maximize ecosystem functioning in European forests. *Journal of Applied Ecology* 56: 733–44.

Bàrberi, P. (2013). Functional agrobiodiversity: the key to sustainability? In: Bhullar, G. S. and Bhullar, N. K. (Eds), *Agricultural Sustainability: Progress and Prospects in Crop Research*. Elsevier, Amsterdam, the Netherlands, pp. 3–20.

Bàrberi, P. (2015). Functional biodiversity in organic systems: the way forward? *Sustainable Agriculture Research* 4: 26–31.

Bàrberi, P. (2018). Functional agrobiodiversity: approach, examples and applicability to banana cropping systems. *Acta Horticulturae* 1196: 55–62.

Bàrberi, P., G. Burgio, G. Dinelli, A. C. Moonen, S. Otto, et al. (2010). Functional biodiversity in the agricultural landscape: relationships between weeds and arthropod fauna. *Weed Research* 50: 388–401.

Bàrberi, P., G. Bocci, S. Carlesi, L. Armengot, J. M. Blanco-Moreno, et al. (2018). Linking species attributes to agroecosystem services: a functional analysis of weed communities. *Weed Research* 58: 76–88.

Barbosa, P. (Ed.). (1998). *Conservation Biological Control*. Academic Press, San Diego, CA.

Barea, J. M., M. J. Pozo, R. Azcón and C. Azcón-Aguilar (2005). Microbial co-operation in the rhizosphere. *Journal of Experimental Botany* 56(417): 1761–78.

Batáry, P., L. V. Dicks, D. Kleijn and W. J. Sutherland (2015). The role of agri-environment schemes in conservation and environmental management. *Conservation Biology* 29(4): 1006–16.

Baumann, D. T., L. Bastiaans and M. J. Kropff (2004). Analysis and design of a leek-celery intercropping system using mechanistic and descriptive models. *Acta Horticulturae* 638: 59–68.

Bengtsson, J. (1998). Which species? What kind of diversity? Which ecosystem function? Some problems in studies of relations between biodiversity and ecosystem function. *Applied Soil Ecology* 10: 191–9.

Bengtsson, J., J. Ahnström and A. C. Weibull (2005). The effects of organic agriculture on biodiversity and abundance: a meta-analysis. *Journal of Applied Ecology* 42(2): 261–9.

Benton, T. G., J. A. Vickery and J. D. Wilson (2003). Farmland biodiversity: Is habitat heterogeneity the key? *Trends in Ecology and Evolution* 18(4): 182–8.

Bianchi, F. J. J. A., V. Mikos, L. Brussaard, B. Delbaere and M. M. Pulleman (2013). Opportunities and limitations for functional agrobiodiversity in the European context. *Environmental Science and Policy* 27, 223–31.

Blouin, M., M. E. Hodson, E. A. Delgado, G. Baker, L. Brussaard, et al. (2013). A review of earthworm impact on soil function and ecosystem services. *European Journal of Soil Science* 64(2): 161–82.

Bommarco, R., D. Kleijn and S. Potts (2013). Ecological intensification: harnessing ecosystem services for food security. *Trends in Ecology and Evolution* 28(4): 230–8.

Bondy, M. and D. C. Cole (2019). Change as a double-edged sword: ecological farmers' stressors and responses to changes in farming in Grey County, Ontario. *The Journal of Rural and Community Development* 14(4): 114–31.

Bowman, M. S. and D. Zilberman (2013). Economic factors affecting diversified farming systems. *Ecology and Society* 18(1): 33.

Brooker, R. W., A. E. Bennett, W. F. Cong, T. J. Daniell, T. S. George, et al. (2015). Improving intercropping: a synthesis of research in agronomy, plant physiology and ecology. *New Phytologist* 206(1): 107–17.

Brown, M. J. F. and R. J. Paxton (2009). The conservation of bees: a global perspective. *Apidologie* 40(3): 410–16.

Bruner, A. G., R. E. Gullison, R. E. Rice and G. A. B. Da Fonseca (2001). Effectiveness of parks in protecting tropical biodiversity. *Science* 291(5501): 125–8.

Brush, S. B., J. E. Taylor and M. R. Bellon (1992). Technology adoption and biological diversity in Andean potato agriculture. *Journal of Development Economics* 39(2): 365–87.

Brussaard, L., P. C. de Ruiter and G. G. Brown (2007). Soil biodiversity for agricultural sustainability. *Agriculture, Ecosystems and Environment* 121(3): 233–44.

Butler, S. J., J. A. Vickery and K. Norris (2007). Farmland biodiversity and the footprint of agriculture. *Science* 315(5810): 381–4.

Carpenter, S. R., N. F. Caraco, D. L. Correll, R. W. Howarth, A. N. Sharpley, et al. (1998). Nonpoint pollution of surface waters with phosphorus and nitrogen. *Ecological Applications* 8(3): 559–68.

Carranza-Gallego, G., G. I. Guzmán, R. Garcia-Ruíz, M. G. de Molina and E. Aguilera (2019). Addressing the role of landraces in the sustainability of Mediterranean agroecosystems. *Sustainability* 11(21): 6029.

Carrié, R. J. G., D. R. George and F. L. Wäckers (2012). Selection of floral resources to optimise conservation of agriculturally-functional insect groups. *Journal of Insect Conservation* 16: 635–40.

Chamberlain, D. E., R. J. Fuller, R. G. H. Bunce, J. C. Duckworth and M. Shrubb (2000). Changes in the abundance of farmland birds in relation to the timing of agricultural intensification in England and Wales. *Journal of Applied Ecology* 37(5): 771–88.

Chamorro, L., R. M. Masalles and F. X. Sans (2016). Arable weed decline in Northeast Spain: does organic farming recover functional biodiversity? *Agriculture, Ecosystems and Environment* 223: 1–9.

Chan, K. M. A., M. R. Shaw, D. R. Cameron, E. C. Underwood and G. C. Daily (2006). Conservation planning for ecosystem services. *PLoS Biology* 4(11): 2138–52.

Chaplin-Kramer, R., M. E. O'Rourke, E. J. Blitzer and C. Kremen (2011). A meta-analysis of crop pest and natural enemy response to landscape complexity. *Ecology Letters* 14(9): 922–32.

Chazdon, R. L., C. A. Harvey, O. Komar, D. M. Griffith, B. G. Ferguson, M. Martínez-Ramos, H. Morales, R. Nigh, L. Soto-Pinto, M. Van Breugel, and S. M. Philpott (2009). Beyond reserves: A research agenda for conserving biodiversity in human-modified tropical landscapes. Biotropica 41(2): 142–53.

Clergue, B., B. Amiaud, F. Pervanchon, F. Lasserre-Joulin and S. Plantureux (2005). Biodiversity: function and assessment in agricultural areas – a review. *Agronomy for Sustainable Development* 25: 1–15.

Colbach, N., H. Darmency, A. Fernier, S. Granger, V. Le Corre, et al. (2017). Simulating changes in cropping practices in conventional and glyphosate-resistant maize. II. Weed impacts on crop production and biodiversity. *Environmental Science and Pollution Research* 24: 13121–35.

Cook, J. H., J. Beyea and K. H. Keeler (1991). Preserving biological diversity in the face of large-scale demands for biofuels. Symposium Papers – Energy from Biomass and Wastes, Washington DC.

Cook, S. M., Z. R. Khan and J. A. Pickett (2007). The use of push-pull strategies in integrated pest management. *Annual Review of Entomology* 52: 375–400.

Costa, C. A., M. C. Godinho, S. Duarte, C. Mateus, E. Figueiredo, et al. (2012). Functional biodiversity and farming techniques: How to measure impacts? *Acta Horticulturae* 933: 455-62.

Costanzo, A. and P. Bàrberi (2014). Functional agrobiodiversity and agroecosystem services in sustainable wheat production. A review. *Agronomy for Sustainable Development* 34: 327-48.

Costanzo, A. and P. Bàrberi (2016). Field scale functional agrobiodiversity in organic wheat: effects on weed reduction, disease susceptibility and yield. *European Journal of Agronomy* 76: 1-16.

Cresswell, J. E. (2011). A meta-analysis of experiments testing the effects of a neonicotinoid insecticide (imidacloprid) on honey bees. *Ecotoxicology* 20(1): 149-57.

Crossley Jr, D. A., B. R. Mueller and J. C. Perdue (1992). Biodiversity of microarthropods in agricultural soils: relations to processes. *Agriculture, Ecosystems and Environment* 40(1-4): 37-46.

Crutsinger, G. M., M. D. Collins, J. A. Fordyce and N. J. Sanders (2008). Temporal dynamics in non-additive responses of arthropods to host-plant genotypic diversity. *Oikos* 117: 255-64.

Cycoń, M., A. Markowicz and Z. Piotrowska-Seget (2013). Structural and functional diversity of bacterial community in soil treated with the herbicide napropamide estimated by the DGGE, CLPP and r/K-strategy approaches. *Applied Soil Ecology* 72: 242-50.

Dale, V. H. and S. Polasky (2007). Measures of the effects of agricultural practices on ecosystem services. *Ecological Economics* 64(2): 286-96.

Davey, J. S., I. P. Vaughan, R. Andrew King, J. R. Bell, D. A. Bohan, et al. (2013). Intraguild predation in winter wheat: Prey choice by a common epigeal carabid consuming spiders. *Journal of Applied Ecology* 50: 271-9.

Davis, A. S., J. D. Hill, C. A. Chase, A. M. Johanns and M. Liebman (2012). Increasing cropping system diversity balances productivity, profitability and environmental health. *PLoS ONE* 7: 10.

DeClerck, F. A. J., F. Jessica, P. Cheryl and R. Roseline (2011). Ecological approaches to human nutrition. *Food and Nutrition Bulletin* 32: S41-S50.

Delbaere, B., V. Mikos and M. Pulleman (2014). European Policy Review: Functional agrobiodiversity supporting sustainable agriculture. *Journal for Nature Conservation* 22: 193-4.

Delmont, T. O., E. Prestat, K. P. Keegan, M. Faubladier, P. Robe, et al. (2012). Structure, fluctuation and magnitude of a natural grassland soil metagenome. *ISME Journal* 6: 1677-87.

de Miranda, E. E. and C. Mattos (1992). Brazilian rain forest colonization and biodiversity. *Agriculture, Ecosystems and Environment* 40(1-4): 275-96.

Dennis, P. and G. L. A. Fry (1992). Field margins: can they enhance natural enemy population densities and general arthropod diversity on farmland? *Agriculture, Ecosystems and Environment* 40(1-4): 95-115.

Donald, P. F., R. E. Green and M. F. Heath (2001). Agricultural intensification and the collapse of Europe's farmland bird populations. *Proceedings of the Royal Society B: Biological Sciences* 268(1462): 25-9.

Duflot, R., R. Georges, A. Ernoult, S. Aviron and F. Burel (2014). Landscape heterogeneity as an ecological filter of species traits. *Acta Oecologica* 56: 19-26.

Duru, M., O. Therond and M. Fares (2015). Designing agroecological transitions: a review. *Agronomy for Sustainable Development* 35: 1237-57.

European Commission (2015). Food and farming. *Focus on Land*, 8p. Available: https://ec.europa.eu/agriculture/sites/agriculture/files/events/2015/outlook-conference/brochure-land_en.pdf

European Environment Agency (2004). *High Nature Value Farmland - Characteristics, Trends, and Policy Challenges*. European Environment Agency, Copenhagen, Denmark.

Evans, M. C. (2019). Revisiting the role of science in biodiversity conservation. *Proceedings Biodiversity Revisited Symposium 'Seeds of Change. Provocations for a New Research Agenda'*, Vienna (AT), 11–13 September, pp. 28–31.

Farber, S., R. Costanza, D. L. Childers, J. Erickson, K. Gross, et al. (2006). Linking ecology and economics for ecosystem management. *BioScience* 56(2): 121–33.

Fiedler, A. K., D. A. Landis and S. D. Wratten (2008). Maximizing ecosystem services from conservation biological control: the role of habitat management. *Biological Control* 45(2): 254–71.

Fischer, J., D. J. Abson, V. Butsic, M. J. Chappell, J. Ekroos, et al. (2014). Land sparing versus land sharing: moving forward. *Conservation Letters* 7(3): 149–57.

Flowers, T. J. (2004). Improving crop salt tolerance. *Journal of Experimental Botany* 55(396): 307–19.

Foley, J. A., R. DeFries, G. P. Asner, C. Barford, G. Bonan, et al. (2005). Global consequences of land use. *Science* 309(5734): 570–4.

Fornara, D. A. and D. Tilman (2008). Plant functional composition influences rates of soil carbon and nitrogen accumulation. *Journal of Ecology* 96(2): 314–22.

France Stratégie (2019). Faire de la politique agricole commune un levier de la transition agroécologique. Rapport, 106p. Available at: www.strategie.gouv.fr/publications/faire-de-politique-agricole-commune-un-levier-de-transition-agroecologique

French, D. D. and N. Picozzi (2002). 'Functional Groups' of bird species, biodiversity and landscapes in Scotland. *Journal of Biogeography* 29(2): 231–59.

Frith, U. (2020). Fast lane to slow science. *Trends in Cognitive Research* 24(1): 1–2.

Furtak, K., K. Gawryjołek, A. M. Gajda and A. Gałązka (2017). Effects of maize and winter wheat grown under different cultivation techniques on biological activity of soil. *Plant, Soil and Environment* 63: 449–54.

Gabriel, D., S. M. Sait, J. A. Hodgson, U. Schmutz, W. E. Kunin, et al. (2010). Scale matters: the impact of organic farming on biodiversity at different spatial scales. *Ecology Letters* 13(7): 858–69.

Gadgil, M. (1992). Conserving biodiversity as if people matter: a case study from India. *Ambio* 21(3): 266–70.

Gall, G. A. E. and M. Staton (1992). Conclusions. *Agriculture, Ecosystems and Environment* 42(1-2): 217–30.

Gallai, N., J. M. Salles, J. Settele and B. E. Vaissière (2009). Economic valuation of the vulnerability of world agriculture confronted with pollinator decline. *Ecological Economics* 68(3): 810–21.

Gardiner, M. M., D. A. Landis, C. Gratton, C. D. DiFonzo, M. O'Neal, et al. (2009). Landscape diversity enhances biological control of an introduced crop pest in the north-central USA. *Ecological Applications* 19(1): 143–54.

Garibaldi, L. A., I. Steffan-Dewenter, C. Kremen, J. M. Morales, R. Bommarco, et al. (2011). Stability of pollination services decreases with isolation from natural areas despite honey bee visits. *Ecology Letters* 14(10): 1062–72.

Garnett, T., M. C. Appleby, A. Balmford, I. J. Bateman, T. G. Benton, et al. (2013). Sustainable intensification in agriculture: premises and policies. *Science* 341(6141): 33–4.

Garnier, E. and M.-L. Navas (2012). A trait-based approach to comparative functional plant ecology: concepts, methods and applications for agroecology. A review. *Agronomy for Sustainable Development* 32: 365–99.

Garnier, E., M.-L. Navas and K. Grigulis (2015). *Plant Functional Diversity: Organism Traits, Community Structure, and Ecosystem Properties*. Oxford University Press, New York, NY, 256p.

Geiger, F., J. Bengtsson, F. Berendse, W. W. Weisser, M. Emmerson, et al. (2010). Persistent negative effects of pesticides on biodiversity and biological control potential on European farmland. *Basic and Applied Ecology* 11(2): 97–105.

Gianinazzi, S., A. Gollotte, M. N. Binet, D. van Tuinen, D. Redecker, et al. (2010). Agroecology: the key role of arbuscular mycorrhizas in ecosystem services. *Mycorrhiza* 20(8): 519–30.

Gill, R. J., O. Ramos-Rodriguez and N. E. Raine (2012). Combined pesticide exposure severely affects individual-and colony-level traits in bees. *Nature* 491(7422): 105–8.

Gkisakis, V., N. Volakakis, D. Kollaros, P. Bàrberi and E. M. Kabourakis (2016). Soil arthropod community in the olive agroecosystem: determined by environment and farming practices in different management systems and agroecological zones. *Agriculture, Ecosystems and Environment* 218: 178–89.

Goldstein, J. H., G. Caldarone, T. K. Duarte, D. Ennaanay, N. Hannahs, et al. (2012). Integrating ecosystem-service tradeoffs into land-use decisions. *Proceedings of the National Academy of Sciences of the United States of America* 109(19): 7565–70.

Gordon, L. J., C. M. Finlayson and M. Falkenmark (2010). Managing water in agriculture for food production and other ecosystem services. *Agricultural Water Management* 97(4): 512–19.

Green, R. E., S. J. Cornell, J. P. W. Scharlemann and A. Balmford (2005). Farming and the fate of wild nature. *Science* 307(5709): 550–5.

Gregory, R. D., A. van Strien, P. Vorisek, A. W. Gmelig Meyling, D. G. Noble, et al. (2005). Developing indicators for European birds. *Philosophical Transactions of the Royal Society B: Biological Sciences* 360(1454): 269–88.

Gurr, G. M., S. D. Wratten and J. M. Luna (2003). Multi-function agricultural biodiversity: pest management and other benefits. *Basic and Applied Ecology* 4: 107–16.

Hendrix, P. F., D. C. Coleman and D. A. Crossley Jr (1992). Using knowledge of soil nutrient cycling processes to design sustainable agriculture. *Journal of Sustainable Agriculture* 2(3): 63–82.

Henle, K., D. Alard, J. Clitherow, P. Cobb, L. Firbank, et al. (2008). Identifying and managing the conflicts between agriculture and biodiversity conservation in Europe – a review. *Agriculture, Ecosystems & Environment* 124: 60–71.

Holden, C. (1989). Entomologists wane as insects wax. *Science* 246(4931): 754–6.

Hole, D. G., A. J. Perkins, J. D. Wilson, I. H. Alexander, P. V. Grice, et al. (2005). Does organic farming benefit biodiversity? *Biological Conservation* 122(1): 113–30.

Holzschuh, A., I. Steffan-Dewenter, D. Kleijn and T. Tscharntke (2007). Diversity of flower-visiting bees in cereal fields: effects of farming system, landscape composition and regional context. *Journal of Applied Ecology* 44(1): 41–9.

Huston, M. A. (1992). Biological diversity and human resources. *Impact of Science on Society* 166: 121–30.

Iannetta, P. P. M., G. Begg, C. Hawes, M. Young, J. Russell, et al. (2007). Variation in Capsella (shepherd's purse): an example of intraspecific functional diversity. *Physiologia Plantarum* 129: 542–54.

IPCC (2014). *Climate Change 2014: Synthesis Report. Contribution of Working Groups I, II and III to the Fifth Assessment Report of the Intergovernmental Panel on Climate Change* [Core Writing Team, Pachauri, R. K. and Meyer, L. A. (eds)]. IPCC, Geneva, Switzerland, 151p.

IPES-Food (2016). From uniformity to diversity: a paradigm shift from industrial agriculture to diversified agroecological systems. Louvain-la-Neuve, Belgium, 96p.

Isaacs, R., J. Tuell, A. Fiedler, M. Gardiner and D. Landis (2009). Maximizing arthropod-mediated ecosystem services in agricultural landscapes: the role of native plants. *Frontiers in Ecology and the Environment* 7(4): 196–203.

Jeffries, P., S. Gianinazzi, S. Perotto, K. Turnau and J. M. Barea (2003). The contribution of arbuscular mycorrhizal fungi in sustainable maintenance of plant health and soil fertility. *Biology and Fertility of Soils* 37(1): 1–16.

Jensen, E. S., M. B. Peoples, R. M. Boddey, P. M. Gresshoff, H. N. Henrik, et al. (2012). Legumes for mitigation of climate change and the provision of feedstock for biofuels and biorefineries. A review. *Agronomy for Sustainable Development* 32(2): 329–64.

Johns, R. J. (1992). The influence of deforestation and selective logging operations on plant diversity in Papua New Guinea. In: Whitmore, T. C. and Sayer, J. A. (Eds), *Tropical Deforestation and Species Extinction*. Chapman & Hall, London, UK, pp. 143–7.

Jose, S. (2009). Agroforestry for ecosystem services and environmental benefits: an overview. *Agroforestry Systems* 76(1): 1–10.

Kareiva, P., S. Watts, R. McDonald and T. Boucher (2007). Domesticated nature: shaping landscapes and ecosystems for human welfare. *Science* 316(5833): 1866–9.

Kay, S., J. Crous-Duran, N. Ferreiro-Domínguez, S. García de Jalón, A. Graves, et al. (2018). Spatial similarities between European agroforestry systems and ecosystem services at the landscape scale. *Agroforestry Systems* 92: 1075–89.

Kennedy, A. C. and K. L. Smith (1995). Soil microbial diversity and the sustainability of agricultural soils. *Plant and Soil* 170(1): 75–86.

Kennedy, C. M., E. Lonsdorf, M. C. Neel, N. M. Williams, T. H. Ricketts, et al. (2013). A global quantitative synthesis of local and landscape effects on wild bee pollinators in agroecosystems. *Ecology Letters* 16(5): 584–99.

Kennish, M. J. (2002). Environmental threats and environmental future of estuaries. *Environmental Conservation* 29(1): 78–107.

Kibblewhite, M. G., K. Ritz and M. J. Swift (2008). Soil health in agricultural systems. *Philosophical Transactions of the Royal Society B: Biological Sciences* 363(1492): 685–701.

Kleijn, D. and W. J. Sutherland (2003). How effective are European agri-environment schemes in conserving and promoting biodiversity? *Journal of Applied Ecology* 40(6): 947–69.

Kleijn, D., R. Winfree, I. Bartomeus, L. G. Carvalheiro, M. Henry, et al. (2015). Delivery of crop pollination services is an insufficient argument for wild pollinator conservation. *Nature Communications* 6: 7414.

Klink, C. A. and R. B. Machado (2005). Conservation of the Brazilian Cerrado. *Conservation Biology* 19(3): 707–13.

Krebs, J. R., J. D. Wilson, R. B. Bradbury and G. M. Siriwardena (1999). The second silent spring? *Nature* 400(6745): 611–12.

Kremen, C. and A. Miles (2012). Ecosystem services in biologically diversified versus conventional farming systems: benefits, externalities, and trade-offs. *Ecology and Society* 17(4): 40.

Kremen, C., N. M. Williams and R. W. Thorp (2002). Crop pollination from native bees at risk from agricultural intensification. *Proceedings of the National Academy of Sciences of the United States of America* 99(26): 16812-16.

Kremen, C., A. Iles and C. Bacon (2012). Diversified farming systems: an agroecological, systems-based alternative to modern industrial agriculture. *Ecology and Society* 17(4): 44.

Kruess, A. and T. Tscharntke (1994). Habitat fragmentation, species loss, and biological control. *Science* 264(5165): 1581-4.

Kunz, T. H., E. B. de Torrez, D. Bauer, T. Lobova and T. H. Fleming (2011). Ecosystem services provided by bats. *Annals of the New York Academy of Sciences* 1223: 1-38.

Lal, R., R. F. Follett, B. A. Stewart and J. M. Kimble (2007). Soil carbon sequestration to mitigate climate change and advance food security. *Soil Science* 172(12): 943-56.

Landis, D. A., M. M. Gardiner, W. Van Der Werf and S. M. Swinton (2008). Increasing corn for biofuel production reduces biocontrol services in agricultural landscapes. *Proceedings of the National Academy of Sciences of the United States of America* 105(51): 20552-7.

Larson, L. R., C. B. Cooper, S. Futch, D. Singh, N. J. Shipley, et al. (2020). The diverse motivations of citizen scientists: does conservation emphasis grow as volunteer participation progresses? *Biological Conservation* 242: 108428.

Lawler, J. J., D. J. Lewis, E. Nelson, A. J. Plantinga, S. Polasky, et al. (2014). Projected land-use change impacts on ecosystem services in the United States. *Proceedings of the National Academy of Sciences of the United States of America* 111(20): 7492-7.

Lazzaro, M., A. Costanzo and P. Bàrberi (2018). Single vs multiple agroecosystem services provided by common wheat cultivar mixtures: weed suppression, grain yield and quality. *Field Crops Research* 221: 277-97.

Lazzaro, M., P. Bàrberi, M. Dell'Acqua, M. E. Pè, M. Limonta, et al. (2019). Unraveling diversity in wheat competitive ability traits can improve integrated weed management. *Agronomy for Sustainable Development* 39: 6.

Letourneau, D. K., J. A. Jedlicka, S. G. Bothwell and C. R. Moreno (2009). Effects of natural enemy biodiversity on the suppression of arthropod herbivores in terrestrial ecosystems. *Annual Review of Ecology, Evolution, and Systematics* 40: 573-92.

Lockie, S. (2009). Agricultural biodiversity and neoliberal regimes of agri-environmental governance in Australia. *Current Sociology* 57: 407-26, 470-1.

MacDonald, D., J. R. Crabtree, G. Wiesinger, T. Dax, N. Stamou, et al. (2000). Agricultural abandonment in mountain areas of Europe: environmental consequences and policy response. *Journal of Environmental Management* 59(1): 47-69.

Mäder, P., A. Fließbach, D. Dubois, L. Gunst, P. Fried, et al. (2002). Soil fertility and biodiversity in organic farming. *Science* 296(5573): 1694-7.

Malmqvist, B. and S. Rundle (2002). Threats to the running water ecosystems of the world. *Environmental Conservation* 29(2): 134-53.

Mannion, A. M. (1992). Sustainable development and biotechnology. *Environmental Conservation* 19(4): 297-306.

Marshall, E. J. P. and A. C. Moonen (2002). Field margins in northern Europe: their functions and interactions with agriculture. *Agriculture, Ecosystems and Environment* 89: 5-21.

Martin, E. A., M. Dainese, Y. Clough, A. Báldi, R. Bommarco, et al. (2019). The interplay of landscape composition and configuration: new pathways to manage functional biodiversity and agroecosystem services across Europe. *Ecology Letters* 22: 1083–94.

Matson, P. A., W. J. Parton, A. G. Power and M. J. Swift (1997). Agricultural intensification and ecosystem properties. *Science* 277: 504–9.

McNeely, J. A. and R. B. Norgaard (1992). Developed country policies and biological diversity in developing countries. *Agriculture, Ecosystems and Environment* 42(1–2): 194–204.

Meyling, N. V. and J. Eilenberg (2007): Ecology of the entomopathogenic fungi *Beauveria bassiana* and *Metarhizium anisopliae* in temperate agroecosystems: potential for conservation biological control. *Biological Control* 43(2): 145–55.

Millennium Ecosystem Assessment (2005). *Ecosystems and Human Well-being: Synthesis*. Island Press, Washington DC.

Moonen, A. C. and P. Bàrberi (2008). Functional biodiversity: an agroecosystem approach. *Agriculture, Ecosystems and Environment* 127(1–2): 7–21.

Mulder, Ch, D. De Zwart, H. J. Van Wijnen, A. J. Schouten and A. M. Breure (2003). Observational and simulated evidence of ecological shifts within the soil nematode community of agroecosystems under conventional and organic farming. *Functional Ecology* 17(4): 516–25.

Muller, A., C. Schader, N. El-Hage Scialabba, J. Brüggemann, A. Isensee, et al. (2017). Strategies for feeding the world more sustainably with organic agriculture. *Nature Communications* 8: 1290.

Ndebele-Murisa, M. R., C. P. Mubaya, L. Pretorius, R. Mamombe, K. Iipinge, et al. (2020). City to city learning and knowledge exchange for climate resilience in southern Africa. *PLoS ONE* 15(1): e0227915.

Nereu, M., R. H. Heleno, F. Lopez-Núñez, M. Agostinho and J. A. Ramos (2018). Effects of native biodiversity on grape loss of four castes: testing the biotic resistance hypothesis. *Web Ecology* 18(1): 15–27.

Nilsson, L., Y. Clough, H. G. Smith, J. Alkan Olsson, M. V. Brady, J. Hristov, P. Olsson, K. Skantze, D. Ståhlberg, and J. Dänhardt (2019). A suboptimal array of options erodes the value of CAP ecological focus areas. Land Use Policy 85, 407–18.

Njeru, E. M., G. Bocci, L. Avio, C. Sbrana, A. Turrini, et al. (2017). Functional identity has a stronger effect than diversity on mycorrhizal symbiosis and productivity of field grown organic tomato. *European Journal of Agronomy* 86: 1–11.

Norgaard, R. B. (1992). Coordinating disciplinary and organizational ways of knowing. *Agriculture, Ecosystems and Environment* 42(1–2): 205–16.

Orians, G. H. and P. Lack (1992). Arable lands. *Agriculture, Ecosystems and Environment* 42(1–2): 101–24.

Oro, D., M. Genovart, G. Tavecchia, M. S. Fowler and A. Martínez-Abraín (2013). Ecological and evolutionary implications of food subsidies from humans. *Ecology Letters* 16(12): 1501–14.

Pardon, P., J. Mertens, B. Reubens, D. Reheul, T. Coussement, et al. (2019). *Juglans regia* (walnut) in temperate arable agroforestry systems: effects on soil characteristics, arthropod diversity and crop yield. *Renewable Agriculture and Food Systems* 1–17.

Parris, K. (2001). OECD Agri-biodiversity indicators: background paper. *Proceedings OECD Expert Meeting on Agri-biodiversity Indicators*. Zurich, Switzerland, p. 42.

Patz, J. A., T. K. Graczyk, N. Geller and A. Y. Vittor (2000). Effects of environmental change on emerging parasitic diseases. *International Journal for Parasitology* 30(12–13): 1395–405.

Pearce, D. and D. Moran (1994). *The Economic Value of Biodiversity*. Earthscan, London, UK.

Perfecto, I. and J. Vandermeer (2010). The agroecological matrix as alternative to the land-sparing/agriculture intensification model. *Proceedings of the National Academy of Sciences of the United States of America* 107(13): 5786–91.

Petit, S., N. Munier-Jolain, V. Bretagnolle, C. Bockstaller, S. Gaba, et al. (2015). Ecological intensification through pesticide reduction: weed control, weed biodiversity and sustainability in arable farming. *Environmental Management* 56: 1078–90.

Phalan, B., M. Onial, A. Balmford and R. E. Green (2011). Reconciling Food Production and Biodiversity conservation: land sharing and land sparing compared. *Science* 333: 1289–91.

Pham, D. (2016). Public engagement is key for the future of science research. *NPJ Science of Learning* 1: 16010.

Potts, S. G., J. C. Biesmeijer, C. Kremen, P. Neumann, O. Schweiger, et al. (2010). Global pollinator declines: trends, impacts and drivers. *Trends in Ecology and Evolution* 25(6): 345–53.

Power, A. G. (2010). Ecosystem services and agriculture: tradeoffs and synergies. *Philosophical Transactions of the Royal Society B: Biological Sciences* 365(1554): 2959–71.

Powlson, D. S., P. J. Gregory, W. R. Whalley, J. N. Quinton, D. W. Hopkins, et al. (2011). Soil management in relation to sustainable agriculture and ecosystem services. *Food Policy* 36(Suppl. 1): S72–87.

Rai, P. K. and J. S. Singh (2020). Invasive alien plant species: their impact on environment, ecosystem services and human health. *Ecological Indicators* 111: 106020.

Ranaldo, M., S. Carlesi, A. Costanzo and P. Bàrberi (2020). Functional diversity of cover crop mixtures enhances biomass yield and weed suppression in a Mediterranean agroecosystem. *Weed Research* 60: 96–108.

Raudsepp-Hearne, C., G. D. Peterson and E. M. Bennett (2010). Ecosystem service bundles for analyzing tradeoffs in diverse landscapes. *Proceedings of the National Academy of Sciences of the United States of America* 107(11): 5242–7.

Reganold, J. P. and J. M. Wachter (2016). Organic agriculture in the twenty-first century. *Nature Plants* 2: 15221.

Renting, H., W. A. H. Rossing, J. C. J. Groot, J. D. Van der Ploeg, C. Laurent, et al. (2009). Exploring multifunctional agriculture. A review of conceptual approaches and prospects for an integrative transitional framework. *Journal of Environmental Management* 90: S112–23.

Requier, F., A. Fournier, Q. Rome and E. Darrouzet (2020). Science communication is needed to inform risk perception and action of stakeholders. *Journal of Environmental Management* 257: 109983.

Ricketts, T. H., J. Regetz, I. Steffan-Dewenter, S. A. Cunningham, C. Kremen, et al. (2008). Landscape effects on crop pollination services: are there general patterns? *Ecology Letters* 11(5): 499–515.

Rinaudo, V., P. Bàrberi, M. Giovannetti and M. G. A. van der Heijden (2010). Mycorrhizal fungi suppress aggressive agricultural weeds *Plant and Soil* 333: 7–20.

Rundlöf, M. and H. G. Smith (2006). The effect of organic farming on butterfly diversity depends on landscape context. *Journal of Applied Ecology* 43: 1121–7.

Sasaki, T., Y. Yoshihara, Y. Suyama and T. Nakashizuka (2011). Clipping stimulates productivity but not diversity in improved and semi-natural pastures in temperate Japan. *Agriculture, Ecosystems and Environment* 142: 428–31.

Sattenspiel, L. (2000). Tropical environments, human activities, and the transmission of infectious diseases. *Yearbook of Physical Anthropology* 43: 3-31.

Sayer, J., T. Sunderland, J. Ghazoul, J. -L. Pfund, D. Sheil, et al. (2013). Ten principles for a landscape approach to reconciling agriculture, conservation, and other competing land uses. *Proceedings of the National Academy of Sciences of the United States of America* 110(21): 8349–56.

Scherr, S. J. and J. A. McNeely (2008). Biodiversity conservation and agricultural sustainability: towards a new paradigm of 'ecoagriculture' landscapes. *Philosophical Transactions of the Royal Society B: Biological Sciences* 363(1491): 477–94.

Schorpp, Q. and S. Schrader (2016). Earthworm functional groups respond to the perennial energy cropping system of the cup plant (*Silphium perfoliatum* L.). *Biomass and Bioenergy* 87: 61–8.

Schröter, D., W. Cramer, R. Leemans, I. C. Prentice, M. B. Araújo, et al. (2005). Ecology: ecosystem service supply and vulnerability to global change in Europe. *Science* 310(5752): 1333-7.

Science for Environment Policy (2017). Agri-environmental schemes: how to enhance the agriculture-environment relationship. Thematic Issue 57. Issue produced for the European Commission DG Environment by the Science Communication Unit, UWE, Bristol, 44p. Available at: http://ec.europa.eu/science-environmentpolicy.

Serra, A. A., A. Miqueau, F. Ramel, I. Couée, C. Sulmon, et al. (2019). Species- and organ-specific responses of agri-environmental plants to residual agricultural pollutants. *Science of the Total Environment* 694, 133661.

Sharafatmandrad, M. and A. K. Mashizi (2020). Investigating distribution of ecosystem services in rangeland landscapes: an approach based on weighted key functional traits. *Ecological Indicators* 111: 105971.

Sloots, K. and A. W. van der Vlies (2007). Emission reduction by multipurpose buffer strips on arable fields. *Water Science and Technology* 56: 81–8.

Spence, N., L. Hill and J. Morris (2019). How the global threat of pests and diseases impacts plants, people, and the planet. *Plants, People, Planet* 2: 5–13.

Spurgeon, D. J., A. M. Keith, O. Schmidt, D. R. Lammertsma and J. H. Faber (2013). Land-use and land-management change: relationships with earthworm and fungi communities and soil structural properties. *BMC Ecology* 13: 46.

Stoate, C., N. D. Boatman, R. J. Borralho, C. R. Carvalho, G. R. de Snoo, et al. (2001). Ecological impacts of arable intensification in Europe. *Journal of Environmental Management* 63(4): 337–65.

Strobel, G. and B. Daisy (2003). Bioprospecting for microbial endophytes and their natural products. *Microbiology and Molecular Biology Reviews* 67(4): 491–502.

Sutter, L., M. Albrecht and P. Jeanneret (2018). Landscape greening and local creation of wildflower strips and hedgerows promote multiple ecosystem services. *Journal of Applied Ecology* 55: 612-20.

Swanson, T. M. (1992). Economics of a biodiversity convention. *Ambio* 21(3): 250–8.

Swinton, S. M., F. Lupi, G. P. Robertson and S. K. Hamilton (2007). Ecosystem services and agriculture: cultivating agricultural ecosystems for diverse benefits. *Ecological Economics* 64: 245–52.

Tamiru, A., T. J. A. Bruce, C. M. Woodcock, J. C. Caulfield, C. A. O. Midega, et al. (2011). Maize landraces recruit egg and larval parasitoids in response to egg deposition by a herbivore. *Ecology Letters* 14: 1075–83.

Tanksley, S. D. and S. R. McCouch (1997). Seed banks and molecular maps: unlocking genetic potential from the wild. *Science* 277(5329): 1063-6.

Tilman, D. (1999). Global environmental impacts of agricultural expansion: the need for sustainable and efficient practices. *Proceedings of the National Academy of Sciences of the United States of America* 96(11): 5995-6000.

Tilman, D., P. B. Reich, J. Knops, D. Wedin, T. Mielke, et al. (2001). Diversity and productivity in a long-term grassland experiment. *Science* 294(5543): 843-5.

Tilman, D., K. G. Cassman, P. A. Matson, R. Naylor and S. Polasky (2002). Agricultural sustainability and intensive production practices. *Nature* 418: 671-7.

Tittonell, P. (2014). Ecological intensification of agriculture - sustainable by nature. *Current Opinion in Environmental Sustainability* 8: 53-61.

Topp, E. (2003). Bacteria in agricultural soils: diversity, role and future perspectives. *Canadian Journal of Soil Science* 83(3): 303-9.

Trivellone, V., N. Schoenenberger, B. Bellosi, M. Jermini, F. de Bello, et al. (2014). Indicators for taxonomic and functional aspects of biodiversity in the vineyard agroecosystem of Southern Switzerland. *Biological Conservation* 170: 103-9.

Tscharntke, T., A. M. Klein, A. Kruess, I. Steffan-Dewenter and C. Thies (2005). Landscape perspectives on agricultural intensification and biodiversity - ecosystem service management. *Ecology Letters* 8(8): 857-74.

Tscharntke, T., Y. Clough, S. A. Bhagwat, D. Buchori, H. Faust, et al. (2011). Multifunctional shade-tree management in tropical agroforestry landscapes - a review. *Journal of Applied Ecology* 48(3): 619-29.

Tscharntke, T., Y. Clough, T. C. Wanger, L. Jackson, I. Motzke, et al. (2012). Global food security, biodiversity conservation and the future of agricultural intensification. *Biological Conservation* 151(1): 53-9.

Tylianakis, J. M., T. Tscharntke and O. T. Lewis (2007). Habitat modification alters the structure of tropical host-parasitoid food webs. *Nature* 445(7124): 202-5.

United Nations (1992). Convention on biological diversity, 30p. Available at: www.cbd.int/doc/legal/cbd-en.pdf.

van Bueren, E. T. L., P. C. Struik and E. Jacobsen (2002). Ecological concepts in organic farming and their consequences for an organic crop ideotype. *Netherlands Journal of Agricultural Science* 50: 1-26.

Van Mele, P. and H. Van Chien (2004). Farmers, biodiversity and plant protection: developing a learning environment for sustainable tree cropping systems. *International Journal of Agricultural Sustainability* 2: 67-76.

van Rijn, P. C. J. and F. L. Wäckers (2016). Nectar accessibility determines fitness, flower choice and abundance of hoverflies that provide natural pest control. *Journal of Applied Ecology* 53: 925-33.

Verburg, P. H., J. van de Steeg, A. Veldkamp and L. Willemen (2009). From land cover change to land function dynamics: a major challenge to improve land characterization. *Journal of Environmental Management* 90(3): 1327-35.

Verheyen, K., M. Buggenhout, P. Vangansbeke, A. De Dobbelaere, P. Verdonckt, et al. (2014). Potential of short rotation coppice plantations to reinforce functional biodiversity in agricultural landscapes. *Biomass and Bioenergy* 67: 435-42.

Vickery, J. A., J. R. B. Tallowin, R. E. Feber, E. J. Asteraki, P. W. Atkinson, et al. (2001). The management of lowland neutral grasslands in Britain: effects of agricultural practices on birds and their food resources. *Journal of Applied Ecology* 38(3): 647-64.

Violle, C., M.-L. Navas, D. Vile, E. Kazakou, C. Fortunel, et al. (2007). Let the concept of trait be functional! *Oikos* 116: 882-92.

Walter, D. E. and S. Winterton (2007). Keys and the crisis in taxonomy: extinction or reinvention? *Annual Review of Entomology* 52: 193-208.

Weisser, W. W., C. Roscher, S. T. Meyer, A. Ebeling, G. Luo, et al. (2017). Biodiversity effects on ecosystem functioning in a 15-year grassland experiment: patterns, mechanisms, and open questions. *Basic and Applied Ecology* 23: 1-73.

Wezel, A., S. Bellon, T. Doré, C. Francis, C. Vallod, et al. (2009). Agroecology as a science, a movement and a practice. A review. *Agronomy for Sustainable Development* 29: 503-15.

Willett, W., J. Rockström, B. Loken, M. Springmann, T. Lang, et al. (2019). Food in the Anthropocene: the EAT-Lancet Commission on healthy diets from sustainable food systems. *The Lancet* 393, 447-92. doi:10.1016/S0140-6736(18)31788-4.

Winfree, R., N. M. Williams, J. Dushoff and C. Kremen (2007). Native bees provide insurance against ongoing honey bee losses. *Ecology Letters* 10(11): 1105-13.

Winfree, R., J. W. Fox, N. M. Williams, J. R. Reilly and D. P. Cariveau (2015). Abundance of common species, not species richness, drives delivery of a real-world ecosystem service. *Ecology Letters* 18(7): 626-35.

Winqvist, C., J. Bengtsson, E. Öckinger, T. Aavik, F. Berendse, et al. (2014). Species' traits influence ground beetle responses to farm and landscape level agricultural intensification in Europe. *Journal of Insect Conservation* 18: 837-46.

Yachi, S. and M. Loreau (1999). Biodiversity and ecosystem productivity in a fluctuating environment: the insurance hypothesis. *Proceedings of the National Academy of Sciences of the United States of America* 96: 1463-8.

Zhang, W., T. H. Ricketts, C. Kremen, K. Carney and S. M. Swinton (2007). Ecosystem services and dis-services to agriculture. *Ecological Economics* 64(2): 253-60.

Zhao, N., W. Lin, K. Cai and J. Wang (2010). Impacts of livestock waste compost on tomato bacterial wilt, soil enzyme activity and soil microbial functional diversity. *Shengtai Xuebao/Acta Ecologica Sinica* 30(19): 5327-37.

Zhu, Y., H. Chen, J. Fan, Y. Wang, Y. Li, et al. (2000). Genetic diversity and disease control in rice. *Nature* 406(6797): 718-22.

Zimmerer, K. S. (1992). The loss and maintenance of native crops in mountain agriculture. *GeoJournal* 27(1): 61-72.

Part 2

Management practices to support agroecosystem services

Chapter 5

The role of field margins in biodiversity conservation in agroecosystems

Alicia Cirujeda and Gabriel Pardo, Centro de Investigación y Tecnología Agroalimentaria de Aragón (CITA-Universidad de Zaragoza), Spain

1 Introduction

There is increasing evidence of serious declines in the abundance and diversity of fauna such as invertebrates, including pollinators, in many regions, in the United States (van Engelsdorp et al., 2009) and Europe (Potts et al., 2010). Research in Germany, for example, estimated a 76% seasonal decline of flying insect biomass between 1989 and 2016 (Vogel, 2017; Hallmann et al., 2017). A recent review of over 70 studies has confirmed that insect loss is happening across the planet (Sánchez-Bayo and Wyckhuys, 2019). Public concern has focussed on the importance of pollination for crop production, though other insects are also important to preserve natural biodiversity, for example, as a food source for birds, bats and reptiles. Multiple anthropogenic pressures seem to be responsible for insect pollinator declines, including intensification in land use for agriculture and other purposes (leading to habitat loss and fragmentation), intensive farming practices (reducing the availability of food), pollution, climate change and the spread of alien species and diseases (Tscharntke et al., 2005; Vanbergen and the Insect Pollinators Initiative, 2013). The seriousness of the problem has been recently highlighted by a group of leading scientists (Harvey

http://dx.doi.org/10.19103/AS.2020.0071.08

et al., 2020). They have called for measures to halt this decline such as reversing recent trends in agricultural intensification (including the reduced application of synthetic pesticides) and increasing landscape heterogeneity, for example, through the establishment and maintenance of natural areas.

This chapter reviews research on ways of modifying the agricultural landscape to reverse the decline in a range of fauna and flora. The chapter starts by looking at the span of options for promoting biodiversity in agricultural landscapes. It then focusses on field margins, their characteristics and types, as well as the role of field margins in agroecosystems. The chapter then reviews research on managing field margins to promote insect biodiversity. It also looks at managing field margins to promote rare arable plants (RAPs), reptiles, amphibians, birds and mammals.

2 Options for promoting biodiversity in agricultural landscapes

A key initial issue is how far to promote biodiversity 'inside' or 'outside' cultivated areas, that is, whether to prioritise preserving or creating dedicated uncultivated areas as wildlife habitats or to focus on changing on-farm practices. Most agronomists and farmers prefer to promote biodiversity outside the farms. In fact, retaining non-cultivated areas has been a common practice in official land consolidation programmes, for example, in Spain since the 1950s (Bielza de Ory, 1989). The main goal of these programmes has been to develop land for more intensive agriculture, for example, to increase the field size to facilitate irrigation and mechanisation. However, an environmental impact assessment is now compulsory in these programmes to identify key areas of natural habitat which need to be preserved near to the fields as well as field boundaries separating terraced fields in steep areas. These areas can be used to create a network of suitably diverse and connected habitats through Natural Resource Restoration Projects, which have become an established part of many land consolidation programmes (Ramírez del Palacio and Hernández, 2013).

Many insects need landscape complexity/heterogeneity offered, for example, by non-crop areas or perennial vegetation located close to a crop food source where they can live throughout the year (Benton et al., 2003; Paredes et al., 2015). Permanent set-aside land where natural vegetation can recover (towards the state represented by climax vegetation in non-cultivated areas) should be interconnected by corridors such as field margins to optimise species biodiversity (Tischendorf and Wissel, 1997; Rands and Whitney, 2011; Landis, 2017). Thomas et al. (2001) have shown the different spatial and seasonal use of varying habitats by beetles, with some species living the whole year within the field, others in the margin and some in both at different times of the year. A review of over 30 studies has highlighted that different species respond in distinct ways to diverse scales of management and that the preservation of multiple taxonomic groups requires multiple scales of conservation combining both 'outside' and 'inside' field approaches (Gonthier et al., 2014).

There are a number of options for promoting plant and habitat biodiversity 'inside' fields, that is, by changing agronomic practices. However, this is a difficult task that requires maintaining diverse weed populations without significant yield reduction. Intensive farming practices favour crop monocultures with low levels of biodiversity. These low levels of biodiversity are reinforced by insecticides to control insect pests and herbicides to control weed populations. These pesticides may also suppress non-target species, whether beneficial insects such as pollinators or plants on which they rely for food (Fried et al., 2009). One approach to promote greater floral diversity (which then benefits local fauna) is to modify herbicide regimes to target only the most competitive weed species while preserving other weed species in a way that still maintains crop yields (Ulber et al., 2010; Nicholls and Altieri, 2013; Davies, 2018). Some studies have shown that increasing species richness inside a field by tolerating some weeds may not damage yields but can promote biodiversity (Storkey and Neve, 2018; Mahaut et al., 2019). Targeting species to suppress, however, can be challenging since some species (e.g. *Papaver rhoeas* L. and *Centaurea cyanus* L.) can be considered as problem weeds in some countries (e.g. Spain) but endangered species in others (e.g. Germany) (Ulber et al., 2010).

Another approach is to modify cropping practices to increase biodiversity by introducing a greater range of crops which, for example, bloom at differing times to provide a more continuous and diverse source of food for insect populations. This can be achieved, for example, through more frequent rotations (Chiatante and Meriggi, 2016). An alternative is to grow a mix of crops simultaneously through techniques such as intercropping and the use of cover crops (Brooker et al., 2015; Wittwer et al., 2017). Incorporating flowering crops such as oilseed rape, sunflowers or legume pastures into cereal cropping systems can promote pollinators. This has been promoted by EU policies based on the concept of 'greening' to construct mosaic landscapes instead of monocultures. As an example, one EU agri-environmental scheme (AES) in north-eastern Spain promotes seeding of sainfoin (*Onobrychis viciifolia* (Scop.)) in rotation with cereals. No herbicides are used during the 3–4 years of crop establishment, which is mowed 1–2 times a year. It has been demonstrated that this scheme leads to higher weed species richness in the sainfoin fields compared to the nearby cereal fields, providing richer wildlife habitat (Cirujeda et al., 2019a).

3 Field margins: characteristics and types

A key tool in promoting biodiversity is field margins in their different versions (sometimes known as buffer strips). A field margin has been defined as 'the whole of the crop edge, any margin strip present and the semi-natural habitat associated with the boundary' (Greaves and Marshall, 1987; Marshall and Moonen, 2002) (Fig. 1). The boundary comprises the barrier that separates one crop field from another. Boundaries may include lines of trees, hedges, walls, fences, ditches and watercourses. The margin strip is any strip of uncultivated land between the edge of the crop and the boundary and generally exists only for practical reasons (e.g. to allow access or because cultivation is not feasible right up to the boundary), or it may be planned for conservation purposes. Field margins are typically non-continuous (e.g. to allow access to the field for farm machinery) (Photo 1).

Many different types of margins have been promoted by AESs (Hacket and Lawrence, 2014), including 'conservation headlands', which are parts of fields seeded with a crop but not treated with fertilisers, insecticides or herbicides, resulting in the growth of a number of broad-leaved weed species which increases insect abundance (SAC, 2002). In 'uncropped annual cultivation' margins (sometimes referred to as fallow or set-aside areas), a single 15 cm depth cultivation is conducted annually, but no crop is seeded. This encourages a wide diversity of annual plant species, in particular (Holland et al., 2007, 2013). Other options are grass-sown margins, wildflower-sown margins, pollen and nectar mix or wild bird seed mixes. Though they may be part of the field, they can be considered as the crop edge part of a field margin despite the fact that they generally exist only for some months in a year. 'Natural regeneration'

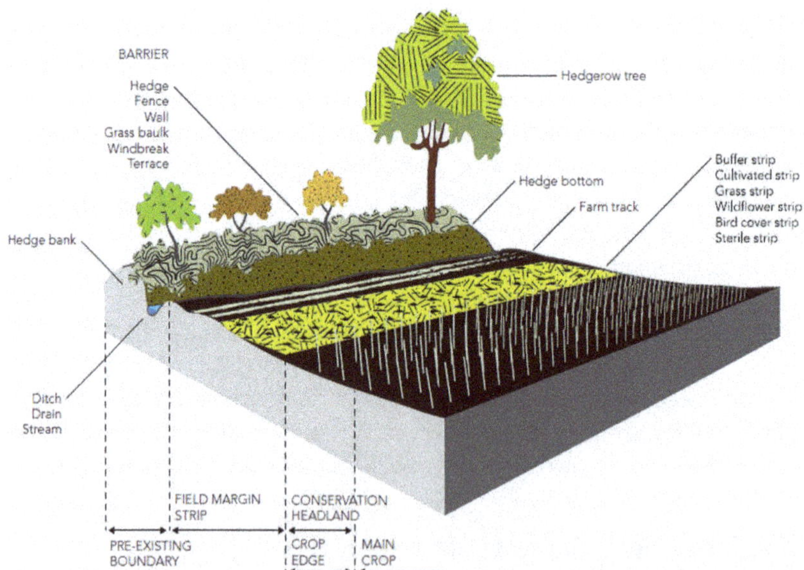

Figure 1 The principal components of an arable field margin (from Hackett and Lawrence (2014) after Greaves and Marshall (1987) and Marshall and Moonen (2002)).

Photograph 1 Non-continuous field boundaries facilitating the mechanisation of the agricultural plots.

margin management implies that woody growth may be cut after the first 12–24 months. Bushes or trees are targeted in these strips to avoid trees becoming established and boundaries evolving to hedges and lineal woods in short periods of time. A recent tendency is to plant margin strips with perennial species (e.g. in Switzerland, see Tschumi et al., 2016a) as annuals do not offer continuous habitats for birds and other wildlife animals (Hacket and Lawrence, 2014; Tschumi et al., 2016a).

Whatever the particular features of a field margin, it is a space where spontaneous vegetation may establish or planting can be undertaken specifically to promote biodiversity (e.g. grass or wildflower strips) (Hackett and Lawrence, 2014). Field margins may be relatively fixed and long-established with mature, perennial vegetation or may be more recent and variable in width and composition. In moist and nutrient-rich areas, boundaries tend to be colonised very rapidly with trees and bushes. In drier areas, boundaries may be covered with small perennial plants and bushes (Morrison et al., 2017; Cirujeda et al., 2019b).

The characteristics and pattern of field margins are determined by a range of factors such as topography and the nature of agricultural production in a region. Increasing intensification in agriculture in many areas has led to larger fields with fewer boundaries that are more amenable to mechanisation, as well as the push to exploit every square metre to maximise yields, leading to the reduction in the size of boundaries and reducing the width of field margin strips or even eliminating them altogether. Given that this is a very widespread development in Europe, most AESs promoted by the EU focus on the crop edge (forming part of the fields) and not on boundaries between fields, because they have often been eliminated. It is not by chance that all the countries considered in the review by Hackett and Lawrence (2014), except Greece, are located in central and northern Europe. Due to the topography and climatic conditions prevalent in Mediterranean countries, boundaries are still frequent there in many areas.

Cirujeda et al. (2019b) have used five key descriptors as a way of classifying field boundaries ranging from very simple boundaries (i.e. flat, narrow, dominated by annual species) up to complex boundaries (e.g. banked with a width of over 3 m and dominated by woody and evergreen species). Following this classification, different boundaries can be identified. In flat areas of northern France or in the Netherlands, for example, agriculture is very intensive, with practically no separation between highly productive fields (Kleijn and Verbeek, 2000) (Photo 2). Only some natural vegetation may be found, for example, in drainage channels. The absence of separations between fields in these landscapes is similar to what occurs in irrigated valleys in some Mediterranean countries where field margins are also minimal (Photo 3) (Cirujeda et al., 2019b). Margins are narrow, and no boundary structures are found. Shrubs or trees are non-existent, and the vegetation in these margins, when present, is probably very similar to the weeds found in fields (Cirujeda et al., 2019b).

Photograph 2 In huge extensions of highly productive flat areas no field separation at all makes the establishment of wildlife very difficult.

Photograph 3 Very narrow margins separate levelled fields in irrigated areas. These margins receive herbicide and fertiliser drifts from the nearby fields and are in occasions even treated with herbicides on purpose.

In hilly areas, land devoted to pasture typically requires few boundaries, as in areas around the Alps and Pyrenees (Photo 4). In contrast, other crops grown in hilly semi-arid or dry environments in Southern Europe (e.g. Spain, Portugal, Italy, Greece and Southern France) still commonly have extensive networks of boundaries, sometimes associated with terracing to create a series of level fields on a slope, including margins in more arid areas where they may help reduce erosion and assure rainfall infiltration (Photos 5 and 6). In these areas the need for sowing flower strips is not always considered necessary because those field margins may offer a well-established and diverse flora with perennial grasses and small bushes dotted with some annuals (Photo 6), providing similar ecosystem services to especially sown margins in other countries.

When rainfall allows the establishment of hedge trees (e.g. in Britain or also in the foothills of the Pyrenees and other mountains), certain tree species, for example, *Fraxinus excelsior* L. (ash), are retained as hedgerow trees in field margins both as windbreaks and as a source of livestock feed in mixed farming systems (Montserrat, 2009) (Photo 7). These boundaries host the most complex vegetation and are also used to smooth slopes and reduce erosion (Bassa et al., 2012; Solé-Senan et al., 2014, Marshall and Arnold, 1995, Cirujeda et al., 2019b). In many areas these margins act as a barrier between agricultural

Photograph 4 Pastures in alpine regions where margins and boundaries are absent despite the steep conditions because the soil is covered all year round.

Photograph 5 Field boundaries in a dryland landscape with a bank. In this case, mainly annual species grow on the boundary.

Photograph 6 Field boundaries in a dryland landscape with banks where bushes and other perennial vegetation is allowed to grow.

Photograph 7 Boundaries with a bank, including trees.

Photograph 8 Natural vegetation establishes below or on the stonewalls when fields are used for agricultural purposes. When stonewalls separate pastured meadows, there is generally less natural vegetation left.

areas and woodland or forest (Marshall and Arnold, 1995) in countries such as Norway (Hovd and Skogen, 2005) or the United States (Reberg-Horton et al., 2011).

Finally, in some areas, fields are separated from each other by stone walls, which have an equivalent role to hedges or fences (Photo 8). These structures are typical in many European regions, especially in Mediterranean countries and sheep or cattle farming in parts of northern England, Scotland and Ireland. Perennial evergreen plants such as *Rubia tinctorum* L., *Smilax aspera* L. or different *Sedum* species and so on may grow between the stones which can also be a refuge for wildlife such as lizards, spiders and insects. Some vegetation may also become established on and under the walls, and almond and other tree species are sometimes planted as part of agroforestry systems next to stonewalls (e.g. in areas of Italy, Greece, Portugal and Spain) (Cirujeda et al., 2019b) (Photo 8). In the case of pasture systems, animals typically graze the whole field and limited natural vegetation strips are found adjacent to walls.

4 The role of field margins in agroecosystems

Despite plenty of evidence that field margins in their different ways are generally beneficial in attracting biodiversity, it is still the case that too few agricultural landscapes contain these structures (Walker et al., 2018). There are a number of reasons why field boundaries are not widespread or evenly distributed throughout the agricultural landscapes and why most AESs focus on annual crop edges and less on promoting the presence of perennial species. These include intensive land use by farmers trying to profit from using as much land for crop production as possible, as well as the concentration of land ownership among fewer owners, which has led to increasing mechanisation that pushes farmers to destroy such physical barriers as crop boundaries.

Another reason is the possible negative interaction between a field margin and the adjacent crop. Margins may be occupied by plant species which compete for resources with the adjacent crop. In areas with high rainfall and fertile soils, trees may develop, shading the crop (so depriving it of light) and taking up water and nutrients from the field. In irrigated areas, perennial tall-growing species like *Arundo donax* L., *Phragmites communis* Trin. may become established. In other cases, smaller but very competitive perennials may grow such as *Elytrigia repens* (L.) Desv. ex. B.D. Jacks in the United Kingdom, the Netherlands and Belgium (Marshall and Arnold, 1995; Kleijn and Verbeek, 2000; de Cauwer et al., 2008), *E. repens* and *Cirsium arvense* L. in France (Cordeau et al., 2012), *Sonchus arvensis* L. and *C. arvense* and *Urtica dioica* L. in Belgium (de Cauwer et al., 2008). Because of this, farmers frequently consider field boundaries to be a source of weeds and pests (Kleijn and Verbeek, 2000; Marshall and Arnold, 1995; Mante and Gerowitt, 2009; Reberg-Horton et al., 2011).

As well as being a problem in themselves by competing with crops for nutrients and other resources, weeds can also host pests which damage crops. As an example, a study conducted to assess the growth of two aphid species on various weed species showed that *Veronica hederifolia* L. and *Solanum nigrum* L. were the most suitable hosts for *Myzus persicae*, whereas *Amsynckia intermedia* Fish. and Mey and *Amaranthus retroflexus* L. were the most suitable weed hosts for *Aphis fabae* (Fernandez-Quintanilla et al., 2002). *Philaenus spumarius*, one of the most common carriers of the *Xylella fastidiosa*, is commonly found on *Sonchus* spp., *Malva* spp. and other weeds (Dongiovanni et al., 2019). As a result, field margins can be perceived as a source of weeds and pests, which negatively impact on crop production (Kleijn and Verbeek, 2000; Marshall and Arnold, 1995; Mante and Gerowitt, 2009; Reberg-Horton et al., 2011). However, some authors consider field margins more as a refuge for arable field plants (including weeds) escaping disturbance by agricultural management rather than as a potential infestation source (Marshall and Arnold, 1995; Poggio et al., 2010).

Apart from serving as a clear physical separation of plots owned by different people, field margins can deliver an important range of ecosystem service benefits to farming and the wider community (Hackett and Lawrence,

2014; Syngenta, Arcadis and Bioversity International, 2018). These services include the following:

- Preventing soil erosion,
- Optimising crop nutrition by helping to retain fertilisers in the field,
- Reducing runoff or spray drift from fertilisers and pesticides, including protecting neighbouring water courses from contamination,
- Improving water retention in soils, especially in hilly areas where moisture is a limiting factor or helping to prevent flood risk,
- Providing shade for crops and livestock (where too much sunlight may be damaging),
- When margins include shrubs or trees, they provide forage, fodder and wood,
- Acting as a windbreak, and
- Promoting beneficial flora and fauna, often serving as ecological corridors, including biota promoting soil health (such as earthworms, woodlice and beetles), pollinators and natural predators of pest species.

Broader ecosystem services include reducing air pollution, improving carbon sequestration and enhancing landscape aesthetics, which may then benefit farms that have diversified into tourism and recreational activities (Wratten et al., 2012).

5 Managing field margins to promote insect biodiversity

Because of the concern that spontaneous vegetation might bring pests into the crops, it is easier to convince farmers and advisors to promote the establishment of certain plant species stating clearly the practical purpose of introducing these species (e.g. attracting natural enemies of existing pests). Since the 1980s there have been lists of plant species, which attract predator species for the biological control of pests (especially in fruit orchards) (e.g. Altieri and Letourneau, 1982). Some of the most popular plants which have been shown to attract natural enemies and which are easy to seed and to maintain are the annual forbs *Tagetes patula* L. (Rahat et al., 2005), *Calendula arvensis* L. (Martinez-Una et al., 2013), *Diplotaxis erucoides* (L.) D.C. (Araj and Wratten, 2015), *Fagopyron esculentum* Moench (Jado et al., 2019; Barloggio et al., 2019), *Phacelia tanacetifolia* Benth. (Baggen et al., 1999), *Centaurea cyanus* L. (Balmer et al., 2014) and several Fabaceae such as, for example, *Vicia sativa* L. (Barloggio et al., 2019). Within the perennial forbs some species stand out, including *Dittrichia viscosa* (L.) Greuter (Kavallieratos et al., 2002) or *Malva sylvestris* L. (hosting different *Orius* spp. species, Bosco and Tavella, 2013). However, it is important to note that the relationships between plants, pests and their natural enemies are complex, and in certain occasions a beneficial plant can also host pests (e.g. *F. esculentum*., Baggen et al., 1999).

Mixing functional traits of the plants (e.g. flowering periods, flower size, flower height) increases the availability of natural enemies and pollinators throughout the crop season (e.g. Balzan et al., 2014). As an example, it is important to take account of the preference of pollinating insects for flying at particular heights in order to conserve energy (Dafni and Potts, 2004). Other research has highlighted the importance of sward architectural complexity for invertebrate habitats (Westbury et al., 2017). Given that nesting pollinators travel relatively small distances from nests to forage, it is important to link field margins, wild areas and other food sources (e.g. crops) as closely as possible (Rands and Whitney, 2011; Stanley and Stout, 2014). Guidelines have been established for selecting perennial floral mixtures to optimise those functional traits (Uyttenbroeck et al., 2015; Cresswell et al., 2019). A more recent focus has been on planting more perennial species since they can provide more continuous habitats for wildlife (Hacket and Lawrence, 2014; Tschumi et al., 2016a).

Management of field margins needs to reconcile different roles, for example, promoting biodiversity of a range of flora and fauna while maintaining or enhancing crop production (Olson and Wackers, 2007). In the review by Hackett and Lawrence (2014) comparing a range of different types of field margin management and their relative environmental benefits, their general conclusion is that longer-term and less-disturbed field margins, such as natural regeneration, grass- or wildflower-sown margins provide the broadest and most consistent environmental benefits for a range of flora and fauna, including insects. The same authors also review factors affecting optimal width and length of margins in particular locations, as well as the nature and timing of interventions such as mowing to control grass and weeds in favour of wild flowering plants. Good management practices for field margins are also outlined by Natural England (2013).

Less research has been conducted on the attractiveness of natural field margins to insects compared to other wildlife. These margins are often more diverse than floral strips since they are spontaneously colonised by a wide range of vegetation types. A study of natural field margins in north-eastern Spain in boundaries >1 m width found that the plant species most visited by bees were *M. sylvestris*, *Ligustrum vulgare* L., *Vicia villosa* Roth., *Rubus* spp. L., *P. rhoeas* and *Torilis arvensis* Huds. (Morrison et al., 2017). The four most visited species were perennial forbs, while the fifth and sixth species most visited were annual weeds within nearby crop fields. The perennial forbs probably attract bees to a greater extent than annual species due to their vigour and provision of more flowers at the same level. However, in a study on organic tomato, Balzan and Moonen (2014) found that, while semi-natural habitats were associated with higher vegetation diversity, natural enemies of crop pests were more strongly associated with sown strips during flowering than with semi-natural habitats.

Grass strips have been found to have beneficial effects, despite providing less nectar than wildflower mixes, depending on factors such as landscape structure (Marshall et al., 2006). Holland et al. (2012) have shown that sown grass strips in England can have a positive role in sheltering natural enemies of cereal aphids when these enemies were released in the area. However, from the point of view of promoting pollination, it has been shown that uncropped margins sown with mixtures containing nectar and pollen-producing plants were more effective in providing bumblebee forage than margins sown with a grass mix (Carvell et al., 2007).

A range of research suggests that, while grass-sown margins are also beneficial, wildflower-sown margins can be particularly effective in promoting a range of insects, including pollinators (Scheper et al., 2013; Blaauw and Isaacs, 2014; Morrison et al., 2017), butterflies (Blake et al., 2011), spiders (Blake et al., 2013) and beetles (Woodcock et al., 2005, 2008) as well as combinations of these species (Marshall et al., 2006; Pywell et al., 2011; Dicks et al., 2013; Holland et al., 2013). Through AESs the EU has promoted sowing flower strips in field margins and set-aside land in countries such as Belgium, Czech Republic, Estonia, Finland, Germany, Greece, Ireland, the Netherlands, Sweden, Switzerland and United Kingdom, particularly to benefit pollinators (Rundlof and Bommarco, 2011; Hackett and Lawrence, 2014). Flower mixtures sown with annual plants in field margins offer abundant nectar and pollen, which attracts natural enemies of pests and thus decreases insect pest infestations on nearby crops (Westbury et al., 2008). As an example, an increase of natural enemies of aphids (hoverflies, lacewings and ladybirds) has been demonstrated in flower mixes adjacent to potato (Tschumi et al., 2016b), and natural enemy densities were also found to be higher within and close to wildflower strips in oilseed rape and wheat (Pollier et al., 2019). However, findings of Ganser et al. (2019) demonstrated that the annuals wildflower strips ploughed during the overwintering period are poor overwintering habitats and may even act as ecological traps for arthropods.

Perennial flower strips were also found to promote parasitoids of coleopteran pests of oilseed rape (Hatt et al., 2018). A significant reduction of the density of cereal leaf beetle *Oulema* spp. was observed in species-rich wildflower strips next to winter cereals (Tschumi et al., 2016a). When perennial flower strips were sown covering as much as 10% of a conventionally managed agricultural landscape, the richness of wild bees and butterflies increased three- to five-fold after more than 2 years, while other species needed 3 years to increase substantially (Buhk et al., 2018).

6 Managing field margins to promote rare arable plants, reptiles, amphibians, birds and mammals

6.1 Rare arable plants

It has been widely demonstrated that RAPs establish best in non-fertilised crop edges or, more generally, under low-input or organic farming (Albrecht et al., 2016; Walker et al., 2007). The main cause of their decline has been attributed to nitrification (from fertiliser use), high crop densities, herbicides and crop seed cleaning (reviewed by Fried et al., 2009). According to Hacket and Lawrence (2014), 'conservation headlands' and 'uncropped annual cultivation' margin types offer the best habitats for RAP. Studies of plant distributions at the edges of arable fields indicate that RAPs are more likely to persist in the seedbank of crop edges rather than in field centres (Marshall, 1989; Wilson and Aebischer, 1995).

However, there are regional differences in the performance of these ecological structures. A survey conducted in Finland on 193 boundaries found that RAP flora may not establish in conventional boundaries at all when no specific measures are taken (Tarmi et al., 2002). A study conducted in north-eastern Spain showed that RAPs were most common on boundaries with the presence of a bank and with <60% of woody and evergreen perennials (excluding trees) or on boundaries with a stonewall (Cirujeda et al., 2019b). In contrast, lower density and abundance of RAP were found in flat and narrow boundaries or in those covered with abundant woody and evergreen perennials. Another study has shown that low-input margins in more complex landscapes are the most suitable habitat for RAP (Solé-Senan et al., 2014).

6.2 Reptiles and amphibians

While they are not known as a crop pest, many reptiles are insectivorous, which means they can contribute to pest management by reducing harmful insect populations. Moreover, reptiles are a food source for some birds. However, despite their potential benefits, herpetofauna are seriously threatened by agricultural intensification worldwide (Biaggini and Corti, 2015). Studies on reptile distribution in agroecosystems have shown that few lizard species live inside crop fields though some generalists manage to survive even after a crop has been harvested (Biaggini et al., 2011; Hansen et al., 2019). Research on the Sonaran Desert in the southwestern United States has compared undisturbed mesquite shrubland (*Prosopis juliflora*) with irregularly shaped clearings and mesquite-free areas (Germano and Hungerford, 1981). Results clearly showed that the total reptile sightings were higher in the first two ecosystems. The authors suggested that, because of the smaller size and lower mobility of reptiles, numbers are closely related to vegetation diversity, more than large mammals and birds. Biaggini and Corti (2015) studied reptile abundance and diversity in eight locations, including crop fields, semi-natural areas and vegetated buffer strips interspersed among fields. They found that most reptiles were present in the buffer strips while intensive crops and pastures hosted just one lizard species. Numbers of individuals and species increased when strips were connected to semi-natural areas, independent of their width and vegetation structure. There has also been some work showing that stonewalls serving as field boundaries are an appropriate habitat for reptiles and their prey (Foster and Baker, 2010).

In the case of amphibians, it is known that habitat connectivity is crucial for their survival (Cushman, 2006). However, few studies have been conducted in agroecosystems and, more specifically, in field margins and boundaries (Hansen et al., 2019). Since this group of animals requires specific moisture conditions for food and breeding, buffer zones around water sources are more useful than field boundaries. A study of Dutch ditch systems in grazing meadows for dairy cattle found that ditches managed in a more nature-friendly way through an AES (e.g. by promoting vegetative cover) had a significantly higher frog abundance than when ditches are managed conventionally (Maes et al., 2008). Field boundaries may also provide corridors linking these habitats.

6.3 Birds

The negative impact of modern farming techniques on farmland birds has been known since at least the 1960s in the United States, and Rachel Carson's *Silent Spring* (Carson, 1962) is considered one of the most important environmental books of the twentieth century. In the United Kingdom, for example, some species of farmland birds were found to have declined by over 50% between the 1960s and 1990s (Fuller et al., 1995; Gregory and Marchant, 1996). A recent meta-analysis of 122 US studies concluded that pesticides and habitat loss were the predominant factors affecting farmland birds (reviewed by Stanton et al., 2018). They emphasised the importance of reducing pesticide inputs and maintaining or restoring uncultivated field margins and native habitats. This has been confirmed by other research which has identified the main negative consequence of agricultural intensification as the loss of ecological heterogeneity at multiple spatial and temporal scales (Benton et al., 2003; Green et al., 2005). Recent findings in Spain, for example, suggest the reduction in fallow land as the main driver in reducing farmland birds (Traba and Morales, 2019).

Rands (1986) reported that the length of the permanent field boundary was closely related to the breeding density of two species of partridge. This author found that the amount of dead grass present in the base of the boundaries or hedges was the best predictor of grey partridge breeding density within farms. Potts (1980) has shown that the decrease of partridges in Great Britain was associated with the removal of hedges or similar nesting sites. Vickery et al. (2009) found that sown grass strips offer limited food resources compared to adding perennial forbs to grass mixtures. This created a structurally and floristically more diverse sward that was able to provide a food supply in both summer and winter. Hacket and Lawrence (2014) have also concluded that the combination of interconnected boundaries and differently sown field margins to provide diverse vegetation offers the best sources of food and habitat for birds and other species. Vickery et al. (2009), as well as Hacket and Lawrence (2014), conclude that a structurally and floristically diverse sward would be the best food supply for summer, whereas structures that offer seed-rich habitats are needed in winter.

It has been found that habitat designed for birds can also increase insect populations (Ouvrard and Jacquemart, 2018). A recent review of buffer strip management to deliver plant and invertebrate resources for farmland birds is provided by Westbury et al. (2017). It is important to note, however, that, in the United Kingdom, where AESs have been in operation since the late 1980s, farmland bird is still declining (Vickery and Tayleur, 2018). The results of one of these studies show that current efforts are not sufficient and that much more of the agricultural landscape needs to come under AES to be able to halt the ongoing bird declines (Walker et al., 2018). Moreover, in most cases the detailed relationships between bird populations and agricultural practices are still unknown (Fuller et al., 1995). This suggests that more research is needed to assess the effectiveness of current approaches (Walker et al., 2018; Stanton et al., 2018).

6.4 Mammals

Like birds, it has long been known that populations of small mammals such as the hare (*Lepus europaeus*) have suffered from the loss of habitat and food sources related to more intensive farming systems, which has then had a knock-on effect on predators such as birds of prey and owls (Tapper and Barnes, 1986). Macdonald et al. (2007) showed that sown grass field margins in the United Kingdom can support communities of smaller mammals, while hedgerows may act as important commuting and hunting routes. However, as observed for birds, these grass field margins might offer limited habitat, which can be improved by enriching them with forbs. Various factors such as cutting management and length of time the rests are left in situ will affect their potential. However, as highlighted for birds, effective conservation of farmland mammals must seek solutions at the landscape scale, addressing such issues as habitat connectivity between farms (Macdonald et al., 2007). Similarly, Rodríguez-Pastor et al. (2016) found much higher concentrations of mammals such as the common vole (*Microtus arvalis*) in field margins compared to adjacent fields, particularly when planted with diverse species and managed (e.g. by cutting) to maintain this diversity by controlling more dominant grass and weed species.

7 Conclusions

A key theme emerging from the research is the importance of restoring complexity to landscapes simplified by more intensive, monoculture-focussed methods of agricultural production (Tscharntke et al., 2005; Morrison et al., 2017). Landscape complexity has a positive influence on the distribution of insects (Woodcock et al., 2007), birds, amphibians, reptiles and mammals as well as RAP (Solé-Senan et al., 2014). Permanent field margins sown with a mixture of annual and perennial species of grasses and wildflowers (with perennials encouraging naturally regenerated boundaries) are a key environment where fauna can shelter, for example, after harvest in monocrop areas. Ideally, these should be combined with several crops planted nearby in a mosaic to provide a more varied and continuous food source, as well as to connect networks of field margin wildlife corridors to areas of uncultivated vegetation. In general terms, the greater the degree of spatial and temporal complexity, the greater the potential benefit for biodiversity, particularly for larger species such as mammals and birds.

A key challenge is to persuade farmers to sacrifice some potential areas of crop cultivation to help achieve this complexity, to put time and resources into creating or regenerating field margins as well as to adopt more agroecological practices, which reduce reliance on synthetic fertilisers and pesticides and result in more diverse cropping patterns (e.g. through rotations, cover crops or intercropping) and greater in-field diversity (e.g. by tolerating less competitive weed species). These are significant changes in terms of time, resources, expertise and potential impact on yields. A starting point has been planting grass and flower mixes in margin strips through AES. There is also a growing body of research that, given the ecosystem services they provide, measures to promote biodiversity such as field margins need not have a negative impact on crop yields and may, indeed, contribute to more sustainable production in the long term (Sutter et al., 2018).

8 Where to look for further information

The classic basic work on this topic was published as early as in 1987 in Great Britain including the article of Greaves and Marshall who designed a complete figure with the different components of a field margin (see Greaves and Marshall, 1987 in the References section for full details).

Later, new concepts were included such as the functional services of the field margins (see Pywell et al., 2011; Balzan and Moonen, 2014; Uyttenbroeck et al., 2015).

An interesting comparison of the effectivity of different margin types in Great Britain is documented in the report written by Hackett and Lawrence (2014): Multifunctional Role of Field Margins in Arable Farming (see Hacket and Lawrence, 2014 in the References section for full details).

Less information is available for this subject in Mediterranean habitats, but most publications have been edited by the Agroecology Research Group of the Plant Biology Department of the University of Barcelona (see Bassa et al., 2012). In 2019 a classification of field boundaries in the Mediterranean area has been published (Cirujeda et al., 2019b).

Despite the topic of biodiversity loss in the landscape that affects the whole humanity, there are not many private initiatives dealing with this topic. However there are some that can be recommended for further information.

The company Syngenta has launched a project to demonstrate that certain annual flower species attract pollinators, which can help farmers to improve the yields in pollinator-dependent crops such as some vegetables and fruits. The project started more than 10 years ago and has been established in 17 different countries including the United States, Brazil, Canada, Chile and several European countries. https://www.syngenta.com/sustainability/operation-pollinator.

Different bird protection organisations raise their voices to defend the habitats for the agricultural-related birds. https://www.birdlife.org/europe-and-central-asia/news/vanishing-europe%E2%80%99s-farmland-birds.

An interesting private experience in north-eastern Spain won an award of the regional Agriculture and Environment Department. In that case 360 ha land in the location of Monreal del Campo were concentrated but 33 km of margins were planted with 167 000 trees and 55 000 shrubs apart from 136 vegetated islands. The details are published in: http://naturaxilocae.blogspot.com/2014/06/el-premio-aragon-de-medio-ambiente-para.html.

9 References

Albrecht, H., Cambecedes, J., Lang, M. and Wagner, M. 2016. Management options for the conservation of rare arable plants in Europe. *Bot. Lett.* 163(4), 389–415. doi:10.1080/23818107.2016.1237886.

Altieri, M. A. and Letourneau, D. K. 1982. Vegetation management and biological control in agroecosystems. *Crop Protection* 1(4), 405–30. doi:10.1016/0261-2194(82)90023-0.

Araj, S. E. and Wratten, S. D. 2015. Comparing existing weeds and commonly used insectary plants as floral resources for a parasitoid. *Biol. Control* 81, 15–20. doi:10.1016/j.biocontrol.2014.11.003.

Baggen, L. R., Gurr, G. M. and Meats, A. 1999. Flowers in tri-trophic systems: mechanisms allowing selective exploitation by insect natural enemies for conservation biological control. *Entomol. Exp. Appl.* 91(1), 155–61. doi:10.1046/j.1570-7458.1999.00478.x.

Balmer, O., Geneau, C. E., Belz, E., Weishaupt, B., Förderer, G., Moos, S., Ditner, N., Juric, I. and Luka, H. 2014. Wildflower companion plants increase pest parasitation and yield in cabbage fields: experimental demonstration and call for caution. *Biol. Control* 76, 19-27. doi:10.1016/j.biocontrol.2014.04.008.

Balzan, M. V. and Moonen, A. C. 2014. Field margin vegetation enhances biological control and crop damage suppression from multiple pests in organic tomato fields. *Entomol. Exp. Appl.* 150(1), 45–65. doi:10.1111/eea.12142.

Balzan, M. V., Bocci, G. and Moonen, A. C. 2014. Augmenting flower trait diversity in wildflower strips to optimise the conservation of arthropod functional groups for multiple agroecosystem services. *J. Insect Conserv.* 18(4), 713–28. doi:10.1007/s10841-014-9680-2.

Barloggio, G., Tamm, L., Nagel, P. and Luka, H. 2019. Selective flowers to attract and enhance *Telenomus laeviceps* (Hymenoptera: Scelionidae): a released biocontrol agent of *Mamestra brassicae* (Lepidoptera: Noctuidae). *Bull. Entomol. Res.* 109(2), 160–8. doi:10.1017/S0007485318000287.

Bassa, M., Chamorro, L., José-María, L., Blanco-Moreno, J. M. and Sans, F. X. 2012. Factors affecting plant species richness in field boundaries in the Mediterranean region. *Biodivers. Conserv.* 21(4), 1101–14. doi:10.1007/s10531-012-0245-y.

Benton, T. G., Vickery, J. A. and Wilson, J. D. 2003. Farmland biodiversity: is habitat heterogeneity the key? *Trends Ecol. Evol.* 18(4), 182–8. doi:10.1016/S0169-5347(03)00011-9.

Biaggini, M., Bazzoffi, P., Gentile, R. and Corti, C. 2011. Effectiveness of the GAEC cross compliance standards Rational management of set aside, Grass strips to control soil erosion and Vegetation buffers along watercourses on surface animal diversity and biological quality of soil. *Ital. J. Agron.* 6(1s), 100–6. doi:10.4081/ija.2011.6.s1.e14.

Biaggini, M. and Corti, C. 2015. Reptile assemblages across agricultural landscapes: where does biodiversity hide? *Anim. Biodiv. Conserv.* 38, 163–74.

Bielza de Ory, V. 1989. *Territorio y Sociedad en España. Vol. 2, Geografía* Humana. Taurus, Madrid.

Blaauw, B. R. and Isaacs, R. 2014. Flower plantings increase wild bee abundance and the pollination services provided to a pollination-dependent crop. *J. Appl. Ecol.* 51(4), 890–8. doi:10.1111/1365-2664.12257.

Blake, R. J., Woodcock, B. A., Westbury, D. B., Sutton, P. and Potts, S. G. 2011. New tools to boost butterfly habitat quality in existing buffer strips. *J. Insect Conserv.* 15(1–2), 221–32. doi:10.1007/s10841-010-9339-6.

Blake, R. J., Woodcock, B. A., Westbury, D. B., Sutton, P. and Potts, S. G. 2013. Novel management to enhance spider diversity in existing Grass buffer strips. *Agric. Forest Entomol.* 15(1), 77–85. doi:10.1111/j.1461-9563.2012.00593.x.

Bosco, L. and Tavella, L. 2013. Distribution and Abundance of Species of the Genus Orius in Horticultural Ecosystems of Northwestern Italy. *Bull. Insectol* 66, 297–307.

Brooker, R. W., Bennett, A. E., Cong, W.-F., Daniell, T. J., George, T. S., Hallett, P. D., Hawes, C., Iannetta, P. P. M., Jones, H. G., Karley, A. J., Li, L., McKenzie, B. M., Pakeman, R. J., Paterson, E., Schöb, C., Shen, J., Squire, G., Watson, C. A., Zhang, C., Zhang, F.,

Zhang, J. and White, P. J. 2015. Improving intercropping: a sunthesis of research in agronomy, plant physiology and ecology. *New Phytologist* 206(1), 107–17. doi:10.1111/nph.13132.

Buhk, C., Oppermann, R., Schanowski, A., Bleil, R., Ludemann, J. and Maus, C. 2018. Flower strip networks offer promising long term effects on pollinator species richness in intensively cultivated agricultural areas. *BMC Ecol.* 18(1), 55. doi:10.1186/s12898-018-0210-z.

Carson, R. 1962. *The Silent Spring*. Houghton Mifflin Harcourt, Boston.

Carvell, C., Meek, W. R., Pywell, R. F., Goulson, D. and Nowakowski, M. 2007. Comparing the efficacy of agri-environment schemes to enhance bumble bee abundance and diversity on arable field margins. *J. Appl. Ecol.* 44(1), 29–40. doi:10.1111/j.1365-2664.2006.01249.x.

Chiatante, G. and Meriggi, A. 2016. The importance of rotational crops for biodiversity conservation in Mediterranean areas. *PLoS ONE* 11(2), e0149323.doi:10.1371/journal.pone.0149323.

Cirujeda, A., Marí, A. I., Murillo, S., Aibar, J., Pardo, G. and Solé-Senan, X. O. 2019a. May the inclusión of a legume crop change weed composition in cereal fields? Example of sainfoin in Aragon (Spain). *Agronomy* 9(3), article number 134. doi:10.3390/agronomy9030134.

Cirujeda, A., Pardo, G., Marí, A. I., Aibar, J., Pallavicini, Y., González-Andújar, J. L., Recasens, J. and Solé-Senan, X. O. 2019b. The structural classification of field boundaries in Mediterranean arable cropping systems allows the prediction of weed abundances in the boundary and in the adjacent crop. *Weed Res.* 59(4), 300–11. doi:10.1111/wre.12366.

Cordeau, S., Petit, S., Reboud, X. and Chauvel, B. 2012. Sown grass strips harbour high weed diversity but decrease weed richness in adjacent crops. *Weed Res.* 52(1), 88–97. doi:10.1111/j.1365-3180.2011.00892.x.

Cresswell, C. J., Cunnigham, H. M., Wicox, A. and Randall, N. P. 2019. A trait-based approach to plant species selection to increase functionality of farmland vegetative strips. *Ecol. Evol.* 9(8), 4532–43. doi:10.1002/ece3.5047.

Cushman, S. A. 2006. Effects of habitat loss and fragmentation on amphibians: a review and prospectus. *Biol. Conserv.* 128(2), 231–40. doi:10.1016/j.biocon.2005.09.031.

Dafni, A. and Potts, S. G. 2004. The role of flower inclination, depth, and height in the preferences of a pollinating beetle (Coleoptera: Glaphyridae). *J. Insect Behav.* 17(6), 823–34. doi:10.1023/B:JOIR.0000048991.45453.73.

Davies, J. 2018. EU herbicide Regulation: protection goals for non-target plants and in-crop protection of beneficial weeds, 18th European Weed Research Society Symposium Ljubljana, Slovenia (vol. 239).

de Cauwer, B., Reheul, D., Nijs, I. and Milbau, A. 2008. Management of newly established field margins on nutrient-rich soil to reduce weed spread and seed rain into adjacent crops. *Weed Res.* 48(2), 102–12. doi:10.1111/j.1365-3180.2007.00607.x.

Dicks, L. V., Ashpole, J. E., James, J., Jönsson, A., Randall, N., Showler, D. A., Smith, R. K., Turpie, S., Williams, D., Sutheland, W. J. 2013. *Farmland Conservation: Evidence for the Effects of Interventions in Northern and Western Europe*. Pelagic Publishing, Exeter.

Dongiovanni, C., Cavalieri, V., Bodino, N., Tauro, D., Di Carolo, M., Fumarola, G., Altamura, G., Lasorella, C. and Bosco, D. 2019. Plant selection and population trend of spittlebug immatures (Hemiptera: Aphrophoridae) in olive Groves of the Apulia region of Italy. *J. Econ. Entomol.* 112(1), 67–74. doi:10.1093/jee/toy289.

Fernandez-Quintanilla, C., Fereres, A., Godfrey, L. and Norris, R. F. 2002. Development and reproduction of *Myzus persicae* and *Aphis fabae* (Hom., Aphididae) on selected weed species surrounding sugar beet fields. *J. Appl. Ent.* 126, 1-7.

Foster, E. P. J. and Baker, J. 2010. Reptile habitat management handbook. *Amphibian Reptile Conserv., Bournemouth*. Available at: http://www.arc-trust.org/downloads/RHMH.pdf (accessed 23 August 2019).

Fried, G., Petit, S., Dessaint, F. and Reboud, X. 2009. Arable weed decline in Northern France: crop edges as refugia for weed conservation? *Biol. Conserv.* 142(1), 238-43. doi:10.1016/j.biocon.2008.09.029.

Fuller, R. J., Gregory, R. D., Gibbons, D. W., Marchant, J. H., Wilson, J. D., Baillie, S. R. and Carter, N. 1995. Population declines and range contractions among lowland farmland birds in Britain. *Conserv. Biol.* 9(6), 1425-41. doi:10.1046/j.1523-1739.1995.09061425.x.

Ganser, D., Knop, E. and Albrecht, M. 2019. Sown wildflower strips as overwintering habitat for arthropods: effective measure or ecological trap? *Agr. Ecosyst. Environ.* 275, 123-31. doi:10.1016/j.agee.2019.02.010.

Germano, D. J. and Hungerford, C. R. 1981. Reptile population changes with manipulation of Sonoran Desert shrub. *Gr Basin Nat.* 41, 129-38.

Gonthier, D. J., Ennis, K. K., Farinas, S., Hsieh, H. Y., Iverson, A. L., Batáry, P., Rudolphi, J., Tscharntke, T., Cardinale, B. J. and Perfecto, I. 2014. Biodiversity conservation in agriculture requires a multi-scale approach. *Proc. Biol. Sci.* 281(1791), 20141358. doi:10.1098/rspb.2014.1358.

Greaves, M. P. and Marshall, E. J. P. 1987. Field margins: definitions and statistics. In: Way, J. M. and Greig-Smith, P. J. (Eds), *Field Margins*. Monograph No. 35. British Crop Protection Council, Thornton Heath, pp. 3-10.

Green, R. E., Cornell, S. J., Scharlemann, J. P. and Balmford, A. 2005. Farming and the fate of wild nature. *Science* 307(5709), 550-5. doi:10.1126/science.1106049.

Gregory, R. D. and Marchant, J. H. 1996. Population trends of jays, magpies, jackdaws and carrioncrows in the UK. *Bird Study* 43(1), 28-37. doi:10.1080/00063659609460993.

Hackett, M. and Lawrence, A. 2014. Multifunctional role of field margins in arable farming. Report for European Crop Protection Association by Cambridge Environmental Assessments Adas Uk Ltd. Report number CEA 1118. Available at: https://www.ecpa.eu/sites/default/files/Field%20Margins%20Arable%20Farming_V02%20%281%29.pdf (accessed 7 July 2020).

Hallmann, C. A., Sorg, M., Jongejans, E., Siepel, H., Hofland, N., Schwan, H., Stenmans, W., Müller, A., Sumser, H., Hörren, T., Goulson, D. and de Kroon, H. 2017. More than 75 percent decline over 27 years in total flying insect biomass in protected areas. *PLoS ONE* 12(10), e0185809. doi:10.1371/journal.pone.0185809.

Hansen, N. A., Scheele, B. C., Driscoll, D. A. and Lindenmayer, D. B. 2019. Amphibians in agricultural landscapes: the habitat value of crop areas, linear plantings and remnant woodland patches. *Anim. Conserv.* 22(1), 72-82. doi:10.1111/acv.12437.

Harvey, J. A., Heinen, R., Armbrecht, I., Basset, Y., Baxter-Gilbert, J. H., Bezemer, T. M., Böhm, M., Bommarco, R., Borges, P. A. V., Cardoso, P., Clausnitzer, V., Cornelisse, T., Crone, E. E., Dicke, M., Dijkstra, K. B., Dyer, L., Ellers, J., Fartmann, T., Forister, M. L., Furlong, M. J., Garcia-Aguayo, A., Gerlach, J., Gols, R., Goulson, D., Habel, J. C., Haddad, N. M., Hallmann, C. A., Henriques, S., Herberstein, M. E., Hochkirch, A., Hughes, A. C., Jepsen, S., Jones, T. H., Kaydan, B. M., Kleijn, D., Klein, A. M., Latty, T., Leather, S. R., Lewis, S. M., Lister, B. C., Losey, J. E., Lowe, E. C., Macadam, C. R., Montoya-Lerma, J., Nagano, C. D., Ogan, S., Orr, M. C., Painting, C. J., Pham, T. H.,

Potts, S. G., Rauf, A., Roslin, T. L., Samways, M. J., Sanchez-Bayo, F., Sar, S. A., Schultz, C. B., Soares, A. O., Thancharoen, A., Tscharntke, T., Tylianakis, J. M., Umbers, K. D. L., Vet, L. E. M., Visser, M. E., Vujic, A., Wagner, D. L., WallisDeVries, M. F., Westphal, C., White, T. E., Wilkins, V. L., Williams, P. H., Wyckhuys, K. A. G., Zhu, Z. R. and de Kroon, H. 2020. International scientists formulate a roadmap for insect conservation and recovery. *Nat. Ecol. Evol.* 4(2), 174–6. Available at: https://www.nature.com/articles/s41559-019-1079-8. doi:10.1038/s41559-019-1079-8.

Hatt, S., Uyttenbroeck, R., Lopes, T., Chen, J. L., Piqueray, J., Monty, A. and Francis, F. 2018. Effect of flower traits and hosts on the abundance of parasitoids in perennial multiple species wildflower strips sown within oilseed rape (*Brassica napus*) crops. *Arthropod-Plant Inte.* 12(6), 787–97. doi:10.1007/s11829-017-9567-8.

Holland, J. M., Orson, J., Powell, W., Storkey, J. and Chamberlain D. 2007. Managing uncropped land in order to enhance biodiversity benefits of the arable farmland landsacape. *Aspects Appl. Biol.* 81, 1–10.

Holland, J. M., Oaten, H., Moreby, S., Birkett, T., Simper, J., Southway, S. and Smith, B. M. 2012. Agri-environment scheme enhancing ecosystem services: a demonstration of improved biological control in cereal crops. *Agr. Ecosyst. Environ.* 155, 147–52. doi:10.1016/j.agee.2012.04.014.

Holland, J. M., Storkey, J., Lutman, P. J. W., Henderson, I. and Orson, J. 2013. Managing uncropped land in order to enhance biodiversity benefits of the arable farmed landscape: the farm4bio project. Report No. 506. Agriculture and Horticulture Development Board (AHDB), Kenilworth.

Hovd, H. and Skogen, A. 2005. Plant species in arable field margins and road verges of central Norway. *Agr. Ecosyst. Environ.* 110(3–4), 257–65. doi:10.1016/j.agee.2005.04.013.

Jado, R. H., Araj, S. E., Abu-Irmaileh, B., Shields, M. W. and Wratten, S. D. 2019. Floral resources to enhance the potential of the parasitoid *Aphidius colemani* for biological control of the aphid *Myzus persicae*. *J. Appl. Entomol.* 143(1–2), 34–42. doi:10.1111/jen.12556.

Kavallieratos, N. G., Stathas, G. J., Athanassious, C. G. and Papadoulis, G. T. 2002. *Dittrichia viscosa* and *Rubus ulmifolius* as reservoirs of aphid parasitoids (Hymenoptera: Braconidae: Aphidiinae) and the role of certain coccinellid species. *Phytoparasitica* 30(3), 231–42. doi:10.1007/BF03039992.

Kleijn, D. and Verbeek, M. 2000. Factors affecting the species composition of arable field boundary vegetation. *J. Appl. Ecol.* 37(2), 256–66. doi:10.1046/j.1365-2664.2000.00486.x.

Landis, D. A. 2017. Designing agricultural landscapes for biodiversity-based ecosystem services. *Basic Appl. Ecol.* 18, 1–12. doi:10.1016/j.baae.2016.07.005.

Macdonald, D. W., Tattersall, F. H., Service, K. M., Firbank, L. G. and Feber, R. E. 2007. Mammals, agri-environment schemes and set-aside - What are the putative benefits? *Mamm. Rev.* 37(4), 259–77. doi:10.1046/j.1440-1770.2002.00172_37_4.x.

Maes, J., Musters, C. J. M. and de Snoo, G. R. 2008. The effect of agri-environment schemes on amphibian diversity and abundance. *Biol. Conserv.* 141(3), 635–45. doi:10.1016/j.biocon.2007.12.018.

Mahaut, L., Gaba, S. and Fried, G. 2019. A functional diversity approach of crop sequences reveals that weed diversity and abundance show different responses to environmental variability. *J. Appl. Ecol.* 56(6), 1400–9. doi:10.1111/1365-2664.13389.

Mante, J. and Gerowitt, B. 2009. Learning from farmers' needs: identifying obstacles to the successful implementation of field margin measures in intensive arable regions. *Landsc. Urban Plan.* 93(3–4), 229–37. doi:10.1016/j.landurbplan.2009.07.010.

Marshall, E. J. P. 1989. Distribution patterns of plants associated with arable field edges. *J. Appl. Ecol.* 26(1), 247–57. doi:10.2307/2403665.

Marshall, E. J. P. and Arnold, G. M. 1995. Factors affecting field weed and field margin flora on a farm in Essex, UK. *Landsc. Urban Plan.* 31(1–3), 205–16. doi:10.1016/0169-2046(94)01047-C.

Marshall, E. J. P. and Moonen, A. C. 2002. Field margins in northern Europe. Their functions and interactions with agriculture. *Agr. Ecosyst. Environ.* 89(1–2), 5–21. doi:10.1016/S0167-8809(01)00315-2.

Marshall, E. J. P., West, T. M., Kleijn, D. 2006. Impacts of an agri-environment field margin prescription on flora and fauna of arable farmland in different landscapes. *Agric. Ecosys Environ.* 113, 236–44.

Martinez-Una, A., Martin, J. M., Fernandez-Quintanilla, C. and Dorado, J. 2013. Provisioning floral resources to attract aphidophagous hoverflies (Diptera: Syrphidae) useful for pest management in central Spain. *J. Econ. Entomol.* 106(6), 2327–35. doi:10.1603/EC13180.

Montserrat, P. 2009. *La cultura que hace el paisaje*. La Fertilidad de la Tierra Ediciones, Estella (Navarra), Spain.

Morrison, J., Izquierdo, J., Hernandez Plaza, E. H. and Gonzalez-Andujar, J. L. 2017. The role of field margins in supporting wild bees in Mediterranean cereal agroecosystems: which biotic and abiotic factors are important? *Agr. Ecosyst. Environ.* 247, 216–24. doi:10.1016/j.agee.2017.06.047.

Natural England 2013. *Entry Level Stewardship – Environmental Stewardship Handbook* (4th edn). Natural England. http://publications.naturalengland.org.uk/publication/2798159.

Nicholls, C. I. and Altieri, M. A. 2013. Plant biodiversity enhances bees and other insect pollinators in agroecosystems. A review. *Agron. Sustain. Dev.* 33(2), 257–74. doi:10.1007/s13593-012-0092-y.

Olson, D. and Wackers, F. 2007. Management of field margins to maximise multiple ecological services. *J. Appl. Ecol.* 44, 13-21.

Ouvrard, P. and Jacquemart, A.-L. 2018. Agri-environment schemes targeting farmland bird populations also provide food for pollinating insects. *Agric. For. Entomol.* 20(4), 558–74. doi:10.1111/afe.12289.

Paredes, D., Cayuela, L., Gurr, G. M. and Campos, M. 2015. Is ground cover vegetation an effective biological control enhancement strategy against olive pests? *PLOS ONE* 10(2), e0117265. doi:10.1371/journal.pone.0117265.

Poggio, S. L., Chaneton, E. J. and Ghersa, C. M. 2010. Landscape complexity differentially affects alpha, beta, and gamma diversities of plants occurring in fencerows and crop fields. *Biol. Conserv.* 143(11), 2477–86. doi:10.1016/j.biocon.2010.06.014.

Pollier, A., Tricault, Y., Plantegenest, M. and Bischoff, A. 2019. Sowing of margin strips rich in floral resources improves herbivore control in adjacent crop fields. *Agric. For. Entomol.* 21(1), 119–29. doi:10.1111/afe.12318.

Potts, G. R. 1980. The effects of modern agriculture, nest predation and game management on the population ecology of partridge (*Perdix perdix* and *Alectoris rufa*). *Adv. Ecol. Res.* 11, 1–79. doi:10.1016/S0065-2504(08)60266-4.

Potts, S. G., Roberts, S. P. M., Dean, R., Marris, G., Brown, M. A., Jones, R., Neumann, P. and Settele, J. 2010. Declines of managed honey bees and beekeepers in Europe. *J. Apicult. Res.* 49(1), 15–22. doi:10.3896/IBRA.1.49.1.02.

Pywell, R., Meek, W.R., Loxton, R.G., Nowakowski, M., Carvell, C., Woodcock, B.A. 2011. Ecological restoration on farmland drive beneficial functional responses in plant and invertebrate communities. *Agric. Ecosys Environ.* 140, 36–44.

Rahat, S., Gurr, G. M., Wratten, S. D., Mo, J. H. and Neeson, R. 2005. Effect of plant nectars on adult longevity of the stinkbug parasitoid, *Trissolcus basalis*. *Int. J. Pest Manag.* 51(4), 321–4. doi:10.1080/09670870500312778.

Ramírez del Palacio, O. J. and Hernández Navarro, S. 2013. Ecological connectivity in agricultural spaces. Contributions from land consolidation projects. *Dossier Ciudades* 1, 269–98. http://iuu.uva.es/DOSSIER/Dossier%2001/Dossier%2001%20269-010%20RAMIREZ%20DEL%20PALACIO.pdf (accessed 24 October 2019).

Rands, M. R. W. 1986. Effect of hedgerow characteristics on partridge breeding densities. *J. Appl. Ecol.* 23(2), 479–87. doi:10.2307/2404030.

Rands, S. A. and Whitney, H. M. 2011. Field margins, foraging distances and their impacts on nesting pollinator success. *PLoS ONE* 6(10), e25971. doi:10.1371/journal.pone.0025971.

Reberg-Horton, S. C., Mueller, J. P., Mellange, S. J., Creamer, N. G., Brownie Cbell, M. and Burton, M. G. 2011. Influence of field margin type on weed species richness and abundance in conventional crop fields. *Renew. Agr. Food Syst.* 26, 127–36.

Rodríguez-Pastor, R., Luque-Larena, J. J., Lambin, X. and Mougeot, F. 2016. "Living on the edge": the role of field margins for common vole (*Microtus arvalis*) populations in recently colonised Mediterranean farmland. *Agr. Ecosyst. Environ.* 231, 206–17. doi:10.1016/j.agee.2016.06.041.

Rundlof, M. and Bommarco, R. 2011. *STEP Deliverable 4.1: Report on the Uptake of Mitigation Strategies Counteracting Pollinator Loss Across Europe*. Swedish University of Agricultural Sciences, Uppsala, Sweden.

SAC. 2002. *Conservation Headlands: Technical Note*. SRUC: https://www.sruc.ac.uk/downloads/file/1337/tn515_conservation_headlands.

Sánchez-Bayo, F. and Wyckhuys, K. A. G. 2019. Wordlwide decline of the entomofauna: a review of its drivers. *Biol. Conserv.* 232, 8–27. doi:10.1016/j.biocon.2019.01.020.

Scheper, J., Holzschuh, A., Kuussaari, M., Potts, S. G., Rundlöf, M., Smith, H. G. and Kleijn, D. 2013. Environmental factors driving the effectiveness of European agri-environmental measures in mitigating pollinator loss - a meta-analysis. *Ecol. Lett.* 16(7), 912–20. doi:10.1111/ele.12128.

Solé-Senan, X. O., Juarez-Escario, A., Conesa, J. A., Torra, J., Royo-Esnal, A. and Recasens, J. 2014. Plant diversity in Mediterranean cereal fields: unraveling the effect of landscape complexity on rare arable plants. *Agr. Ecosyst. Environ.* 185, 221–30. doi:10.1016/j.agee.2014.01.003.

Stanley, D. A. and Stout, J. C. 2014. Pollinator sharing between mass-flowering oilseed rape and co-flowering wild plants: implications for wild plant pollination. *Plant Ecol.* 215(3), 315–25. doi:10.1007/s11258-014-0301-7.

Stanton, R. L., Morrissey, C. A. and Clark, R. G. 2018. Analysis of trends and agricultural drivers of farmland bird declines in North America: a review. *Agr. Ecosyst. Environ.* 254, 244–54. doi:10.1016/j.agee.2017.11.028.

Storkey, J. and Neve, P. 2018. What good is weed diversity? *Weed Res.* 58(4), 239–43. doi:10.1111/wre.12310.

Sutter, L., Albrecht, M. and Jeanneret, P. 2018. Landscape greening and local creation of wildflower strips and hedgerows promote multiple ecosystem services. *J. Appl. Ecol.* 55(2), 612–20. doi:10.1111/1365-2664.12977.

Syngenta, Arcadis and Bioversity International 2018. Multifunctional field margins: assessing the benefits for nature, society and business. Available at: https://www.syngenta.com/sites/syngenta/files/sustainability/the-good-growth-plan/operation-pollinator/MFFM-Assessing-the-benefits-for-nature-society-and-business.pdf.

Tapper, S. C. and Barnes, R. F. W. 1986. Influence of farming practices on the ecology of the brown hare Lepus europaeus. *J. Appl. Ecol.* 23, 39–52.

Tarmi, S., Tuuri, H. and Helenius, J. 2002. Plant communities of field boundaries in Finnish farmland. *Agr. Food Sci. Finland* 11(2), 121–35. doi:10.23986/afsci.5717.

Thomas, C. F. G., Parkinson, L., Griffiths, G. J. K., Garcia, A. F. and Marshall, E. J. P. 2001. Aggregation and temporal stability of carabid beetle distributions in field and hedgerow habitats. *J. Appl. Ecol.* 38(1), 100–16. doi:10.1046/j.1365-2664.2001.00574.x.

Tischendorf, L. and Wissel, C. 1997. Corridors as conduits for small animals: attainable distances depending on movement pattern, boundary reaction and corridor width. *Oikos* 79(3), 603–11. doi:10.2307/3546904.

Traba, J. and Morales, M. B. 2019. The decline of farmland birds in Spain is strongly associated to the loss of fallowland. *Sci. Rep.* 9(1), 9473. doi:10.1038/s41598-019-45854-0.

Tscharntke, T., Klein, A. M., Kruess, A., Steffan-Dewenter, I. and Thies, C. 2005. Landscape perspectives on agricultural intensification and biodiversity-ecosystem service management. *Ecol. Lett.* 8(8), 857–74. doi:10.1111/j.1461-0248.2005.00782.x.

Tschumi, M., Albrecht, M., Baertschi, C., Collatz, J., Entling, M. H. and Jacot, K. 2016a. Perennial, species-rich wildflower strips enhance pest control and crop yield. *Agr. Ecosyst. Environ.* 220, 97–103. doi:10.1016/j.agee.2016.01.001.

Tschumi, M., Albrecht, M., Collatz, J., Dubsky, V., Entling, M. H., Najar-Rodriguez, A. J. and Jacot, K. 2016b. Tailored flower strips promote natural enemy biodiversity and pest control in potato crops. *J. Appl. Ecol.* 53(4), 1169–76. doi:10.1111/1365-2664.12653.

Ulber, L., Steinmann, H. H. and Klimek, S. 2010. Using selective herbicides to manage beneficial and rare weed species in winter wheat. *J. Plant Dis. Protect.* 117(5), 233–9. doi:10.1007/BF03356366.

Uyttenbroeck, R., Hatt, S., Piqueray, J., Paul, A., Bodson, B., Francis, F. and Monty, A. 2015. Creating perennial flower strips: think functional! *Agric. Agric. Sci. Procedia* 6, 95-101. doi:10.1016/j.aaspro.2015.08.044.

Vanbergen, A. J. and the Insect Pollinators Initiative 2013 Threats to an ecosystem service: pressures on pollinators. *Front. Ecol. Environ.* 11(5), 251-9. doi:10.1890/120126.

van Engelsdorp, D., Evans, J. D., Saegerman, C., Mullin, C., Haubruge, E., Nguyen, B. K., Frazier, M., Frazier, J., Cox-Foster, D., Chen, Y., Underwood, R., Tarpy, D. R. and Pettis, J. S. 2009. Colony collapse disorder: a descriptive study. *PLOS ONE* 4(8), e6481. doi:10.1371/journal.pone.0006481.

Vickery, J. A., Feber, R. E. and Fuller, R. J. 2009. Arable field margins managed for biodiversity conservation: a review of food resource provision for farmland birds. *Agr. Ecosyst. Environ.* 133(1–2), 1–13. doi:10.1016/j.agee.2009.05.012.

Vickery, J. A. and Tayleur, C. 2018. Stemming the decline of farmland birds: the need for interventions and evaluations at a large scale. *Anim. Conserv.* 21(3), 195–6. doi:10.1111/acv.12425.

Vogel, G. 2017. Available at: https://www.sciencemag.org/news/2017/05/where-have-all-insects-gone (accessed 16 July 2019).

Walker, K. J., Critchley, C. N. R., Sherwood, A. J., Large, R., Nutall, P., Hulmes, S., Rose, R. and Mountford, J. O. 2007. The conservation of arable plants on cereal field margins: an assessment of new agri-environment scheme options in England, UK. *Biol. Conserv.* 136(2), 260–70. doi:10.1016/j.biocon.2006.11.026.

Walker, L. K., Morris, A. J., Cristinacee, A., Dadam, D., Grice, P. V. and Peach, W. J. 2018. Effects of higher-tier agri-environment scheme on the abundance of priority farmland birds. *Anim. Conserv.* 21(3), 183–92. doi:10.1111/acv.12386.

Westbury, D. B., Woodcock, B. A., Harris, S. J., Brown, V. K. and Potts, S. G. 2008. The effects of seed mix and management on the abundance of desirable and pernicious unsown species in arable buffer strip communities. *Weed Res.* 48(2), 113–23. doi:10.1111/j.1365-3180.2007.00614.x.

Westbury, D. B., Woodcock, B. A., Harris, S. J., Brown, V. K. and Potts, S. G. 2017. Buffer strip management to deliver plant and invertebrate resources for farmland birds in agricultural landscapes. *Agric. Ecosyst. Environ.* 240, 215–23. doi:10.1016/j.agee.2017.02.031.

Wilson, P. J. and Aebischer, N. J. 1995. The distribution of dicotyledonous arable weeds in relation to distance from the field edge. *J. Appl. Ecol.* 32(2), 295–310. doi:10.2307/2405097.

Wittwer, R. A., Dorn, B., Jossi, W. and van der Heijden, M. G. 2017. Cover crops support ecological intensification of arable cropping systems. *Sci. Rep.* 7, 41911. doi:10.1038/srep41911.

Woodcock, B. A., Westbury, D. B., Potts, S. G., Harris, S. J. and Brown, V. K. 2005. Establishing field margins to promote beetle conservation in arable farms. *Agric. Ecosyst. Environ.* 107(2–3), 255–66. doi:10.1016/j.agee.2004.10.029.

Woodcock, B. A., Potts, S. G., Westbury, D. B., Ramsay, A. J., Lambert, M., Harris, S. J. and Brown, V. K. 2007. The importance of sward architectural complexity in structuring predatory and phytophagous invertebrate assemblages. *Ecol. Entomol.* 32(3), 302–11. doi:10.1111/j.1365-2311.2007.00869.x.

Woodcock, B. A., Westbury, D. B., Tscheulin, T., Harrison-Cripps, J., Harris, S. J., Ramsey, A. J., Brown, V. K. and Potts, S. G. 2008. Effects of seed mixture and management on beetle assemblages of arable field margins. *Agric. Ecosyst. Environ.* 125(1–4), 246–54. doi:10.1016/j.agee.2008.01.004.

Wratten, S. D., Gillespie, M., Decourtye, A., Mader, E. and Desneux, N. 2012. Pollinator habitat enhancement: benefits to other ecosystem services. *Agr. Ecosyst. Environ.* 159, 112–22. doi:10.1016/j.agee.2012.06.020.

Chapter 6

The role of hedgerows in supporting biodiversity and other ecosystem services in intensively managed agricultural landscapes

Audrey Alignier, Léa Uroy and Stéphanie Aviron, INRAE, France

1 Introduction

Hedgerows are key features of agricultural landscapes, especially in northwestern Europe (e.g. hedgerow networks such as the 'bocage' landscapes in France, England, Ireland, the Netherlands and northern Germany) where they are the dominant type of semi-natural habitat (Burel and Baudry, 1999). These hedgerows play a wide range of roles, including agronomic, ecological and cultural functions (Baudry et al., 2000). They benefit crop growth by acting as windbreak and shelter for crops (Forman and Baudry, 1984) and contribute to the reduction of soil erosion and pesticide drift from neighboring crops (Marshall and Moonen, 2002). They are critical features for biodiversity conservation, agroecosystem functioning and associated ecosystem services (Pywell et al., 2015). They supply habitat, shelter and resources for a wide range of plant and animal species (Marshall and Moonen, 2002; Hannon and Sisk, 2009; Vickery et al., 2009). By providing high levels of floral diversity, they contribute to the maintenance of pollinator populations (Morandin and

http://dx.doi.org/10.19103/AS.2020.0071.09

Kremen, 2013) and enhance the community of natural enemies of crop pests (Bianchi et al., 2006; Chaplin-Kramer and Kremen, 2012). However, hedgerows may also be reservoirs of weeds and pests, which can spread into the fields (Forman and Baudry, 1984; Pollard and Holland, 2006).

At larger scales, hedgerows may contribute to species movement by acting as corridors and ensuring connectivity through the landscape (Baudry et al., 2000; Concepción et al., 2012). Communities of natural enemies and pollinators are often found to be more abundant in complex landscapes where hedgerows are better conserved (Chaplin-Kramer et al., 2011; Kennedy et al., 2013). Based upon the Millennium Ecosystem Assessment (2005), Cole et al. (2009) identified 20 positive ecosystem services[1] delivered by hedgerows including provisioning (e.g. food, fiber), regulating (e.g. air quality, pest regulation, pollination), cultural (e.g. recreation, aesthetics), supporting (e.g. soil fertility, primary production) and biodiversity conservation services (Table 1). Despite this abundant literature, more research is needed to elucidate the role of hedgerows in supporting multiple ecosystem services at different spatial scales.

This chapter will focus predominantly on the role of hedgerows in supporting biodiversity and associated ecosystem services by:

- examining the evidence that hedgerows offer habitat for a wide range of species,
- considering hedgerows as a management unit and part of farming systems,
- considering hedgerows as part of a networked landscape,
- highlighting the role of hedgerows in delivering ecosystem services of agricultural importance, using the examples of pollination and pest regulation, and
- exploring hedgerow planting and bocage restoration and its contribution to agroecosystem multifunctionality.

Each section of this chapter will be illustrated with case studies resulting from different projects carried out on the 'Zone Atelier Armorique', a Long-Term Socio-Ecological Research (LTSER) site in northern Brittany, France (48°360 N, 1°320 W). This study site is characterized by an extensive hedgerow network, with relatively small fields (on average 1.3 ha) being separated by woody habitats (mainly hedgerows and some woodlands; Fig. 1). In this study site, the hedgerows are made of rows of oak (*Quercus robur*) or chestnut (*Castanea sativa*) generally planted on an earthen bank of 0.5–0.8 m high surrounded by ditches and managed as trees by farmers (pruning for firewood every

1 Ecosystem services are the direct and indirect benefits that people obtain from ecosystems (Millennium Ecosystem Assessment, 2005).

Table 1 Ecosystem services delivered by hedgerows as identified by Cole et al. (2009). Examples are from Forman and Baudry (1984), Baudry et al. (2000), Dainese et al. (2017) and Dover (2019b).

Service	Specific service	Examples
Supporting	Soil formation	Soil formation
	Photosynthesis	Production of oxygen
	Primary production	Production of chemical energy, for example, organic matter
Provisioning	Food	Fruits, vegetables, fodder
	Fiber	Timber
	Fuel	Wood (e.g. from pruning trees and cutting shrubs)
Regulating	Air quality	Removal of particulates
	Climate regulation	Radiation interception, windbreak effect
	Water regulation	Runoff and flood control
	Erosion regulation	Soil retention (by roots)
	Water quality/purification	Removal of pollutants
	Pest control	Source of natural enemies
	Pollination	Nectar, pollen and nesting resources provision
	Hazard reduction	Prevention of landslides
Cultural	Recreation (field sports)	Field sports, for example, walking along footpaths
	Cultural heritage	Cultural and historical values
	Education	Outdoor classroom: natural history education
	Aesthetics/inspiration	Source of inspiration
	Sense of place	Regional identity
Biodiversity	Habitat	Living space for plants and animals
	Refuge	Shelter, overwintering site
	Corridor	Facilitation of the movement through the landscape
	Complementary resources	Provision of additional resources

9–12 years). The shrub layer, dominated by hazel (*Corylus avellana*), hawthorn (*Crataegus monogyna*), blackthorn (*Prunus spinosa*), broom (*Cytisus scoparius*), spindle (*Euonymus europaeus*) or gorse (*Ulex europaeus*), is rarely continuous and can even be absent.

2 What is a hedgerow?

Hedgerows can be found throughout the world including Europe, North and South America, Africa and China and have been an important component of

Figure 1 Hedgerow network ('bocage') of the LTSER Zone Atelier Armorique site. Source: © Air Papillon.

managed landscapes for over 2000 years (Deckers et al., 2004; Hannon and Sisk, 2009). Different countries/regions show marked differences in hedgerow structure, species composition and the type of management. The diversity of hedgerows stems from their multiple origins and includes their role in the landscape, as well their contribution to biodiversity and ecosystem functioning (Baudry et al., 2000).

Characterized by lines of trees and/or shrubs, hedgerows also encompass a herbaceous layer, as well as features such as a bank and ditch separating agricultural fields (Forman and Baudry, 1984) (Fig. 2). Hedgerows vary according to the predominance of shrubs, trees, or both, single or multiple species dominance, species composition, height and width, and structures such as ditch, bank, stonewall, wood or wire fence. They represent relatively permanent habitats within a constantly disturbed environment of agricultural production, acting as suitable habitats for many plant and animal species.

Historically, hedgerows constitute limits, physical as well as symbolic. They delineate physical property limits, that is, crop fields, and prevent cattle from straying as in Britain (Baudry et al., 2000). Hedgerows provide wood (e.g. firewood, timber) as well as fodder used as winter forage in several regions in Europe and may provide fruits and medicinal plants. They also play a broad range of agri-environmental roles including microclimate modification, radiation interception, water infiltration, wind protection, soil erosion and runoff prevention (Burel and Baudry, 1995; Kiepe, 1995; Oyarzun et al., 2007; Sánchez

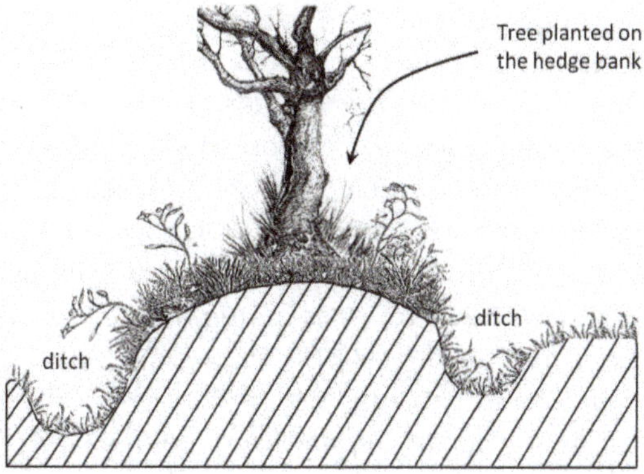

Figure 2 On the left, hedgerow with trees planted on a bank in the Zone Atelier Armorique, France. On the right, schematic representation of a typical hedgerow from Brittany (France) with tree planted on a bank and surrounded by ditches. Source: © A. Alignier.

et al., 2010), although perhaps the greatest volume of scientific literature concerns ecological benefits.

3 The role of hedgerows as habitat for biodiversity

The value of hedgerows as habitat for native plants, butterflies, birds and mammals within intensively managed agricultural landscapes has been widely recognized (Burel, 1996; Baudry et al., 2000; Dover and Sparks, 2000; Hinsley and Bellamy, 2000; Gelling et al., 2007; Dondina et al., 2016). The Royal Society for the Protection of Birds (RSPB) in the UK reports that hedgerows support up to 80% of British woodland birds, 50% of mammals and 30% of butterflies (Anon., 2019).

Hundreds of vascular plant species are known to grow in hedgerows, though they are generally not restricted to hedgerows and are found in other nearby habitats. For example, Wehling and Diekmann (2009) showed that hedgerows contained 77% of all forest species observed in the matrix of neighboring woodlands in Germany. This proportion is comparable to findings from the United States, where about 70% of all forest species were able to colonize hedgerows of the same region (Corbit et al., 1999). By comparing vegetation in forest patches and hedgerows in intensively managed agricultural landscapes in Belgium, Van Den Berge et al. (2019) showed that hedgerow species are typical forest-edge species. Hedgerows include shaded areas, but they are also more exposed to solar radiation and wind. Both light-tolerant plant species as well as species with various degrees of shade tolerance can be found in hedgerows, like ivy (*Hedera helix*), herb-Robert (*Geranium robertianum*) and foxglove (*Digitalis purpurea*).

Hedgerows may also shelter a suite of annual species under disturbed conditions. The perception that weeds spread from hedgerows into crops is prevalent among farmers, leading sometimes to damaging and counter-productive hedgerow management. The removal of perennial species encourages the installation of annual weed species capable of colonizing the field (Marshall and Moonen, 2002). However, studies in Britain have indicated that weed spread from hedgerows/field margins into croplands is limited to a few annual species such as barren brome *Bromus sterilis* and cleavers *Galium aparine* (e.g. Smith et al., 1999).

In natural conditions, the composition of vegetation in hedgerows is a result of soil type, climate, dispersal and regeneration of plant species. In agricultural landscapes, management affects habitats that in turn shape the colonization success of species. By surveying 511 hedgerows in Belgium, Deckers et al. (2004) found that adjacent land use, management, soil conditions and hedgerow structure all affected hedgerow plant communities. Hedgerow structure, partly determined by management, influences light levels and microclimate with a significant influence on the composition of plant communities (McCollin et al., 2000; Van Den Berge et al., 2019).

3.1 Hedgerows as managed elements

Hedgerows are traditionally managed to prevent encroachment on adjacent agricultural land (Baudry et al., 2000). Tree layer management includes infrequent management techniques such as coppicing, applied every 20–40 years, pruning (Fig. 3) and trimming which rejuvenate hedgerows by encouraging them to produce new shoots from the base of woody species. Over 50% of the woody hedge biomass is removed during hedge-laying or coppicing and, between this infrequent rejuvenation management, hedgerows are generally trimmed every few years to maintain their shape (Staley et al., 2013). Over the last 70 years frequent cutting with a mechanized flail has become the dominant form of management across Europe (Croxton et al., 2004).

Herbaceous layer management is much more frequent. Indeed, direct herbicide spraying by farmers at the base of the hedge or on the ditch bank is common to prevent weed invasion into crop fields, for example, spot-wise spraying against species such as creeping thistle *Cirsium arvense*, bramble *Rubus fruticosus agg.* and common nettle *Urtica dioica* (Kleijn and Verbeek, 2000; Alignier and Baudry, 2015). Adjacent agricultural activities (especially plowing or tillage, stock (e.g. cattle, sheep) grazing and herbicide drift; Fig. 4), together with management techniques directly applied to the herbaceous layer of hedgerows result in a high level of disturbance and potential soil nutrient enrichment (Marshall and Moonen, 2002) affecting local environmental heterogeneity, and thus, plant and animal communities within.

Figure 3 Hedgerow with pruned trees in the Zone Atelier Armorique, France. Source: © C. Codet.

3.2 Hedgerows as part of farming systems

The way hedgerows are managed and their structure depends on the farming system (Le Coeur et al., 2002). Cash crop farms mainly have hedgerows with few trees and shrubs. In contrast, diversified productive dairy farms have dense hedgerows. Diversified dairy farms with low production exhibit an intermediate pattern (Le Coeur et al., 2002). In turn, hedgerow structure, by modulating light and moisture, influences plant species composition. Dense hedgerows include plant species requiring forest environment conditions (e.g. false-brome (*Brachypodium sylvaticum*), Solomon's seal (*Polygonatum* multiflorum) and ivy (*H. helix*)). Hygrophilous or meso-hygrophilous species (e.g. common fleabane (*Pulicaria dysenterica*), greater bird's-foot trefoil (*Lotus pedunculatus*) and wild angelica (*Angelica sylvestris*)) have been found to be significantly more frequent in hedgerows with wide and deep banks (Le Coeur et al., 2002).

At the micro-regional scale, Thenail (2002) showed that hedgerow density in bocage landscapes of Brittany (France) can also be related to farming systems. By comparing farms located along a gradient of density of the hedgerow networks in the landscape, the author found that the production, economic levels, productivity and technical means of farms increase from the dense to the open bocage area. Especially, from the dense to the open bocage area, dairy farms contain an increasing proportion of maize used for silage and grain, short

Figure 4 Hedgerow along plowed field in the Zone Atelier Armorique, France. Source: © A. Alignier.

duration grassland and cash crops, which can be seen as a gradient of land-use intensification. The fodder system is based on more maize silage in the open bocage area, and relatively more on multifunctional grassland in the dense bocage area. Complementary studies in the same study site showed that the system of activities implemented for the management of the hedgerows in the farms of the dense bocage landscape, does not only take more time, but is also more complex than in the more open bocage landscapes (Baudry et al., 1998).

Generally considered as semi-natural elements, that is, stable and undisturbed, few ecological studies have investigated the effect of hedgerow management on biodiversity patterns. The management of hedgerows is dependent on types of farming system. However, comparative studies of hedgerows in different farming systems (in a region, but also between countries) are rarely carried out.

3.3 Case study: the importance of taking into account the management of hedgerows and their history for biodiversity conservation

We assessed the relative influence of current and past management practices and landscape context on the local biodiversity of 101 hedgerows in the northern part of the Zone Atelier Armorique. The study focused on the species richness and composition of carabid assemblages. Carabids were trapped partly in 2001 and the other part in 2002. To homogenize, the sampling year of carabids was referred as the year 1. Presence-absence of management practices applied to the herbaceous component of hedgerows was monitored every year between 1995 and 2002, that is, grazing with or without cattle trampling, mechanical management (i.e. mowing, shrub cutting, crushing with or without exportation of plant residues), herbicide spraying and tillage or plowing (Fig. 5). Landscape context of each hedgerow was described annually over the study period in 50 m and 500 m buffer radii with percent cover (%) of woodland, grassland, spring crops and winter crops as landscape composition metrics, and hedge density (total hedge length, in meters) and mean patch size (m^2) as landscape configuration metrics.

We found that carabid species richness and diversity were best explained by past management practices and landscape heterogeneity (4–5 years before carabid surveys) than current management practices (higher values of explained variance; Fig. 6). Overall, the influence of past management practices dedicated to hedgerows was higher than the influence of past landscape heterogeneity except for year 5. This can be explained by the fact that few management operations occurred this particular year. Carabid diversity was higher in grazed or herbicide sprayed hedgerows probably due to increased heterogeneity in habitat conditions. Hedgerows surrounded by grasslands and crops harbored carabid

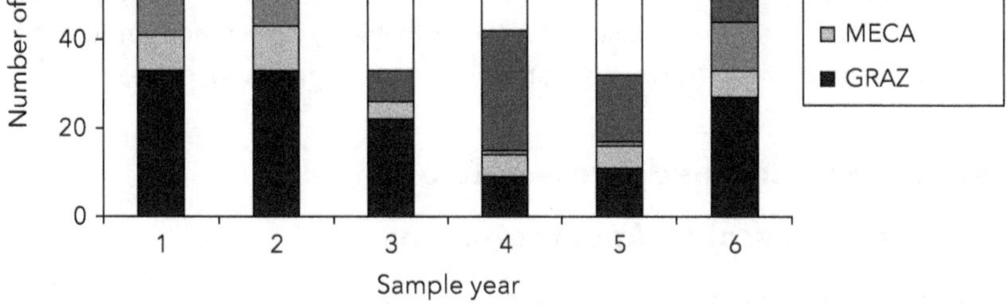

Figure 5 Occurrence frequency of the four management practices over the 101 studied hedgerows, in the sampling year of carabids (year 1) and years before, year 6 being the oldest. NO MANAG. = no management, CHEM = herbicide spraying, TILL = tillage, MECA= mechanical management, GRAZ = cattle grazing. Source: Alignier and Aviron (2017).

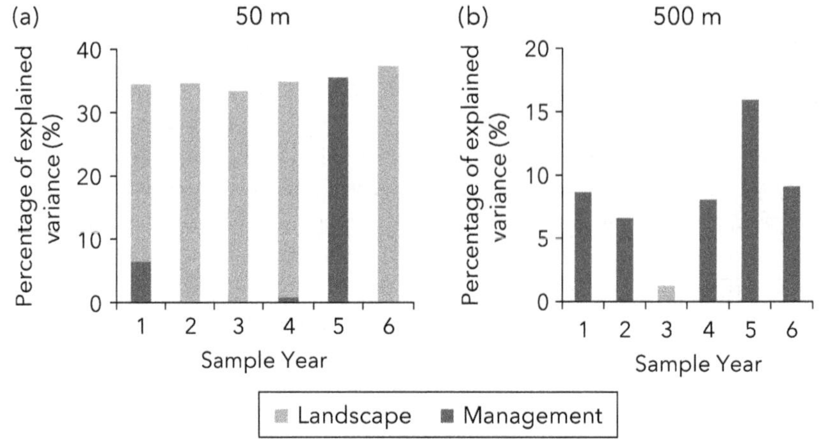

Figure 6 Hierarchical partitioning of explained variance (%) in carabid species richness explained each year (year 1 being the sampling year for carabids and year 6, the eldest) by management practices and landscape metrics at (a) 50 m and (b) 500 m. Source: Alignier and Aviron (2017).

species associated with open habitats while forest carabid species dominated those surrounded by woodland. A landscape effect was scale-dependent and mostly detected at the finest spatial scale, within 50 m around hedgerows. Such distribution of farmland and forest carabid species is consistent with results from previous studies (Millán de la Peña et al., 2003; Aviron et al., 2005).

Our results support the idea that the contribution of hedgerows to the maintenance of biodiversity relies on the way they are managed. Although they are usually considered as semi-natural elements and so, suffering low disturbances, hedgerows should be considered as management units being part of farming activities, as for arable fields. In addition, our results highlight the importance of taking account of time to understand current patterns of carabid assemblages. According to Dover (1996), we therefore suggest that researchers and land managers consider both hedgerow management and history when studying biodiversity patterns in agricultural landscapes.

4 Hedgerows and hedgerow networks

4.1 Brief overview of hedgerow evolution

Since the mid-twentieth century and as a consequence of agricultural intensification, land consolidation programs and urban expansion, hedgerows have suffered from a marked decline in their density in agricultural landscapes. Pointereau (2002) reported that the hedgerow network lost almost 12 000 ha (8%) every year between 1991 and 2000 in France. Bar and Gillespie (2000) indicated a net loss of almost 121 000 km of hedgerows between 1984 and 1990 in the UK. In 2007, the UK Countryside Commission Survey showed a 1.7% decrease in the length of woody linear features between 1998 and 2007 (Carey et al., 2008). This trend has led to severe biodiversity loss and the deterioration of key ecosystem services to agriculture (Tilman et al., 2001). For example, species richness in vegetation associated with all types of landscape boundaries in Great Britain decreased by 4.0% between 1998 and 2007 and by 14.6% between 1978 and 2007. To mitigate the negative effects of agricultural intensification, it has been proposed to focus on the conservation and restoration of hedgerows and hedgerow networks (Batáry et al., 2010; Ponisio et al., 2016; Dainese et al., 2017).

4.2 Hedgerows and landscape connectivity

As linear features of the landscape, hedgerows may serve as a potential corridor for biodiversity and ensure landscape connectivity. Corridors have been proposed as an important conservation tool (Diamond, 1975; Verkaar, 1990), assuming that the connectivity of forest fragments is crucial for the survival of many forest species (Jacquemyn et al., 2003). The efficiency of hedgerows as conservation corridors has mostly been studied in animals (Simberloff and Cox, 1987). For plants, especially animal-dispersed species, corridors may help to maintain viable populations (Corbit et al., 1999; Wehling and Diekmann, 2009).

However, negative effects of hedgerows can also be observed. Corridors can indeed facilitate flows of organisms and matter (e.g. pollen flow) along hedgerows, but impede dispersal across hedgerows (Dover and Settele, 2009). For example, Ouin et al. (2008) showed that hedgerows guided individual butterfly movements in bocage landscapes of Brittany (France). Hedgerows which were oriented toward the receiver forest patch (i.e. parallel hedgerows) clearly promoted butterfly movements, whereas perpendicular hedgerows impeded butterfly movements and acted as barriers. Other negative effects of the hedgerow network are encountered, such as increased spread of contagious diseases, parasites and invasive species (Simberloff et al., 1992; Deckers et al., 2005).

4.3 Case study: increasing landscape connectivity provided by hedgerows increases similarity in plant assemblages for animal-dispersed species in woodlands

We investigated whether and how wooded elements (i.e. hedgerows and woodlots) participated in connectivity and drove composition of plant assemblages according to three dispersal modes: animal dispersal, wind dispersal and unassisted dispersal. We performed detailed surveys of plant assemblages in 25 woodlands, 25 temporary grasslands and 25 winter wheat crops – the three main habitat types of agricultural landscapes in the northern part of the Zone Atelier Armorique. We characterized dispersal modes of each plant species (i.e. animal dispersal, wind dispersal or unassisted dispersal). We assessed landscape connectivity provided by wooded elements (i.e. hedgerows, woodlots and forests) using resistance distance, a measure derived from circuit theory (McRae, 2006; McRae et al., 2008). This method evaluates all possible paths between two habitat patches assuming that dispersal follows random walks (McRae et al., 2008). The resulting connectivity measure is a measure of isolation called resistance distance: the higher the resistance distance, the lower the connectivity. We then evaluated the similarity in composition between pairs of plant assemblages. We found that the similarity in plant assemblages' composition in woodlands was significantly related to the connectivity provided by wooded elements by contrast with temporary grasslands and winter wheat crops. More specifically, only animal-dispersed plant assemblages were more similar in composition than expected by hazard when the connectivity provided by wooded elements increased (Fig. 7).

Figure 7 Partial residual plot denoting the significant effect of the resistance distance (higher the resistance distance, the lower the connectivity) between woodlands provided by wooded elements on the effect size (ES) of similarity values for animal-dispersed plant assemblages in woodlands. The ES of similarity indicates the extent to which the observed value of the similarity between pairs of habitat patches is different from its random expectation based on the regional species pool. Positive ES values indicates a higher similarity than expected, due to a dispersal enhancing effect of the connectivity, whereas negative ES values indicate a lower similarity than expected, due to a dispersal barrier effect of the connectivity. Source: Uroy et al. (2019).

Our results suggest that a high connectivity between patches of woodlands provided by wooded elements (i.e. hedgerows) facilitates the dispersal of animal-dispersal plant species inhabiting these woodlands, hence increasing the similarity in plant composition between patches of woodlands. Overall, this study highlights the fundamental role of wooded habitats as corridor for wood-dwelling animal-dispersed plant species.

4.4 Case study: effective dispersal of plants from woodlands along hedgerows is enhanced by a connected hedgerow network of a good quality

Assessing how connectivity provided by hedgerows affects plant dispersal by studying the established (i.e. above ground) plant assemblages may be difficult, as local factors (e.g. habitat quality) can counterbalance the corridor role of hedgerows. To assess the direct effect of the connectivity provided by hedgerows on seed dispersal, Mony et al. (submitted) analyzed seed dispersal from 16 woodlands along connected and unconnected hedgerow networks in the northern part of the Zone Atelier Armorique. The connected and unconnected hedgerow networks were identified using a least-cost path analysis and their quality (i.e. three-dimensional structure) was also quantified. Using seed traps located along the paths during the peak of seed dispersal, they quantified the species composition, species abundance and richness of trapped seeds after germination in greenhouse conditions. They also analyzed the seed removal by animals of two species, millet (*Panicum miliaceum*) and buckwheat (*Fagopyrum esculentum*), as a proxy of seed dispersal by animals. They reported that connected hedgerow networks displayed both higher abundance and richness compared to unconnected ones. Mony et al. (submitted) also reported that a greater proportion of seeds of *Panicum miliaceum* was removed in connected hedgerow networks compared to unconnected ones (Fig. 8). The connectivity effect was, however, probably partly correlated to the hedgerow quality that was lower in unconnected networks compared to connected ones. This study provides the first empirical evidence that a connected hedgerow network harboring a complex structure enhances plant dispersal from woodlands. Overall, maintaining hedgerow of a good quality should be a major issue of conservation in agricultural landscapes.

4.5 Case study: increasing hedgerow network connectivity contributes to the maintenance of forest carabid species

Hedgerows and other remnant woody habitats are crucial for the dispersal and persistence of forest species with low dispersal abilities in agricultural

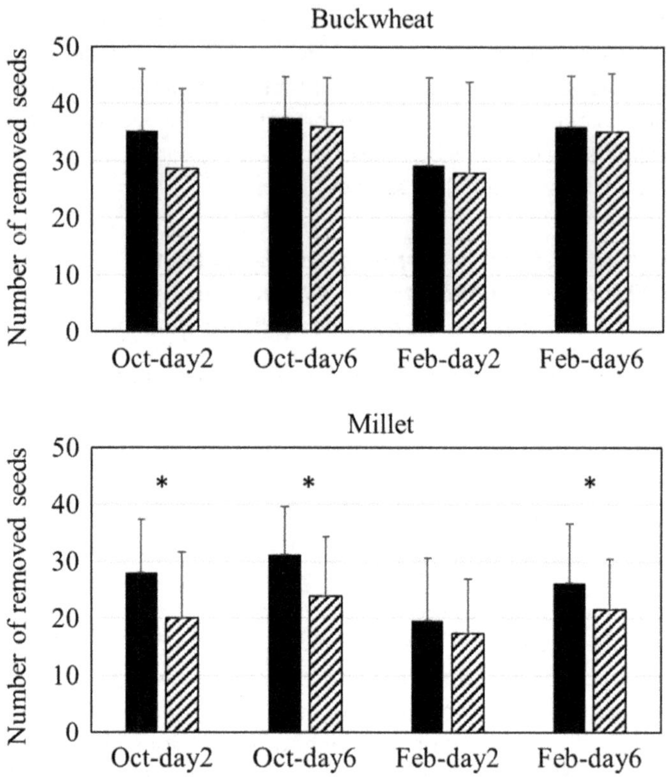

Figure 8 Number of removed seeds of the two studied species, of two species, millet (*Panicum miliaceum*) and buckwheat (*Fagopyrum esculentum*) in connected (solid black bar) and unconnected (hatched bar) hedgerow networks for each sampling period (October and February) and date of survey (day 2 and day 6). Asterisks indicate significant difference in seed removal ($P < 0.05$). Source: Mony et al. (submitted).

landscapes. Among them, forest carabid beetles are low mobile species whose populations are spatially fragmented and located in small woodlots, forests and structurally dense hedgerows (Petit and Burel, 1998; Neumann et al., 2016). By contrast, generalist carabid species, which are also commonly found in hedgerows, are more mobile and less sensitive to the connectivity of the hedgerow network. For this reason, a decrease in hedgerow density and connectivity is expected to cause the replacement of forest species by generalist carabid species. We investigated the contribution of hedgerow density and connectivity for the maintenance of forest carabid species in bocage landscapes. We sampled carabid beetles in 40 hedgerows located in 20 landscapes characterized by varying amount (4.5–28%) and configuration of woody habitats (5–25 km of edges between woody and farmland habitats) in the northern part of the Zone Atelier Armorique. We characterized the amount and spatial connectivity of woody habitats in three radius circles: 50 m in order to account for the immediate context of habitats, as well as 250 m and 500 m to account for the wider landscape context. Spatial connectivity of woody habitats was measured by two metrics: the spatial connectivity index developed by Hanski and Thomas (1994), accounting for woodland area and distance to nearest woody habitat; and the length (in km) of edges between woody habitats and farmland, reflecting the length of the hedgerow network. The results showed that forest carabid species were dominated by two species, *Abax parallelepipedus* and *Pterostichus madidus* (79% of trapped species), which are brachypterous species, that is, species without or with short, unusable wings. Abundances of forest carabid species were significantly and positively related to the length of edges between woody elements and farmland at the finest spatial scale (50 m) (Fig. 9). By contrast, their abundance was reduced in landscapes where annuals crops were directly adjacent, that is, with high edge length between annual crops at the largest spatial scale (500 m). This landscape pattern is typical from intensive agricultural landscapes where hedgerows have been removed and the hedgerow network has been fragmented. In such landscapes, forest carabid species were found to disappear while generalist and highly mobile species, which are adapted to disturbed and open habitats, dominate carabid assemblages (Millán de la Peña et al., 2003). This case study underlines the importance of hedgerow network connectivity for the dispersal and maintenance of forest carabid species in agricultural landscapes.

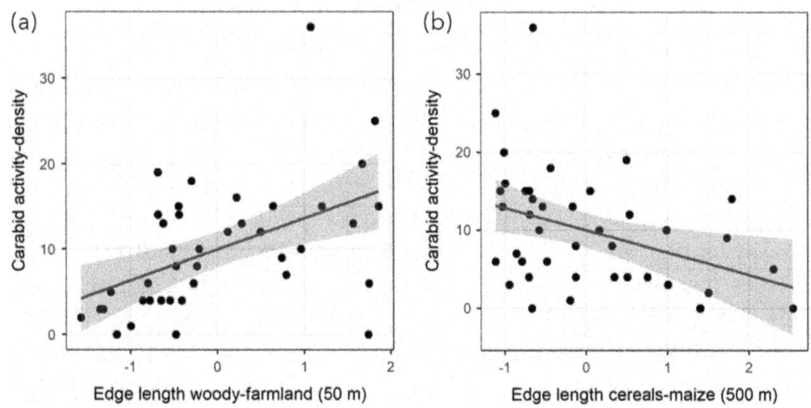

Figure 9 Relationships between activity-density (as proxy of abundance) of forest carabid species in hedgerows and (a) length of edges between woody elements and farmland at 50 m or (b) edge length between annual crops at 500 m. Source: Aviron, unpublished data.

5 Hedgerows and ecosystem services

The ways that ecosystems benefit people are known as ecosystem services. These services can broadly be divided into four categories (Wolton et al., 2014; Table 1):

1. Regulating services - those that regulate processes such as water purification, crop pollination and pest control.
2. Provisioning services - those generating products such as food and energy.
3. Cultural services - those that assist our recreation and well-being, or provide aesthetic experiences and spiritual enrichment or learning opportunities.
4. Supporting services - those that are necessary for the production of all other ecosystem services including soil formation, photosynthesis, primary production and nutrient cycling.

Among regulating services underpinned by biodiversity, pest and weed control and pollination have been demonstrated to have a substantial impact on crop production worldwide. These ecosystem services are also those most at risk from agricultural intensification (Tscharntke et al., 2005). By promoting pest natural enemies (e.g. presence and diversity of predator arthropods) and pollinators, hedgerows sustain these ecosystem services.

5.1 Case study: hedgerow density and configuration promote pest natural enemies in agricultural landscapes

Many generalist insect species in agricultural landscapes use hedgerows and cultivated habitats alternately during their life cycle (Morandin and Kremen, 2013; Garratt et al., 2017; Holland, 2019; Dover et al., 2019a). Among them, natural enemies of crop pests use hedgerows as overwintering sites or refuges when crops do not offer suitable habitat conditions (after harvest for instance), or to find alternative resources (animal preys or nectar resources for adult parasitoids or hoverflies) (Rand et al., 2006; Blitzer et al., 2012; Garratt et al., 2017; Holland, 2019). For this reason, landscape heterogeneity related to the amount of semi-natural habitats including hedgerows is often found to enhance the diversity of natural enemies (Tscharntke et al., 2005; Bianchi et al., 2006) and levels of pest regulation (Thies et al., 2003; Gardiner et al., 2009; Rusch et al., 2011) in agricultural landscapes. This predominant effect of landscape heterogeneity can even be more important than local farming practices to explain the variations in the diversity of natural enemies in crop fields (Rundlöf and Smith, 2006).

We investigated how landscape heterogeneity related to hedgerow networks, how in-field habitat quality related to local farming practices (organic vs. conventional farming) and habitat conditions (crop density, crop height, weed cover), and how farming heterogeneity (surface area of organic vs. conventional crops) influence the diversity of natural enemies in crop fields. Communities of ladybirds, predatory carabid beetles and parasitoids (aphid mummies) were sampled between 2012 and 2013 in 20 pairs of winter wheat fields (one under organic farming, one under conventional farming) in the southern part of the Zone Atelier Armorique. The pairs of studied fields were distributed along two orthogonal landscape gradients in terms of amount of semi-natural habitats (0.5–20%) and farming heterogeneity (6–35% of organic crops). During the study, 657 ladybirds, 54742 carabid beetles and 1120 mummies were collected.

We showed that the diversity of natural enemies was mainly related to the abundance of ladybirds. Abundances of natural enemies were influenced by land-cover heterogeneity and in-field habitat quality (Fig. 10). Specifically, the abundance of natural enemies was increased by increasing hedgerow density and increasing complexity in hedgerow configuration (positive correlation

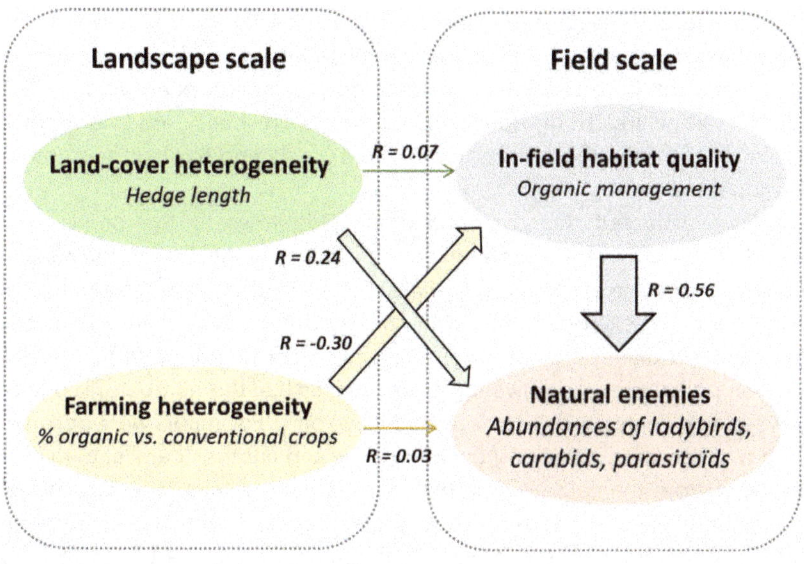

Figure 10 Summary of the relationships between (1) natural enemy abundance in crop fields (related to abundances of ladybirds, carabid beetles and parasitoids) and (2) variables describing in-field habitat quality at the field scale (in pairs organic and conventional fields), (3) metrics describing land-cover heterogeneity and (4) metrics describing farming heterogeneity. Relationships between these four (latent) variables were assessed using PLS-Path modelling. R is the correlation value (path effect) between two latent variables. Depicted exogenous variables (in italics) are variables that are significantly contributed to explain latent variables. Source: modified from Puech et al. (2015).

value: R = 0.24). Local farming type has a stronger and positive effect on the abundances of natural enemies in fields as expressed by the higher, positive correlation value (R = 0.56). These results confirm that the abundances of natural enemies in fields are promoted in complex agricultural landscapes by high density and complex configuration of semi-natural habitats such as hedgerows (Chaplin-Kramer et al., 2011). However, the effects of local crop management are also crucial, and even more important than those of landscape complexity, contrary to results from existing literature (Bengtsson et al., 2005). This result suggests that the relative effects of local farming type and landscape complexity on natural enemies are highly context-dependent. Our study shows, however, that the adoption of low-input farming practices and the restoration of hedgerow networks are both valuable management options to promote natural enemies in agricultural landscapes.

5.2 Case study: hedgerows enhance pollinators in intensive agricultural landscapes

Pollinators play an important functional role in most terrestrial ecosystems and contribute to a key ecosystem service, that is, pollination, that is vital to the maintenance of both wild plant communities and agricultural productivity. Pollinators contribute to 35% of global crop production (Klein et al., 2007). Insects, particularly both domesticated and wild bees, are the primary pollinators of most agricultural crops and wild plants, and so our case study will focus on these insects.

We investigated the contribution of hedgerows to the provision of diversified resources for a wide range of pollinators as well as to the support of plant-pollinator network. Twenty winter cereal fields bordered by at last one hedgerow were selected in the southern part of the Zone Atelier Armorique. We quantified plant-pollinator interactions by recording the number of visits between pollinators and flowering plant species (i.e. the number of contacts between a pollinator and flower's reproductive part). Pollinators were observed along two transects, one along the hedgerow, and another, parallel to the first one, 50 m from the field edge in the field interior. Pollinators were classified into eight groups: domestic bees, solitary bees, bumblebees, syrphids, butterflies, diptera, coleoptera and other hymenoptera. Due to the very few number of observations, domestic bees were excluded from analyses. On the same transects, we estimated the cover (in percentage) of each flowering plant species.

In total, we observed the seven groups of pollinators interacting with 62 flowering plant species. We showed that the cover of flowering plant species in hedgerows was positively related to the abundance of wild bees and bumblebees in hedgerows and, to the abundance of syrphids in both hedgerows and field interiors. Specifically, bumblebees were sensitive to the cover of ligneous flowering plant species, whereas syrphids were sensitive to the cover of herbaceous flowering plant species. Plant-pollinator networks shown in Figure 11 revealed a more complex structure in hedgerows (1008

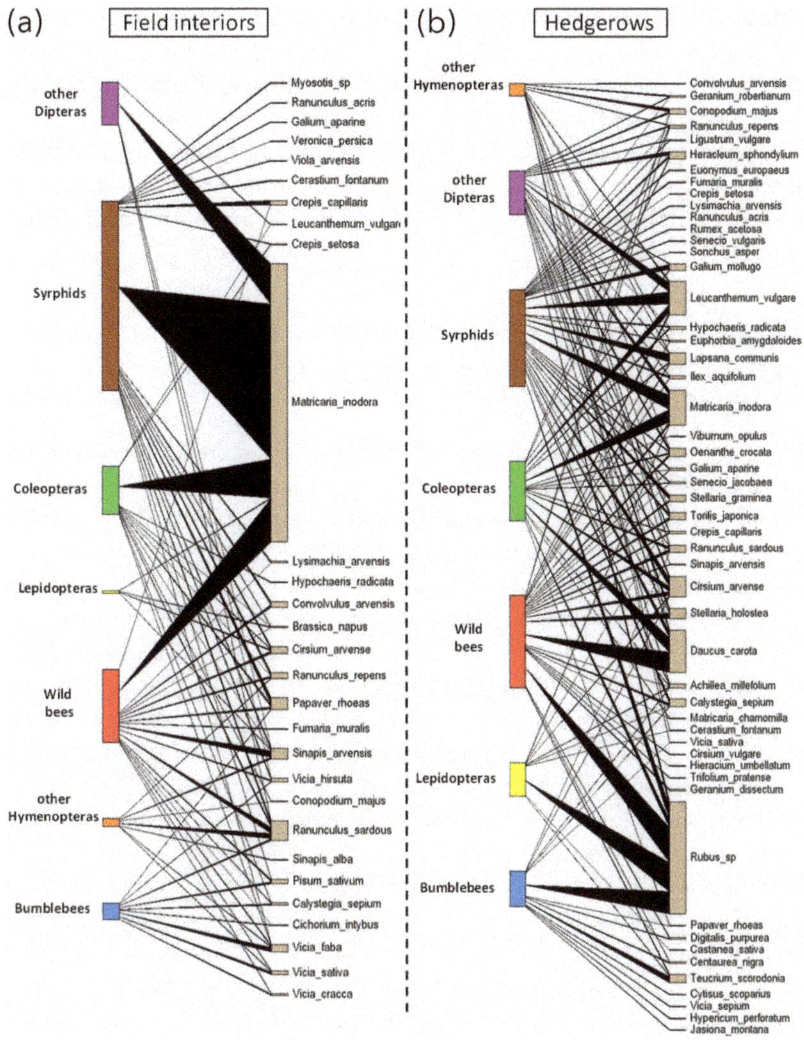

Figure 11 Interaction networks between pollinator groups (N = 7) and flowering plant species in (a) inner fields and (b) hedgerows. The width of links (in black) represents the frequency of interactions. The length of bars (in color) corresponds to the relative abundance of pollinator groups and to the cover of flowering plants (as a proxy of their abundance). Source: Berry, unpublished data.

interaction events involving 55 flowering plant species) than in inner fields (883 interaction events involving 29 flowering plant species).

In fields, mayweed (*Matricaria inodora*) was the most frequently visited plant species (855 visits) and represented a key resource for most pollinator groups. In hedgerows, bramble (*Rubus fruticosus agg.*) was the most frequently visited plant species and was particularly important for wild bees, butterflies and bumblebees. Besides herbaceous species, shrub/tree species like common holly (*Ilex aquifolium*), spindle (*Euonymus europaeus*) and chestnut (*Castanea sativa*) contributed to the interaction network. We showed that hedgerows offer a wide range of resources (through higher diversity of flowering plants than inner fields) as suggested by Venjakob et al. (2016). Our results emphasize the positive role and the need to maintain hedgerows to provide diversified resources for pollinators and support their maintenance in intensive agricultural landscapes.

6 Case study: hedgerow plantation and bocage restoration enhance biodiversity and other ecosystem services

Studies addressing the effect of conservation and restoration of hedgerows have mostly focused on one particular ecosystem service or ecological function like pollination (e.g. Morandin and Kremen, 2013), while there are only a few studies that study the impacts of hedgerows on multiple ecosystem services, simultaneously (e.g. Van Vooren et al., 2018). Here, we assessed the multifunctionality of recently planted hedgerows (i.e. 15 years ago; N = 16) compared to old traditional field margins (old hedgerows: N = 8; grassy margins: N = 8) on several farms pertaining to hedgerow network landscapes in Brittany, northwestern France. Multifunctionality was assessed by considering the ability of field margins to support (i) biodiversity of conservation interest, (ii) functional biodiversity involved in regulation functions (pest control and pollination) and (iii) biodiversity associated with negative disservices for agriculture, related to pest infestations in crops.

We showed that recent hedgerows harbor, 15 years after planting, similar levels of species diversity than old, traditional field margins (Fig. 12). Some recent hedgerows harbor species that are not found in old traditional hedgerows (and vice versa), suggesting that the diversity of field margins contributes to biodiversity conservation at farm/landscape level. Recent hedgerows also had carabid species that are typically associated with forest habitats, showing that these hedgerows fulfilled the suitable microclimatic conditions and vegetation density required by these forest species.

Recent hedgerows also harbored comparable levels of functional biodiversity, 15 years after planting. Recent hedgerows had similar abundance

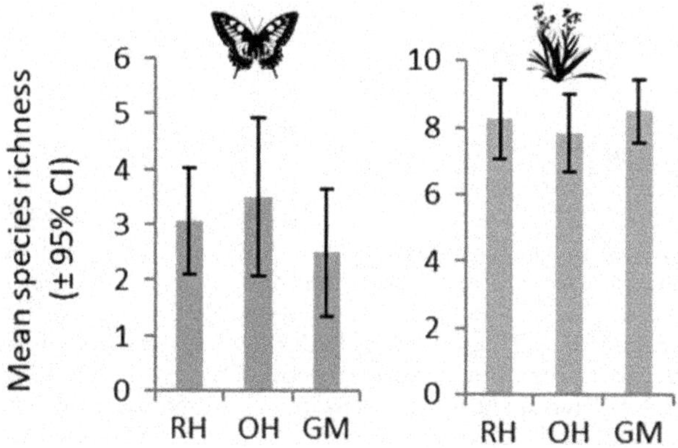

Figure 12 Species richness of biodiversity of conservation interest (butterflies and vascular plants) in recent hedgerows (RH), old hedgerows (OH) and grassy field margins (GM). Differences in species richness between the different types of field margins are not significant (tested in mixed generalized linear models, combined with multi-model inference and model averaging procedures). Source: Aviron, unpublished data.

levels of predatory carabid beetles (Fig. 13) as well as pollinating insects. Annual crops adjacent to recent hedgerows also tended to have higher activity-density of predatory arthropods than field adjacent to traditional margins (Fig. 13). It suggests that recent hedgerows might play a similar role of refuge and

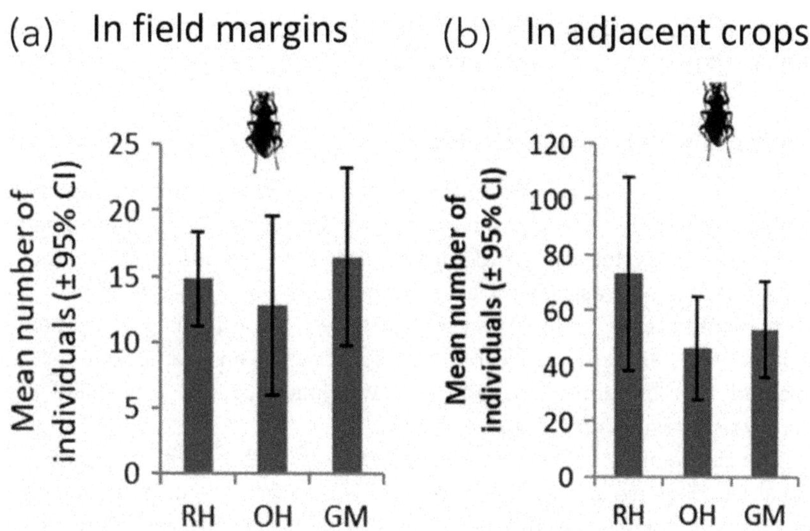

Figure 13 Abundances (activity-density) of predatory carabid beetles observed (a) in field margins and (b) in adjacent fields. RH: recent hedgerows; OH: old hedgerows; GM: grassy field margins. Differences in species richness between the different types of field margins are not significant (tested in mixed generalized linear models, combined with multi-model inference and model averaging procedures). Source: Aviron, unpublished data.

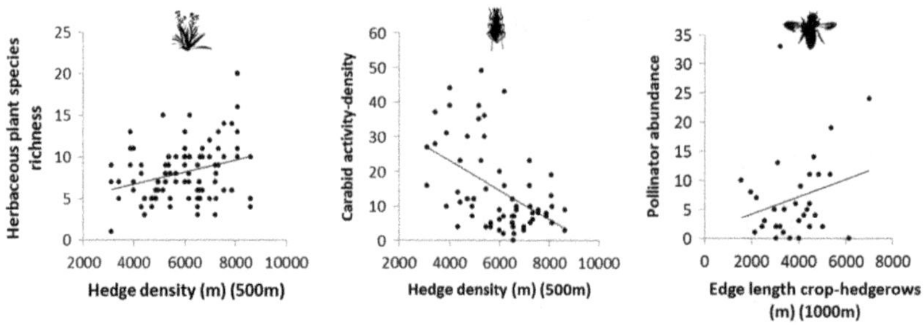

Figure 14 Impact of the configuration of hedgerow networks (hedgerow density or edge length between crops and hedgerows) on local biodiversity. Relationships are all significant at $P < 0.05$ (tested in mixed generalized linear models, combined with multi-model inference and model averaging procedures). Source: Aviron, unpublished data.

reservoir of predatory arthropods to other field margins. They do not increase weed infestation in crops as suggested by comparable levels of weed cover between field margin types.

At the landscape scale, the configuration of hedgerow networks had a strong impact on the different type of biodiversity. But we observed contrasting responses for the different species groups. Hedgerow density had a positive effect on vascular plants and pollinators, but a negative effect on carabid abundance (Fig. 14). These results suggest that enhancing multifunctionality on farms through bocage restoration might not simultaneously enhance contrasted ecological functions and that management measures should be targeted according to required functions.

7 Summary and future trends

The way in which hedgerows are managed as well as their management history can affect their structure and composition. Often considered as semi-natural elements, studies considering the effect of hedgerow management on biodiversity and associated ecosystem services are still rare. There is a dearth of information on the effect of frequency, type and timing of hedgerow management particularly on ecosystem services. As an example, it can be expected that hedgerow management will influence the availability and quantity of floral resources for pollinators.

The connectivity of the landscape, and thereby the ability of plants and animals to disperse across the landscape is also important for biodiversity conservation. But, tracking the movement of organisms through landscapes is difficult. Thus evidence of the value of hedgerow networks in providing effective landscape connectivity remains scarce. If hedgerows promote the abundance and dispersal of pest natural enemies, they may also favor the abundance and dispersal of pests. Whether hedgerows have any impacts on levels of pest regulation will be the outcome of highly complex interactions that are not well understood (Dover, 2019b).

To conclude, conservation and/or restoration of hedgerows appears to be a promising intervention that can promote both biodiversity conservation and associated ecosystem services. However, cost-benefit studies are needed to provide convincing evidence to farmers of the interest of such interventions and over what time-frame benefits might be realized (Garbach and Long, 2017).

8 Where to look for further information

The following articles provide a good introduction of the subject:

- Baudry, J., Bunce, R. G. H. and Burel, F. 2000. Hedgerows: an international perspective on their origin, function and management. *Journal of Environmental Management* 60, 7–22.
- Forman, R. T. and Baudry, J. 1984. Hedgerows and hedgerow networks in landscape ecology. *Environmental Management* 8, 495–510.

Deeper information about the ecology of hedgerows can be found in:

- Dover, J. W. 2019. *The Ecology of Hedgerows and Field Margins*. Routledge, London.

Key conferences:

- IALE congresses: The International Association of Landscape Ecology holds regular meetings on landscape ecology (https://www.landscape-ecology.org/home.html).

9 References

Alignier, A. and Baudry, J. 2015. Changes in management practices over time explain most variation in vegetation of field margins in Brittany, France. *Agriculture, Ecosystems and Environment* 211, 164–72. doi:10.1016/j.agee.2015.06.008.

Alignier, A. and Aviron, S. 2017. Time-lagged response of carabid species richness and composition to past management practices and landscape context of semi-natural field margins. *Journal of Environmental Management* 204(1), 282–90. doi:10.1016/j.jenvman.2017.08.054.

ANON 2019. The value of hedgerows for wildlife. Available at: https://www.rspb.org.uk/our-work/conservation/conservation-and-sustainability/advice/conservation-land-management-advice/farm-hedges/the-value-of-hedgerows-for-wildlife/ (accessed 18 November 2019).

Aviron, S., Burel, F., Baudry, J. and Schermann, N. 2005. Carabid assemblages in agricultural landscapes: impacts of habitat features, landscape context at different spatial scales and farming intensity. *Agriculture, Ecosystems and Environment* 108(3), 205–17. doi:10.1016/j.agee.2005.02.004.

Barr, C. J. and Gillespie, M. K. 2000. Estimating hedgerow length and pattern characteristics in Great Britain using Countryside Survey data. *Journal of Environmental Management* 60(1), 23–32. doi:10.1006/jema.2000.0359.

Batáry, P., Báldi, A., Kleijn, D. and Tscharntke, T. 2010. Landscape-moderated biodiversity effects of agri-environmental management: a meta-analysis. *Proceedings of the Royal Society. Series B: Biological Sciences* 278, 1894–902.

Baudry, J., Jouin, A. and Thenail, C. 1998. La diversité des bordures de champ dans les exploitations agricoles de pays de bocage. *Études et Recherches sur les Systèmes Agraires et le développement* 31, 117–34.

Baudry, J., Bunce, R. G. H. and Burel, F. 2000. Hedgerows: an international perspective on their origin, function and management. *Journal of Environmental Management* 60(1), 7–22. doi:10.1006/jema.2000.0358.

Bengtsson, J., Ahnström, J. and Weibull, A.-C. 2005. The effects of organic agriculture on biodiversity and abundance: a meta-analysis. *Journal of Applied Ecology* 42(2), 261–9. doi:10.1111/j.1365-2664.2005.01005.x.

Bianchi, F. J. J. A., Booij, C. J. H. and Tscharntke, T. 2006. Sustainable pest regulation in agricultural landscapes: a review on landscape composition, biodiversity and natural pest control. *Proceedings of the Royal Society B: Biological Sciences* 273(1595), 1715–27. doi:10.1098/rspb.2006.3530.

Blitzer, E. J., Dormann, C. F., Holzschuh, A., Klein, A.-M., Rand, T. A. and Tscharntke, T. 2012. Spillover of functionally important organisms between managed and natural habitats. *Agriculture, Ecosystems and Environment* 146(1), 34–43. doi:10.1016/j.agee.2011.09.005.

Burel, F. and Baudry, J. 1995. Social, aesthetic and ecological aspects of hedgerows in rural landscapes as a framework for greenways. *Landscape and Urban Planning* 33(1–3), 327–40. doi:10.1016/0169-2046(94)02026-C.

Burel, F. 1996. Hedgerows and their role in agricultural landscapes. *Critical Reviews in Plant Sciences* 15(2), 169–90. doi:10.1080/07352689.1996.10393185.

Burel, F. and Baudry, J. 1999. *Écologie du paysage. Concepts, méthodes et applications.* TEC & DOC, Paris, 362p.

Carey, P. D., Wallis, S., Chamberlain, P. M., Cooper, A., Emmett, B. A., Maskell, L. C., Mccann, T., Murphy, J., Norton, L. R., Reynolds, B., Scott, W. A., Simpson, I. C., Smart, S. M. and Ullyett, J. M. 2008. *Countryside Survey: UK Results from 2007. NERC/Centre for Ecology & Hydrology*, 105pp (CEH Project Number: C03259).

Chaplin-Kramer, R., O'Rourke, M. E., Blitzer, E. J. and Kremen, C. 2011. A meta-analysis of crop pest and natural enemy response to landscape complexity. *Ecology Letters* 14(9), 922–32. doi:10.1111/j.1461-0248.2011.01642.x.

Chaplin-Kramer, R. and Kremen, C. 2012. Pest control experiments show benefits of complexity at landscape and local scales. *Ecological Applications: A Publication of the Ecological Society of America* 22(7), 1936–48. doi:10.1890/11-1844.1.

Cole, L. J., Deane, R. and Rayment, M. 2009. *Provision of Ecosystem Services through the Environmental Stewardship Scheme*. DEFRA, London.

Concepción, E. D., Fernández-González, F. and Díaz, M. 2012. Plant diversity partitioning in Mediterranean croplands: effects of farming intensity, field edge, and landscape context. *Ecological Applications: A Publication of the Ecological Society of America* 22(3), 972–81. doi:10.1890/11-1471.1.

Corbit, M., Marks, P. L. and Gardescu, S. 1999. Hedgerows as habitat corridors for forest herbs in central New York, USA. *Journal of Ecology* 87(2), 220–32. doi:10.1046/j.1365-2745.1999.00339.x.

Croxton, P. J., Franssen, W., Myhill, D. G. and Sparks, T. H. 2004. The restoration of neglected hedges: a comparison of management treatments. *Biological Conservation* 117(1), 19–23. doi:10.1016/S0006-3207(03)00258-1.

Dainese, M., Montecchiari, S., Sitzia, T., Sigura, M. and Marini, L. 2017. High cover of hedgerows in the landscape supports multiple ecosystem services in Mediterranean cereal fields. *Journal of Applied Ecology* 54(2), 380–8. doi:10.1111/1365-2664.12747.

Deckers, B., Hermy, M. and Muys, B. 2004. Factors affecting plant species composition of hedgerows: relative importance and hierarchy. *Acta Oecologica* 26(1), 23–37. doi:10.1016/j.actao.2004.03.002.

Deckers, B., Verheyen, K., Hermy, M. and Muys, B. 2005. Effects of landscape structure on the invasive spread of black cherry *Prunus serotina* in an agricultural landscape in Flanders, Belgium. *Ecography* 28(1), 99–109. doi:10.1111/j.0906-7590.2005.04054.x.

Diamond, J. M. 1975. The island dilemma: lessons of modern biogeographic studies for the design of nature reserves. *Biological Conservation* 7(2), 129–46. doi:10.1016/0006-3207(75)90052-X.

Dondina, O., Kataoka, L., Orioli, V. and Bani, L. 2016. How to manage hedgerows as effective ecological corridors for mammals: a two-species approach. *Agriculture, Ecosystems and Environment* 231, 283–90. doi:10.1016/j.agee.2016.07.005.

Dover, J. and Sparks, T. 2000. A review of the ecology of butterflies in British hedgerows. *Journal of Environmental Management* 60(1), 51–63. doi:10.1006/jema.2000.0361.

Dover, J. and Settele, J. 2009. The influences of landscape structure on butterfly distribution and movement: a review. *Journal of Insect Conservation* 13(1), 3–27. doi:10.1007/s10841-008-9135-8.

Dover, J. W. 1996. Factors affecting the distribution of satyrid butterflies on arable farmland. *Journal of Applied Ecology* 33(4), 723–34. doi:10.2307/2404943.

Dover, J. W. 2019a. The ecology of butterflies and moths in hedgerows and field margins. In: Dover, J. W. (Ed.), *The Ecology of Hedgerows and Field Margins*. Routledge, New York, pp. 186–209.

Dover, J. W. 2019b. *The Ecology of Hedgerows and Field Margins*. Routledge, London.

Forman, R. T. T. and Baudry, J. 1984. Hedgerows and hedgerow networks in landscape ecology. *Environmental Management* 8(6), 495–510. doi:10.1007/BF01871575.

Garbach, K. and Long, R. F. 2017. Determinants of field edge habitat restoration on farms in California's Sacramento Valley. *Journal of Environmental Management* 189, 134–41. doi:10.1016/j.jenvman.2016.12.036.

Gardiner, M. M., Landis, D. A., Gratton, C., Difonzo, C. D., O'Neal, M., Chacon, J. M., Wayo, M. T., Schmidt, N. P., Mueller, E. E. and Heimpel, G. E. 2009. Landscape diversity enhances biological control of an introduced crop pest in the north-central USA. *Ecological Applications: A Publication of the Ecological Society of America* 19(1), 143–54. doi:10.1890/07-1265.1.

Garratt, M. P. D., Senapathi, D., Coston, D. J., Mortimer, S. R. and Potts, S. G. 2017. The benefits of hedgerows for pollinators and natural enemies depends on hedge quality and landscape context. *Agriculture, Ecosystems and Environment* 247, 363–70. doi:10.1016/j.agee.2017.06.048.

Gelling, M., Macdonald, D. W. and Mathews, F. 2007. Are hedgerows the route to increased farmland small mammal density? Use of hedgerows in British pastoral habitats. *Landscape Ecology* 22(7), 1019–32. doi:10.1007/s10980-007-9088-4.

Hannon, L. E. and Sisk, T. D. 2009. Hedgerows in an agri-natural landscape: potential habitat value for native bees. *Biological Conservation* 142(10), 2140–54. doi:10.1016/j.biocon.2009.04.014.

Hanski, I. and Thomas, C. D. 1994. Metapopulation dynamics and conservation: a spatially explicit model applied to butterflies. *Biological Conservation* 68(2), 167–80. doi:10.1016/0006-3207(94)90348-4.

Hinsley, S. A. and Bellamy, P. E. 2000. The influence of hedge structure, management and landscape context on the value of hedgerows to birds: a review. *Journal of Environmental Management* 60(1), 33–49. doi:10.1006/jema.2000.0360.

Holland, J. 2019. Contribution of hedgerows to biological control. In: Dover, J. W. (Ed.), *The Ecology of Hedgerows and Field Margins*. Routledge, New York, pp. 123–46.

Jacquemyn, H., Butaye, J. and Hermy, M. 2003. Influence of environmental and spatial variables on regional distribution of forest plant species in a fragmented and changing landscape. *Ecography* 26(6), 768–76. doi:10.1111/j.0906-7590.2003.03620.x.

Kennedy, C. M., Lonsdorf, E., Neel, M. C., Williams, N. M., Ricketts, T. H., Winfree, R., Bommarco, R., Brittain, C., Burley, A. L., Cariveau, D., Carvalheiro, L. G., Chacoff, N. P., Cunningham, S. A., Danforth, B. N., Dudenhöffer, J. H., Elle, E., Gaines, H. R., Garibaldi, L. A., Gratton, C., Holzschuh, A., Isaacs, R., Javorek, S. K., Jha, S., Klein, A. M., Krewenka, K., Mandelik, Y., Mayfield, M. M., Morandin, L., Neame, L. A., Otieno, M., Park, M., Potts, S. G., Rundlöf, M., Saez, A., Steffan-Dewenter, I., Taki, H., Viana, B. F., Westphal, C., Wilson, J. K., Greenleaf, S. S. and Kremen, C. 2013. A global quantitative synthesis of local and landscape effects on wild bee pollinators in agroecosystems. *Ecology Letters* 16(5), 584–99. doi:10.1111/ele.12082.

Kiepe, P. 1995. No runoff, no soil loss: soil and water conservation in hedgerow barrier systems. Doctoral Thesis. Agricultural University, Wageningen.

Kleijn, D. and Verbeek, M. 2000. Factors affecting the species composition of arable field boundary vegetation. *Journal of Applied Ecology* 37(2), 256–66. doi:10.1046/j.1365-2664.2000.00486.x.

Klein, A. M., Vaissière, B. E., Cane, J. H., Steffan-Dewenter, I., Cunningham, S. A., Kremen, C. and Tscharntke, T. 2007. Importance of pollinators in changing landscapes for world crops. *Proceedings. Biological Sciences* 274(1608), 303–13. doi:10.1098/rspb.2006.3721.

Le Cœur, D., Baudry, J., Burel, F. and Thenail, C. 2002. Why and how we should study field boundary biodiversity in an agrarian landscape context. *Agriculture, Ecosystems and Environment* 89(1–2), 23–40. doi:10.1016/S0167-8809(01)00316-4.

Marshall, E. J. P. and Moonen, A. C. 2002. Field margins in northern Europe: their functions and interactions with agriculture. *Agriculture, Ecosystems and Environment* 89(1-2), 5–21. doi:10.1016/S0167-8809(01)00315-2.

McCollin, D., Jackson, J. I., Bunce, R. G. H., Barr, C. J. and Stuart, R. 2000. Hedgerows as habitat for woodland plants. *Journal of Environmental Management* 60(1), 77–90. doi:10.1006/jema.2000.0363.

McRae, B. H. 2006. Isolation by resistance. *Evolution; International Journal of Organic Evolution* 60(8), 1551–61. doi:10.1111/j.0014-3820.2006.tb00500.x.

McRae, B. H., Dickson, B. G., Keitt, T. H. and Shah, V. B. 2008. Using circuit theory to model connectivity in ecology, evolution, and conservation. *Ecology* 89(10), 2712–24. doi:10.1890/07-1861.1.

Millán De La Peña, N., Butet, A., Delettre, Y., Morant, P. and Burel, F. 2003. Landscape context and carabid beetles (Coleoptera: Carabidae) communities of hedgerows in western France. *Agriculture, Ecosystems and Environment* 94(1), 59–72. doi:10.1016/S0167-8809(02)00012-9.

Millennium Ecosystem Assessment 2005. *Ecosystems and Human Well-Being*. Island Press, Washington, DC.

Mony, C., Landais, Q., Georges, R., Butet, A., Burel, F., Jambon, O., Gouesbet, V. and Ernoult, A. submitted. Effects of connectivity on forest actual plant dispersal in agriculture-dominated landscapes: a community-scale analysis.

Morandin, L. A. and Kremen, C. 2013. Hedgerow restoration promotes pollinator populations and exports native bees to adjacent fields. *Ecological Applications* 23(4), 829–39. doi:10.1890/12-1051.1.

Neumann, J. L., Griffiths, G. H., Hoodless, A. and Holloway, G. J. 2016. The compositional and configurational heterogeneity of matrix habitats shape woodland carabid communities in wooded-agricultural landscapes. *Landscape Ecology* 31(2), 301–15. doi:10.1007/s10980-015-0244-y.

Ouin, A., Martin, M. and Burel, F. 2008. Agricultural landscape connectivity for the meadow brown butterfly (*Maniola jurtina*). *Agriculture, Ecosystems and Environment* 124(3–4), 193–9. doi:10.1016/j.agee.2007.09.010.

Oyarzun, R. A., Stöckle, C. O. and Whiting, M. D. 2007. A simple approach to modeling radiation interception by fruit-tree orchards. *Agricultural and Forest Meteorology* 142(1), 12–24. doi:10.1016/j.agrformet.2006.10.004.

Petit, S. and Burel, F. 1998. Effects of landscape dynamics on the metapopulation of a ground beetle (Coleoptera, Carabidae) in a hedgerow network. *Agriculture, Ecosystems and Environment* 69(3), 243–52. doi:10.1016/S0167-8809(98)00111-X.

Pointereau, P. 2002. Les haies: évolution du linéaire en France depuis quarante ans. *Le Courrier de l'Environnement* 46, 69–73.

Pollard, K. A. and Holland, J. M. 2006. Arthropods within the woody element of hedgerows and their distribution pattern. *Agricultural and Forest Entomology* 8(3), 203–11. doi:10.1111/j.1461-9563.2006.00297.x.

Ponisio, L. C., M'Gonigle, L. K. and Kremen, C. 2016. On-farm habitat restoration counters biotic homogenization in intensively managed agriculture. *Global Change Biology* 22(2), 704–15. doi:10.1111/gcb.13117.

Puech, C., Poggi, S., Baudry, J. and Aviron, S. 2015. Do farming practices affect natural enemies at the landscape scale? *Landscape Ecology* 30(1), 125–40. doi:10.1007/s10980-014-0103-2.

Pywell, R. F., Heard, M. S., Woodcock, B. A., Hinsley, S., Ridding, L., Nowakowski, M. and Bullock, J. M. 2015. Wildlife-friendly farming increases crop yield: evidence for ecological intensification. *Proceedings of the Royal Society B: Biological Sciences* 282(1816), 20151740. doi:10.1098/rspb.2015.1740.

Rand, T. A., Tylianakis, J. M. and Tscharntke, T. 2006. Spillover edge effects: the dispersal of agriculturally subsidized insect natural enemies into adjacent natural habitats. *Ecology Letters* 9(5), 603–14. doi:10.1111/j.1461-0248.2006.00911.x.

Rundlöf, M. and Smith, H. G. 2006. The effect of organic farming on butterfly diversity depends on landscape context. *Journal of Applied Ecology* 43(6), 1121–7. doi:10.1111/j.1365-2664.2006.01233.x.

Rusch, A., Valantin-Morison, M., Sarthou, J.-P. and Roger-Estrade, J. 2011. Multi-scale effects of landscape complexity and crop management on pollen beetle parasitism rate. *Landscape Ecology* 26(4), 473–86. doi:10.1007/s10980-011-9573-7.

Sánchez, I. A., Lassaletta, L., McCollin, D. and Bunce, R. G. H. 2010. The effect of hedgerow loss on microclimate in the Mediterranean region: an investigation in Central Spain. *Agroforestry Systems* 78(1), 13–25. doi:10.1007/s10457-009-9224-z.

Simberloff, D. and Cox, J. 1987. Consequences and costs of conservation corridors. *Conservation Biology* 1(1), 63–71. doi:10.1111/j.1523-1739.1987.tb00010.x.

Simberloff, D., Farr, J. A., Cox, J. and Mehlman, D. W. 1992. Movement corridors: conservation bargains or poor investments? *Conservation Biology* 6(4), 493–504. doi:10.1046/j.1523-1739.1992.06040493.x.

Smith, H., Firbank, L. G. and Macdonald, D. W. 1999. Uncropped edges of arable fields managed for biodiversity do not increase weed occurrence in adjacent crops. *Biological Conservation* 89(1), 107–11. doi:10.1016/S0006-3207(98)00125-6.

Staley, J. T., Bullock, J. M., Baldock, K. C. R., Redhead, J. W., Hooftman, D. A. P., Button, N. and Pywell, R. F. 2013. Changes in hedgerow floral diversity over 70 years in an English rural landscape, and the impacts of management. *Biological Conservation* 167, 97–105. doi:10.1016/j.biocon.2013.07.033.

Thenail, C. 2002. Relationships between farm characteristics and the variation of the density of hedgerows at the level of a micro-region of Bocage landscape. Study case in Brittany, France. *Agricultural Systems* 71(3), 207–30. doi:10.1016/S0308-521X(01)00048-8.

Thies, C., Steffan-Dewenter, I. and Tscharntke, T. 2003. Effects of landscape context on herbivory and parasitism at different spatial scales. *Oikos* 101(1), 18–25. doi:10.1034/j.1600-0706.2003.12567.x.

Tilman, D., Fargione, J., Wolff, B., D'Antonio, C., Dobson, A., Howarth, R., Schindler, D., Schlesinger, W. H., Simberloff, D. and Swackhamer, D. 2001. Forecasting agriculturally driven global environmental change. *Science* 292(5515), 281–4. doi:10.1126/science.1057544.

Tscharntke, T., Klein, A. M., Kruess, A., Steffan-Dewenter, I. and Thies, C. 2005. Landscape perspectives on agricultural intensification and biodiversity - ecosystem service management. *Ecology Letters* 8(8), 857–74. doi:10.1111/j.1461-0248.2005.00782.x.

Uroy, L., Mony, C. and Ernoult, A. 2019. Additive effects of connectivity provided by different habitat types drive plant assembly. *Scientific Reports* 9(1), 13952. doi:10.1038/s41598-019-50184-2.

Van Den Berge, S., Tessens, S., Baeten, L., Vanderschaeve, C. and Verheyen, K. 2019. Contrasting vegetation change (1974–2015) in hedgerows and forests in an intensively used agricultural landscape. *Applied Vegetation Science* 22(2), 269–81. doi:10.1111/avsc.12424.

Van Vooren, L., Reubens, B., Ampoorter, E., Broekx, S., Pardon, P., Van Waes, C. and Verheyen, K. 2018. Monitoring the impact of hedgerows and grass strips on the performance of multiple ecosystem service indicators. *Environmental Management* 62(2), 241–59. doi:10.1007/s00267-018-1043-4.

Venjakob, C., Klein, A. M., Ebeling, A., Tscharntke, T. and Scherber, C. 2016. Plant diversity increases spatio-temporal niche complementarity in plant-pollinator interactions. *Ecology and Evolution* 6(8), 2249–61. doi:10.1002/ece3.2026.

Verkaar, H. 1990. Corridors as a tool for plant species conservation? In: Bunce, R. H. G. and Howard, D. C. (Eds), *Species Dispersal in Agricultural Habitats*. Belhaven Press, London, pp. 82–97.

Vickery, J. A., Feber, R. E. and Fuller, R. J. 2009. Arable field margins managed for biodiversity conservation: a review of food resource provision for farmland birds. *Agriculture, Ecosystems and Environment* 133(1–2), 1–13. doi:10.1016/j.agee.2009.05.012.

Wehling, S. and Diekmann, M. 2009. Importance of hedgerows as habitat corridors for forest plants in agricultural landscapes. *Biological Conservation* 142(11), 2522–30. doi:10.1016/j.biocon.2009.05.023.

Wolton, R. J., Pollard, K., Goodwin, A. and Norton, L. R. 2014. *Regulatory Services Delivered by Hedges: The Evidence Base*. Report of Defra project LM0106, Hedgelink.

Chapter 7

Reconciling production and biodiversity in management of pastures and grasslands

Sylvain Plantureux, Université de Lorraine, INRAE, LAE, France

1 Introduction

Agriculture in the twenty-first century faces two major challenges: to increase food and non-food production to meet the needs of a growing population and, at the same time, to ensure that production will not damage the environment. These needs are often seen as contradictory, with one prioritized at the expense of the other or resolved by subsidizing farmers to compensate them for losses of production from more environmental-friendly management. However, this solution is not sustainable in the long run since it is a drain on the often-constrained finances of governments.

Agroecology is a potential solution to resolve this problem because it proposes to use biodiversity in the production process, substituting energy and chemical inputs (e.g. fertilizers and pesticides) with biological and ecological processes. In this way production and the quality of the environment become

http://dx.doi.org/10.19103/AS.2020.0071.02

linked. Although the principles of agroecology are well-established, their implementation is still limited by many factors, including:

1. the willingness of farmers, extension services and policy makers to change their views on how to assess and manage agricultural systems, and
2. the availability of scientific and technical knowledge about production-biodiversity relationships.

Grasslands are particularly affected by these issues. According to the FAO (2019), 26% of the world's land area and 70% of the world's agricultural land are covered by grasslands, contributing to the livelihoods of over 800 million people. The environmental benefits provided by grasslands are now well-recognized, including:

- preservation of biodiversity (flora, above- and belowground fauna and microflora),
- prevention of chemical pollution,
- soil erosion and water control,
- carbon sequestration, and
- reduction of greenhouse gases.

In Europe half of the region's flora is dependent on grassland habitats (Eckhard et al. in (Veen et al., 2009)). However, in recent decades there has been a significant decline in Europe in the quality of semi-natural grasslands, particularly by conversion to the production of forage crops such as silage maize, the shift to more temporary grasslands (e.g. as part of a rotation) or more intensive cultivation of existing grasslands. Semi-natural grasslands are defined (Peeters et al., 2014) as: 'Low-yielding permanent grasslands, dominated by indigenous, naturally occurring grass communities, other herbaceous species and, in some cases, shrubs and/or trees. These mown and/or grazed ecosystems have not been substantially modified by fertilization, liming, drainage, soil cultivation, herbicide use, introduction of exotic species and (over)sowing.' This increase in intensification in grassland management has resulted in a drastic reduction in the extent and quality of semi-natural grasslands and sometimes even in chemical pollution (such as leaching of nitrates and herbicides used in more intensive grassland management). In most cases intensive grass production is incompatible with biodiversity maintenance (Plantureux et al., 2005). From the environmental point of view, temporary grasslands or intensive permanent grasslands have some benefits (e.g. prevention of soil erosion), but semi-natural grasslands offer the greatest environmental benefits because of their greater biodiversity.

other criteria must be taken into account. Unlike productivity, biodiversity needs to be assessed using a much larger range of criteria. Of these criteria, if species richness is the most commonly used, many others may also be considered, such as specific composition, functional diversity, phylogenetic diversity or evenness.

The various methods used to study the biodiversity-production (BP) relationship have their advantages and disadvantages. Experimentation compares grasslands with differing levels of biodiversity under comparable soil, climate and management conditions. Experimentation makes it possible to control species richness, functional diversity or botanical composition. Controlled experiments have been criticized for their lack of representativeness of actual semi-natural grasslands. They differ, for example, in terms of factors such as the type of sown species, a low number of species, immaturity of grassland communities, lower frequency of harvest and low rates of grazing (Isselstein, 2005).

If observations of grasslands in 'natural conditions' are more realistic, there is a much higher risk of confusion between the effects of biodiversity and the effects of environmental and management factors on productivity ('hidden treatments'). Species removal or oversowing of semi-natural grasslands is an intermediate experimental system to reduce the number of variables to take into account. After having sown a mixture, it is important to wait a few years before beginning to measure BP relationships (Lepš, 2004). While ecologists compare species mixtures with monocultures, agronomists usually compare mixtures with the most productive species or species combination (Schulze et al., 2018).

It is clear that there is no unique BP relationship for grasslands. This relationship varies according to the choice of criteria measuring productivity and biodiversity, methods of study (experiment vs. observation), diversity gradients, or environmental and management conditions. The relationship also depends on the spatial scale (Li et al., 2019). Biodiversity may be favorable or unfavorable or have no effect on the productivity of grasslands, resulting in the absence of consensus on the BP relationship in grasslands. Some important advances have been made over recent decades to better understand the BP relationship. This can help avoid errors like the 'hidden treatments' (or unidentified causes), highlighted by Huston (1997), in the study of complex ecosystems. This understanding also avoids the controversies raised by the assertion that biodiversity always increases the productivity of grasslands or often unsuccessful attempts to extrapolate experimental results to real farm conditions.

3 Biodiversity factors affecting grassland productivity

3.1 Species richness

Many experiments conducted by artificially manipulating species diversity over a long period have shown a very positive effect of biodiversity on the

Intensification or replacement of semi-natural grasslands is, I not the only possible way to improve grassland production. Agroec a potential solution in balancing forage production with the preserv the environmental quality. This requires changing methods of asses agronomic and environmental value of grasslands. This chapter revi concepts underlying this change of paradigm. Based on the current knc of the relationship between the biodiversity of grasslands and producti show how production should not be assessed just on the annual produ dry matter. Factors such as the nutritive value of forage, quality of p such as milk, animal health and profitability also need to be taken into a The chapter then reviews methods to manage grasslands agro-ecol Finally, the experience of a 'flowering meadows competition' will be de to illustrate how farmers can ensure a successful trade-off between pro and biodiversity in grasslands. Though all the types of grassland are dis the main focus is on semi-natural grasslands.

2 Analyzing the relationship between biodiversity and productivity in grasslands

The relationship between biodiversity and productivity of crop agroecos has been studied since the nineteenth century, but research intensified second half of the twentieth century (MacArthur, 1955). This research c all terrestrial and aquatic ecosystems, with some work focused on gras These studies are based on three types of approaches:

- observation: observation in natural, poorly controlled conditi permanent or semi-natural grasslands with various levels of biodive
- experimentation: setting up of experiments where the number a nature of the species are controlled by the sowing pure stands or mi by overseeding an existing sward, by removing species, or by cha the composition of grasslands through management, and
- synthesis: more recently through meta-analyses based on the comp and analysis of results of individual observations or experiments.

In experimental studies, the aboveground net primary productivity (NPI increase of green biomass by the unit of surface and time) is generally ch to measure the productivity of grasslands. In contrast, observation studie on proxy measures of NPP (e.g. annual yield harvested). Productivity is estin on annual, intra- or inter-annual time scales. Comparing annual results ma possible to assess temporal stability in the productivity and resilience of grassl The following sections of this chapter will demonstrate that the evaluation c agronomical value of grasslands should not be limited to NPP and that r productivity of grasslands. These studies include research in North America (Tilman et al., 1996, 2001) (Prieto et al., 2015) and, for example, the Biodepth Project in Europe (Hector et al., 1999). This positive effect of an increasing number of species (or the number of functional groups) is mainly linked to factors such as the complementary use of resources by plant species, niche differentiation and the increasing probability of encountering highly productive species for a particular environment (the 'selection probability effect'). Numerous experiments have demonstrated the particular effect of adding legumes to grass-dominated grasslands. A meta-analysis of 2753 studies carried out over 50 years has concluded that combining legumes with grasses increases biodiversity but has very little effect on productivity on any grassland mixture containing more than two legume species (Ashworth et al., 2018). This positive BP relationship in some experimental situations is in contrast to other experimental results and the observation of the lower productivity of grasslands with high biodiversity in natural conditions (Seither and Wrage, 2014). Grace et al. (2016) have highlighted the need for models to help explain the complex relations between productivity and biodiversity and the various processes driving both productivity and biodiversity.

Many theories and models have been proposed to explain why the BP relationship is not constant (e.g. linear, log-linear, unimodal, decreasing or increasing). On the basis of the 'hump-backed' model proposed by Grime (1973), Schmid (2002) proposed a theoretical framework to resolve the apparent contradiction between the beneficial effect of biodiversity on productivity generally observed in experimental situations (Fig. 1a) and the negative effect observed in actual semi-natural grasslands conditions (Fig. 1b).

The level of soil fertility can be seen as a key element in explaining the different ways the productivity changes with biodiversity (Fig. 2). The lower productivity of semi-natural grasslands with high biodiversity can be seen to be clearly related to lower levels of soil fertility. By assessing soil fertility and species richness, Fridley (2002) showed that the beneficial effect of biodiversity on productivity increases with the level of fertility. However, by analyzing 16 grassland experiments in North America, Craven et al. (2016) showed that the fertility of the soil (as well as the level of water stress) does not alter the biodiversity-productivity relationship. They suggested that this relationship is primarily determined by taxonomical plant species richness. Besides the fertility of the soil, the density of seedlings of species also influences the BP relationship. The increase in the number of species has a greater effect on productivity at low density than at high density because of density-dependent reductions in the growth rate of individuals of different species within the grassland community (He et al., 2005).

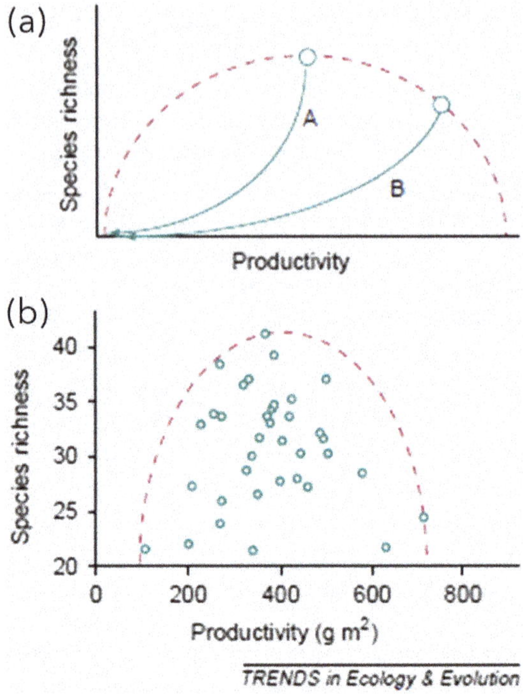

Figure 1 The relationship between plant species richness and productivity in grassland ecosystems (a). The dashed line has been observed in field surveys; the solid lines (A and B) indicate trajectories that are typically obtained in experiments that simulate random species loss. (b) Example of real data from an observational study that included 36 pre-alpine wet grasslands in Switzerland. Each data point is the average of eight sample plots. The dashed curve is an upper envelope over the point cloud. Data taken from D. Pauli, PhD thesis, University of Zurich, 1998.

3.2 Other biodiversity parameters affecting grassland productivity

The link between productivity and species richness is not constant. The effect of particular species and their relative contribution to biomass production also has a very strong effect on productivity, factors which are not taken into account by focusing just on species richness. Grassland diversity should therefore be assessed by parameters other than species richness. Other parameters include genetic and phylogenetic diversity, botanical composition, evenness of species composition (SC) and functional group diversity. Numerous studies have demonstrated that these criteria are better predictors of productivity than species richness.

Ecosystem functioning, including productivity, can be explained by mean plant-trait values, following the so-called mass ratio hypothesis. The 'mass ratio

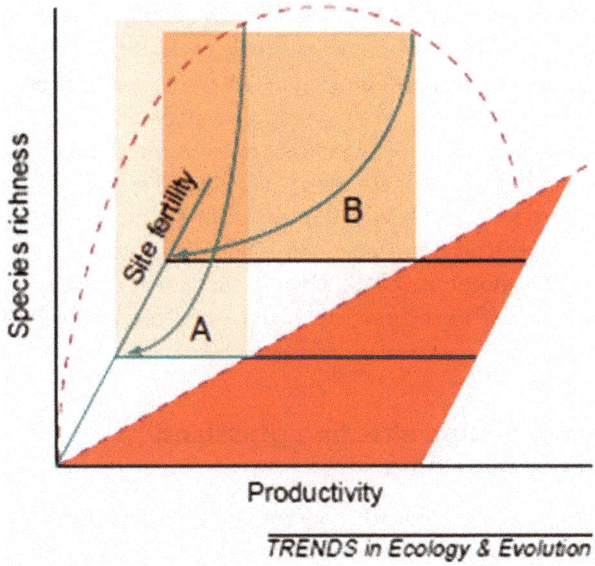

Figure 2 A possible resolution of observational and experimental results. The 3D-representation of the hump-backed relationship indicates the correlation between site fertility and productivity that often occurs in observational studies. The dashed and solid curves correspond to those in Fig. 1a, which can be obtained as a projection of the curves onto the productivity-species richness plane. However, in the 3D-representation, a move out of the rectangular planes framing curves A and B is only possible if site fertility is changed. It is assumed that an increase in productivity from left to right across the diagonal line (shaded dark red) in the productivity-site fertility plane is not possible with any combination of species present in the species pool.

hypothesis' means that the relative abundance of a given species in a community is a better predictor of ecological processes than just the fact of its presence in a grassland community. Duru et al. (2009) demonstrated that grouping species in plant functional types (PFTs) can be used to predict the herbage biomass accumulation rate of grasslands under different management practices for a wide range of plant communities.

Studying 10 native grassland sites in Australia, Mokany et al. (2008) found that aboveground NPP was determined by the traits of the dominant species rather than by grassland diversity. Analyzing the annual biomass production of a national network of 190 semi-natural grasslands, Michaud et al. (2015) found that production was more correlated to functional composition and botanical composition than species richness. In the BIODEPTH project based on a group of European experimental grasslands, Hector et al. (1999) demonstrated that grassland productivity is positively related to functional diversity. Species evenness can also be more predictive of productivity than species richness (Assaf et al., 2011).

Using a mathematical model, Nijs and Roy (2000) found that, while the level of species richness does not correlate with productivity, the response of productivity to species evenness and functional diversity are, respectively, linear and exponential. However, re-examinationof 16 grassland biodiversity studies (Venail et al., 2015) showed non-significant relationships between productivity, phylogenetic diversity, biomass and stability, whereas species richness was found to increase biomass. Other research suggests that plant genetic diversity and species diversity may have different but complementary effects on productivity (Prieto et al., 2015). Overall, functional diversity, in its various forms, appears as the best predictor of grassland productivity compared to species richness.

4 Biodiversity factors affecting grassland productivity over time

Beyond annual productivity or NPP, performance at different times of the year (intra-annual production) and over a number of years (inter-annual production) must also be taken into account in assessing the agronomic value of a grassland. The productivity performance of grasslands is associated with resistance (i.e. the ability to resist biotic and abiotic stresses) and resilience (i.e. the capacity to return quickly to an initial state after stress). Resistance and resilience together define stability. From a farmer's point of view, intra-annual production is related to use. In the case of forage stocks (hay or silage), concentrating production, generally in the spring and early autumn, limits the number of harvests needed. In the case of grazing, it is best to let the grass grow regularly during the whole growing season. The economic sustainability of farming systems is also determined by inter-annual stability of production, that is, the ability to maintain consistent quality and quantity of output. This is becoming more critical with climate change, which tends to increase the frequency of extreme events, especially droughts.

From an ecological point of view, the stability of productivity over time is important to secure ecosystem functioning (Eisenhauer et al., 2018). Intra-annual dynamics and inter-annual stability are both influenced by the level of biodiversity of grasslands. The diversity of responses to environmental change among species is crucial for the resilience of ecosystem functioning (Elmqvist et al., 2003). A high level of biodiversity of grasslands is usually associated with better stability of production, because of overyielding by some species, species asynchrony and portfolio effects (i.e. statistical averaging) (Richardson and Pyšek, 2007) (Caldeira et al., 2005). The portfolio theory was first used by economists to explain why diversification can stabilize returns. In ecology it is used to explain the links between ecosystem performances and diversity. Species richness and evenness may produce this effect (Hillebrand et al., 2008).

Productivity and stability appear to be relatively independent of each other (Cardinale et al., 2013). Küchenmeister et al. (2012) found, for example, that more legumes in temporary grassland mixtures increased yield but decreased yield stability, whereas forbs decreased both yield and stability. Temporal stability also depends on species abundance and growth-related traits (Polley et al., 2013). Studying the production stability of individual species in a species-rich meadow, Lepš (2004) found that dominant species had more impact on stability than the whole community. By combining a legume, two grasses and thee forbs, Küchenmeister et al. (2012) showed that species identity and functional composition affect productivity and stability differently.

Most studies assessing the stability of grassland production have been conducted with experimental plots and have generally tested drought stress and sometimes temperature (heat or frost) stress. Some studies have shown the beneficial effect of grassland diversity (species richness, evenness) on resistance to invasive species (Brett Mattingly et al., 2007), including weed species (Abernathy et al., 2016), though others have not shown this effect (Ganguli et al., 2008).

The link between the stability of grassland production in drought conditions and grassland biodiversity is still an open question. Analyzing 46 experiments studying different types of grasslands composition, Isbell et al. (2015) found that communities with 16–32 species were twice as resilient to extreme wet or dry conditions of varying lengths compared to less diverse communities. In both cases, there was a full recovery of production after the stress event. These results are similar to those of van Ruijven and Berendse (2007), who also found greater resilience from more diverse experimental plots. Re-analyzing 27 experiments (Gross et al., 2014) suggested that interspecific interactions explain the stabilization of productivity in diverse communities. Prieto et al. (2015) demonstrated in experimental conditions (124 plots, 5 species, 10 genotypes) that temporal stability of productivity increased with the number of genotypes but not with the number of species, under both drought and non-drought conditions. Interaction between management and species richness was found by Vogel et al. (2012) when submitting experimental plots (1-60 species, 4 management regimes) to drought. As with NPP, species richness may not be the best predictor of production stability, compared to functional or SC (Xu et al., 2017). Contrasted results were also observed in natural conditions. Comparing intensively managed poor-species grasslands to semi-natural grasslands in the Netherlands, De Keersmaecker et al. (2016) showed that the resistance of semi-natural grasslands was higher than that of intensive grasslands, whereas an opposite trend was observed for resilience. However, these authors note that management may affect resilience, and results should be consequently interpreted with care.

5 Biodiversity factors affecting the quality of forage and animal products

In semi-natural grasslands, conservation of biodiversity and the related ecosystem services generally requires limiting intensification practices (e.g. fertilizer input, grazing intensity or mowing frequency). It is thus difficult, but not impossible, to combine a high level of productivity and biodiversity. Do farmers have to choose between these two objectives? The answer is no because biomass production from species-rich grasslands may be more consistent over time or more resilient in response to an extreme event, which may partly compensate for the loss of productivity. Another advantage of species-rich grasslands is potentially higher forage quality, though this depends on the quality criteria chosen (Farruggia et al., 2008). There are three main types of criteria:

- Feeding value of grass forage (nutritive value, ingestibility and digestibility),
- Medicinal value of grass for animals and for humans (in terms of animal products), and
- Organoleptic value that eating a particular grass gives to animal products.

These types of criteria are influenced by grassland biodiversity. They may result in higher profitability (e.g. in a reduction in production costs, better product quality and thus better-selling prices and market access).

5.1 Biodiversity and the feeding value of grass forage

The feeding value of grass forage encompasses (Baumont, 1996):

- its nutritive content (energy, protein, minerals, vitamins),
- its digestibility (the fraction that will be really assimilated by animals),
- its ingestibililty (quantity of forage voluntarily ingested by the animal), and
- its palatability (i.e. forage characteristics (taste, smell, appearance), which modifies animal feeding behavior.

Each single species in a grassland community has its own characteristics, which may vary with phenological stage, the type of organ (leaf/stem/inflorescence) and environmental conditions (nutritive and water status, temperature, biotic interactions). These values have been measured for a few of the most common species, mainly sown grasses and legumes, which are also found in semi-natural grasslands. However, the number of species described is very low compared to the number of grassland species found in species-rich grasslands. In addition, feeding values differ for various types of animals (cattle, sheep, goat, horse, etc.) because their physiological requirements, feeding behavior and learning processes determining diet choices vary (Duncan et al., 2006). For all these reasons, the relationship between biodiversity and feeding value of forage is complex and not fully understood. The studies that have compared the feeding value of grasslands report either positive or negative effects of biodiversity, depending on the situation (botanical composition, fertility level, type of animal).

Numerous species in a grassland influence the grass feeding value because of differences in digestibility between species at the same phenological stage and differences in maturity between species at the same date (Farruggia et al., 2008). In species-rich grasslands, the proportion of forbs (and sometimes legumes) and the diversity of grasses are generally higher than in species-poor grasslands and are major drivers of forage feeding value. In terms of chemical composition, forbs can increase mineral and secondary metabolites content of forage, and legumes increase the protein content. Based on the analysis of 30 grasslands in the UK, French (2017) found 56% more phosphorus, 106% more potassium, and 183% more calcium in this type of grassland than in conventional hay.

However, the impact of grassland diversity on digestibility can be both positive and negative (Bruinenberg et al., 2002). The lowest fertilization of species-rich grasslands, rather than their higher diversity, explains some results demonstrating lower mineral and protein content. As grasses are more diverse, the shift in their phenology over time makes energy and protein content of grass more stable. Biodiversity also improves palatability and digestibility of forage. However, the ingestibility of forage from species-rich grasslands is frequently lower and there is an increasing risk of consumption of toxic species.

In complex plant mixtures like in semi-natural grasslands, the interaction effects between species influence forage quality. Based on a literature review, Niderkorn and Baumont (2009) demonstrated that feed intake and digestion in ruminants are modified by the interactions between grasses and legumes in the rumen, as well as the impact of bioactive secondary metabolites on digestive processes in the rumen. The lack of knowledge on all species and all interactions between species makes it difficult to directly predict grass-feeding value from grassland plant diversity. However, Jouven et al. (2006) have proposed a model predicting the digestibility of forage from permanent pastures, describing grassland biodiversity through PFT.

Two types of studies have been conducted to estimate the feeding value of grass in relation to biodiversity:

- experiments comparing the feeding value of individual plant species (or genotypes) or simple mixtures, and
- empirical estimations analyzing the forage value of semi-natural grasslands.

As an example of the first type of study, Pontes et al. (2007) measured the crude protein (CP) concentration and pepsin cellulase dry matter digestibility of 13 native grasses cultivated in pure stands. The results show that, if the phenological stage and nitrogen status are the main factors explaining these nutritive criteria, there is a great variability between species, and differences occur between early and late flowering species. Duru et al. (2008) demonstrated that a plant functional traits like leaf dry matter content (LDMC) can be a good predictor of organic matter digestibility (OMD) of 19 grasses cultivated in pure stands. These authors found similar results at the community level in semi-natural grasslands.

In semi-natural grasslands, each grassland community can be considered as a unique mixture of species and phenological stages, and this complexity makes it difficult to assess the effect of biodiversity on feeding value. Analyzing 190 semi-natural and intensive permanent grasslands in France, Michaud et al. (2015) found that grass nutritive value (energy, crude protein, organic matter digestibility, mineral content) was mainly explained by botanical families, plant stage and sward dry matter content. Other plant functional traits than botanical family, and other biodiversity criteria (species richness and evenness), were less predictive.

5.2 Biodiversity and the medicinal value of grass forage for animal and human health

Species in species-rich grasslands, especially forbs and legumes, have a higher secondary metabolite content than most grasses. These molecules belong to several chemical families (Poutaraud et al., 2017): phenolic compounds, terpenoids, carotenoids, alkaloids, vitamins and enzymes. The link between grassland botanical composition and secondary metabolites content of grass forage has been clearly demonstrated by authors like Reynaud et al. (2010). The probability of finding a large diversity of molecules increases with the diversity of species and botanical families in a grassland. French (2017) showed that grass produced by UK-native species-rich grasslands had a greater abundance of medicinal compounds than intensively managed and species-poor grasslands.

Some studies have demonstrated a wide range of anti-inflammatory, antimicrobial, anti-cancer, cardio-protective, neuro-protective, anti-osteo-porotic, estrogenic/anti-estrogenic, anxiolytic, analgesic and anti-oxidant activities of secondary metabolites in forage in laboratory animals and/or herbivores (Poutaraud et al., 2017). Recently attention has focused on the role of condensed tannins, which can prevent bloat in cattle, control parasites like coccidian as well as control flystrike (Lüscher et al., 2014), but which can be toxic above a threshold level (Makkar, 2003). Some species are particularly rich in condensed tannin, like legumes (*Onobrychis* spp., *Lotus* spp., *Acacia* spp.)

and species like *Taraxacum* spp. (Jayanegara et al., 2009). Other studies have focused on the role of plant polyphenols in preventing oxidative stress and improving animal health and production performance (Gessner et al., 2017).

On-farm surveys and experimental results show that there is also an effect of animal-diet composition on human health. The main components involved are fatty acids (FAs), vitamins and polyphenols (Duru, 2016). Research suggests grass in animal diets has a very positive effect on unsaturated FAs (Couvreur et al., 2006) (Coppa et al., 2013). When comparing meat quality of beef cattle grazing intensive or *molinia*-dominated semi-natural grasslands, Fraser et al. (2007) found that there is a positive effect of grassland diversity on unsaturated FA content. FA composition is related both to the proportion of grass in animal diets and to the occurrence of different plant species or families (Collomb et al., 2002). Gorlier et al. (2012) found significant correlations between FA concentration in milk and the contribution of the main botanical families found in alpine grasslands. Ravetto Enri et al. (2017) found the FA profile less dependent on botanical family than on the presence of given species in six types of alpine semi-natural grasslands. As an example, linoleic acid was negatively correlated with SC of grasses and positively correlated with the SC of legumes.

5.3 Biodiversity and the organoleptic value of animal products

There is a significant body of research showing that, among other factors, grassland botanical composition has a major effect on the sensory quality of animal products, as shown in the example of cheese (Fig. 3) (Buchin et al., 1999) (Coulon et al., 2004). There are many mechanisms that help to explain the link between organoleptic product quality and botanical composition of grasslands (Martin et al., 2005) (Coulon et al., 2004):

- direct transfer of molecules from grass to milk and cheese (e.g. aromatic compounds like terpenes and sesquiterpenes), and
- indirect chemical (protein and fat), enzymological (effect on lipolysis and proteolysis) and microbial activity (in rumen, milk and cheese). An indirect effect can, for example, result in the chemical transformation of a molecule present in the forage by a rumen microorganism.

Comparing cheeses made from cows grazing in a mountain area on either species-rich semi-natural grasslands or temporary grasslands, Coppa et al. (2011) found that cheese coming from species-rich grasslands had more intense odor and aroma, and a sourer, more pungent and saltier taste. Buchin et al. (1999) found alpine cheese produced from grasslands with different botanical compositions showed significant differences in aroma type and intensity. Bovolenta et al. (2014) showed that species-rich pastures produced

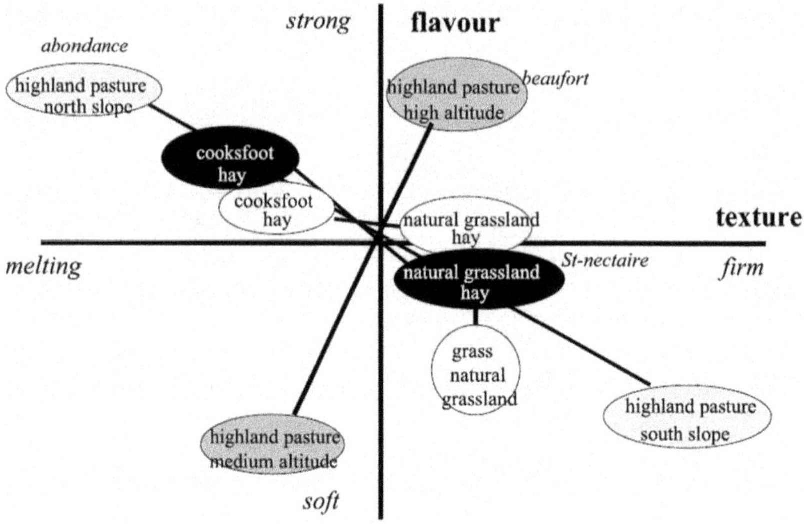

Figure 3 Sensory characteristics differentiation of French mountain cheeses (Beaufort, St-Nectaire, Abondance) according to grass botanical composition (Coulon et al., 2004).

cheese with more volatile compounds (ketones, phenolic compounds and terpenes) than species-poor pastures.

6 Grassland biodiversity and ecosystem services

The 2005 United Nations report on the Millennium Ecosystem Assessment popularized the concept of ecosystem services (ESs). These include provisioning, regulating, cultural, and supporting services. Economists like Costanza et al. (1998) define ES as: 'The benefits human populations derive, directly or indirectly, from ecosystem functions'. However, for ecologists like Daily et al. (1997) ESs are a 'wide range of conditions and processes through which natural ecosystems, and the species that are a part of them, help sustain and fulfill human life'. One definition focuses on (measurable) outcomes, while the other focuses on processes. Daily's conceptualization is that ESs are a part of the ecosystem and, with the addition of anthropogenic capital, humans derive benefits from them. This means that there is a more economically focused definition and a purely ecological definition of ES. This poses a problem because, in order to successfully identify and evaluate ES, whether it be their monetary value or their potential, it is important that studies carried out on the subject use the same definition. The second problem is that the ES definition from the Millennium Ecosystem Assessment does not clearly distinguish conditions and determinants, functions, processes, mechanisms, goods, benefits and profits. The scientific literature about ES delivered by grasslands uses both definitions.

The latest European initiatives refer to the ecologists' definition, especially the 'Mapping of Ecosystems and Ecosystem Services (MAES)' or the 'Common International Classification of Ecosystem Services (CICES)' developed from the work on environmental accounting undertaken by the European Environment Agency (EEA). In the majority of frameworks, biodiversity is considered, by definition, as a key determinant of ES. In the conceptual framework of IBPES (Intergovernmental Science-Policy Platform on Biodiversity and Ecosystem Services), ESs are considered as dependent on biological and ecological processes (Díaz et al., 2015). Biodiversity conservation itself is considered as a key objective to enable ecosystems to provide ES.

Grasslands contribute numerous potential ESs, including pollination (of natural and agricultural species); control of invasive and weed species; pest control; maintenance of soil biological activity; soil stabilization and structure; erosion prevention; flood regulation; mineral recycling; pollution attenuation (pesticides or fertilizers); reduction of cattle methane emissions; water quality regulation; climate regulation; carbon sequestration; biodiversity conservation (flora, soil and aerial fauna, soil microflora); and production of forage, bioenergy, medicinal and aromatic plants (Bengtsson et al., 2019).

Different characteristics of grassland affect different ES. Biodiversity conservation, pollination, control of invasions, cultural services, maintenance of soil biological activity and even reduction of cattle methane emissions clearly depend on the overall level of biodiversity. ESs related to water and physical soil processes (soil stabilization and structure, erosion prevention, flood regulation, pollution prevention) mainly depend on above biomass. Biomass and bioenergy production are generally maximized in intensified and species-poor grasslands, including temporary grasslands (Yang et al., 2019). Analyzing 147 plants species studied in 17 experiments (Isbell et al., 2011) concluded that the conservation of numerous species, including those apparently 'redundant' for ecosystem functioning, is essential to provide multiple ES.

6.1 Biodiversity conservation

Conservation of species-rich grasslands, by definition, conserves floral diversity. Many animal and microbial taxa are reliant on plant diversity for habitat and food. As an example, Kruess and Tscharntke (2002a) have established a link between plant species richness and the diversity of grasshoppers, butterflies, bees and wasps in interaction with grazing activity.

6.2 Pollination and pollinators

There is a close link between grassland plant biodiversity and pollinator abundance and diversity (Potts et al., 2003). As shown by Orford et al. (2016) in a grassland field experiment manipulating species mixtures, even a modest enhancement in plant diversity, by the addition of legumes and forbs, results in better functional diversity of grassland pollinator communities. Beyond the diversity of a single grassland community, the co-occurrence of several plant communities may improve the type, number and frequency of contacts between plant and pollinators (Fantinato et al., 2019).

6.3 Soil biological activity

Grassland plant diversity is correlated with soil microbial diversity (fungi and bacteria) (Spehn et al., 2000) (Benizri and Amiaud, 2005). Similar effects have been observed for other soil taxa like earthworms (Piotrowska et al., 2013). Plant functional traits also improve the structure of the microbial community (de Vries et al., 2012) (Strecker et al., 2015). In a long-term grassland diversity experiment, Eisenhauer et al. (2018) showed that plant diversity also improves root biomass and soil respiration as well as soil microbial biomass.

6.4 Mitigation of cattle methane emissions and reduction of global warming

Assessing the role of grassland biodiversity in mitigating cattle methane emissions is complex for a number of reasons:

1. A realistic evaluation must take into account not only the direct effect of grass composition and diversity on animal physiology but also indirect farm-level effects such as the consumption of concentrates or forage crop cultivation.
2. Findings may change according to the units measured. Examples include greenhouse gas emissions calculated per animal, per kg of meat or milk, or per ha. In an on-farm experiment comparing beef grazing species-poor or species-rich grasslands, Richmond et al. (2015) found that biodiversity reduced daily methane emissions per animal but not per kg of weight gain.
3. It is crucial to consider all fluxes contributing both positively (C sequestration) and negatively (CH_4, CO_2, N_2O) to global warming at the farm level.

Macheboeuf et al. (2014) screened 156 grassland plant species and evaluated their methane emission potential in artificial rumens. They concluded that the botanical families that reduced methane most efficiently were the Rosaceae, Onagraceae, Polygonaceae and Dipsacaceae.

6.5 Control of plant invasions

The resistance of grasslands to invasive plant species generally increases with grassland diversity (Hector et al., 2001). Maintaining a high plant functional diversity also helps prevent invasive species (Pokorny et al., 2005). However, manipulating species evenness within native prairie grassland mesocosms, Brett Mattingly et al. (2007) found no effect on community invasion by exotic plants. The effect of grassland diversity on invasion resistance is generally explained by niche complementarity and better use of resources (light, water and minerals), as demonstrated by Frankow-Lindberg (2012).

6.6 Cultural services: recreation, education and research

Grasslands, and particularly species-rich semi-natural grasslands, are important for the whole of society and not just farmers (Quétier et al., 2010). Society gains a set of 'cultural' benefits that generally increase with grassland diversity and the rarity of species or habitats. The whole of society can appreciate this biodiversity and the contribution of grasslands to the beauty of the landscape (Marzetti et al., 2011).

7 Assessing the economic value of species-rich grasslands

From a macroeconomic point of view, awareness of the consequences of the loss of biodiversity and the potential profits from improving biodiversity has risen all over the world since the Millennium Ecosystem Assessment (2005). In 2007 the governments of the G8 + 5 (bringing together both leading developed economies and leading countries in the developing world) proposed the setting up of TEEB (The Economics of Ecosystems and Biodiversity), a global initiative hosted by the United Nations Environment Programme to analyze the global economic benefit of biological diversity (Balmford et al., 2008). Given the importance of grasslands and rangelands (representing a third of the world's agricultural area), they are at the heart of the debate about the economic value of biodiversity. From a microeconomic point of view, the question is the economic incentive of farmers to promote systems, and in particular grasslands, with high biodiversity.

At the microeconomic (farm or field) level, the interest of conservation of the biodiversity of grasslands can be considered from different perspectives, using the typology of values proposed by Chevassus-Au-Louis et al. (2009):

- Direct uses/value of biodiversity: including forage production (increased productivity, more resilient production (with reduced risk) or better quality of species-rich grasslands), production of medicinal and aromatic plants, better prices for grassland animal products and production of grass biofuel.

- Indirect uses/value of biodiversity: including the prevention of soil erosion, better water retention, prevention of weeds and improved animal health.
- 'Option value', that is, the value of biodiversity for the future, for example, by maintaining the fertility of the soil.

In addition to these three types of values, economists suggest a so-called non-use value, which covers the broader value of biodiversity for humankind, regardless of its direct usefulness, as well as the heritage value of passing on biodiversity to future generations.

In a recent publication named 'Biodiversity Can Pay' Binder et al. (2018) used an economic model to find the optimum level of biodiversity in agricultural systems and concluded that the optimal species richness falls 'between the low levels currently found in commercial grasslands and the high levels found in natural grasslands'. A key element in species-rich grassland profitability lies in the reduction of mid- and long-term risks. Using data from the Jena biodiversity Experiment in Germany (plot experiments with a combination of 16 grasses, 12 legumes, and 20 tall and 16 small herbs), Finger and Buchmann (2015) demonstrated significant insurance cost benefits associated with grassland diversity (e.g. in improving grassland resilience to more extreme weather events). Similar results were found by Schläpfer et al. (2002) when assessing returns from hay cultivation in diverse versus non-diverse grasslands. There is a need for more research on the economic value of species-rich grasslands. This research needs to take into account many factors, such as:

- the differences in quantity and quality of grass produced between species-rich and species-poor grasslands,
- long-term stability of production in the face of climatic change,
- differences in production costs compared to more intensive, forage-based systems (e.g. fertilizer, harvesting and labor costs),
- reduced costs of animal feed and veterinary costs,
- improved feed self-reliance (the level of dependence on concentrate purchases),
- the premium price allowed by grassland production (including protected designation of origin (POD) labeling),
- future proofing in the light of the growing demand for more environmental-friendly animal production with higher standards of animal welfare, and
- the payment of environmental subsidies to support biodiversity conservation.

These studies will help make the economic case for increasing biodiversity in grasslands.

8 Management of grasslands for both biodiversity and production

Is there a management system of grasslands that can both maintain an acceptable level of production for farmers and ensure the conservation of biodiversity? Several answers can be given to this question:

- There is a wealth of research both on grassland management and on promoting grassland biodiversity, which provides a foundation for improving management practices.
- Many ecological studies linking the level of diversity and grassland production provide elements for management solutions (as discussed earlier). Although higher levels of biodiversity are not always compatible with high levels of production, the stability of production over time is frequently associated with a minimum level of biodiversity.
- The previous discussion of economic value also indicates that it is possible to combine a high level of biodiversity and a good level of profitability.
- As noted earlier, the annual production of biomass should not be the only criterion to evaluate the agronomic value of grasslands for farmers; other factors such as quality affect management practices to adopt grass quality.

Because of the huge variety of different situations, a win-win situation is not possible for all grasslands. However, reconciling production and biodiversity is achievable on a case-by-case basis.

A key initial decision is whether to (Green, et al., 2005):

- promote biodiversity across the whole of a grassland (land sharing), or
- dedicate specific space to promote biodiversity (land sparing).

Both solutions have their advantages and limitations (Balmford et al., 2005). From the point of view of biodiversity conservation in grasslands, land sparing can ensure high levels of biodiversity (high level of flora and fauna, taxonomic and functional diversity, preservation of patrimonial or endangered species), though at the expense of forage production. The maintenance of dedicated areas often requires government subsidies to compensate for the loss of production. Because of this cost, the major ecological disadvantage is that the proportion of land allocated to these 'high nature value' purposes may be low. This limits processes such as natural regeneration of grasslands from seed dispersal. Dividing land in this way may have benefits for farmers in allowing them to tailor production grassland precisely to the needs of their animals and optimize grass yields. On the other hand, land sharing helps conserve biodiversity over the whole field and contributes to ecosystem services at the field and landscape scale. Its acceptance by farmers depends on the adoption of an extensive but profitable farming production system.

Sabatier et al. (2014) concluded that heterogeneous landscape management (i.e. land sparing) is the best way to reconcile agricultural and ecological objectives. Loucougaray et al. (2015) have also identified the advantages of a balance between sown and permanent grasslands in optimizing the ecological intensification of alpine grasslands. These results fit with other studies, for example, on bird communities in tropical cloud forests of the Colombian Andes (Cannon et al., 2019), or on trade-offs between overall agricultural production and biodiversity in France (Teillard et al., 2017).

Before screening the main factors explaining the impact of management on grassland production and biodiversity at the field level, it is important to keep in mind two elements:

1 The configuration of the surrounding landscape partly helps to explain grassland biodiversity (Gaujour et al., 2012), but this effect is low compared to local factors (Klimek et al., 2007), and
2 Current management may be sometimes less predictive of current biodiversity level than historical grassland management (Boeuf-Tremblay et al., 1995; Gustavsson et al., 2007) (Yoshihara et al., 2016).

An extensive review of factors influencing grassland biodiversity has been conducted by Isselstein (2018). Several management factors affect grassland biodiversity at field and farm scale, as reviewed by Huguenin-Elie et al. (2012), Gaujour et al. (2012), Tälle et al. (2016), Rosa García et al. (2013) and Marini et al. (2007). These studies suggest that management practices which particularly affect grassland biodiversity include the intensity of grazing, frequency and intensity of cutting as well as fertilization practices (N, P, K and other elements from mineral and organic fertilization as well as animal droppings).

Nutrient inputs generally reduce plant diversity, as shown in either short- or long- term experiments (Schellberg et al., 1999) (Silvertown et al., 2006) or in farm conditions (Isselstein, 2005) (Riesch et al., 2018). Mineral fertilization tends to be more detrimental to species richness than organic manure (Kirkham et al., 2008). Mineral nitrogen application may impact biodiversity even at low rates (30 kg N ha^{-1} $year^{-1}$) (Plantureux, 1996a) (Joyce, 2001), and at even lower rates when considering some patrimonial species that are very sensitive to fertility levels (Grevilliot et al., 1998).

Other management practices may be negative, neutral or positive in their effects on biodiversity (Olff and Ritchie, 1998) (Marriott et al., 2004).

Analyzing 35 studies, Tälle et al. (2016) concluded that grazing was generally better for the conservation of semi-natural grasslands than mowing. A meta-analysis conducted by Diaz et al. (2007) on 197 studies concluded that grazing is favorable to annual species, smaller and stoloniferous plants. Animals do not only graze but also deposit huge amounts of nutrients (60-80% of their mineral intake), and plant diversity can also be depressed by this nutrient load. Medium-intensity grazing (in terms of stocking rate and duration of grazing) generally appears more favorable than very low or very high grazing pressure (Kruess and Tscharntke, 2002b) (Tichit et al., 2005) (Fischer and Wipf, 2002). For the cutting regime, the date of the first cut is crucial, with an early cut excluding non-clonal species (Zechmeister et al., 2003) (Smith et al., 2002).

The two main management factors determining grassland biodiversity – the type of use and nutrient supply – also significantly impact grassland production and quality. When these two types of factors are maximized, annual grass biomass is generally optimized, but potentially at the expense of forage quality and biodiversity. In practice a few competitive grasses (like *Lolium perenne* or *Dactylis glomerate)*, few legumes (*Trifolium repens* and *T. pratense*) and forbs (*Taraxacum* spp., *Ranunculus* spp.) are adapted to these intensive conditions and dominate almost all other grassland species. As previously mentioned, medium intensification is more favorable for biodiversity. This can be made profitable by the limitation of production costs (e.g. less reliance on fertilizers, animal feed and veterinary costs), but also by higher selling prices for animal products with better organoleptic quality, together with potential improvement animal welfare. This usually means not aiming for the highest animal yields (annual quantities of milk or meat per animal or per hectare).

In many regions with significant livestock production, semi-natural grasslands that have not been replaced by forage crops or temporary grasslands are generally highly intensively managed, and their level of biodiversity is generally very low. Based on the French national database of permanent grasslands eflorasys (http://eflorasys.univ-lorraine.fr), Plantureux (1996b) suggest that only 20–25% of French permanent grasslands can be considered as species-rich semi-natural grasslands. Since the 1980s, this proportion has continuously decreased, due in part to practices such as more intensive grazing, early mowing, a greater focus on silage production as well as an increase in mineral and organic fertilization. There now is a move to the 'extensification' of these grasslands, that is, to restore them as semi-natural grasslands.

Marriott et al. (2004) have reviewed a number of studies of extensification. These show that reducing fertilization has a gradual impact on improving biodiversity. Some change in taxonomic composition can appear quite quickly, but species richness and functional composition generally remain stable, with major improvement taking a long time to occur (15–20 years). For example, Isbell et al. (2013) studied grasslands fertilized for 10 years (with 95–270 kg N ha^{-1} $year^{-1}$) and then not fertilized for another 20 years and found no observable recovery in species richness. Grassland response depends significantly on initial site fertility, and especially on phosphorus fertility, which strongly influences the presence of oligotrophic species (Janssens et al., 1998). In an 18-year study in Southern England grasslands, cessation of fertilization decreased biomass by 50% after only two years, while species richness remained stable for five years, before increasing to 300% of the initial richness after 18 years (Pallett et al., 2016).

According to the review by Marriott et al. (2004), later cutting is favorable for plant species using sexual reproduction, and decreasing animal stocking rate increases local heterogeneity of soil resources and, consequently, species biodiversity. It is important to note that improvement of biodiversity by less-intensive management generally results in the reduction in total animal output and, in some cases, in individual animal performance. Abandonment of grassland management (cessation of grazing and harvesting) generally results in a short-term change in grassland composition, with the domination of tall herbs and coarse grasses, and woody plants. Species richness rapidly decreases (over 1–5 years), beginning a secondary succession to forest vegetation. In those situations, the restoration of abandoned grasslands by grazing may efficiently restore biodiversity (Pykala, 2005).

9 Case study: flowering meadows competition

Although it is possible to reconcile biodiversity and production of grasslands, based both on scientific research and on the practical experience of more innovative farmers, the vast majority of farmers and their advisers remain to be convinced. For most farmers, semi-natural grasslands have low productivity and provide poor-quality forage. There is an important need to change minds through education.

In contrast to most agriculture competitions which focused on productivity, a 'Flowered meadows competition' was organized for the first time in France in 2007 to reward farms achieving the best 'agri-ecological equilibrium' in their grasslands (Magda et al., 2015) (Plantureux et al., 2011). It was based on the agri-environmental MEKAII program in Germany in the Federal State of Baden Württemberg (Oppermann, 2003). The first competition was organized in 2007 in collaboration between the Natural Regional Park (PNR) 'Massif des Bauges' in the French Alps and INRA, the French national agronomic research institute. In 2010, the competition was organized at a national level, and, since 2010, about 50 competitions are held around France (Fig. 4) involving

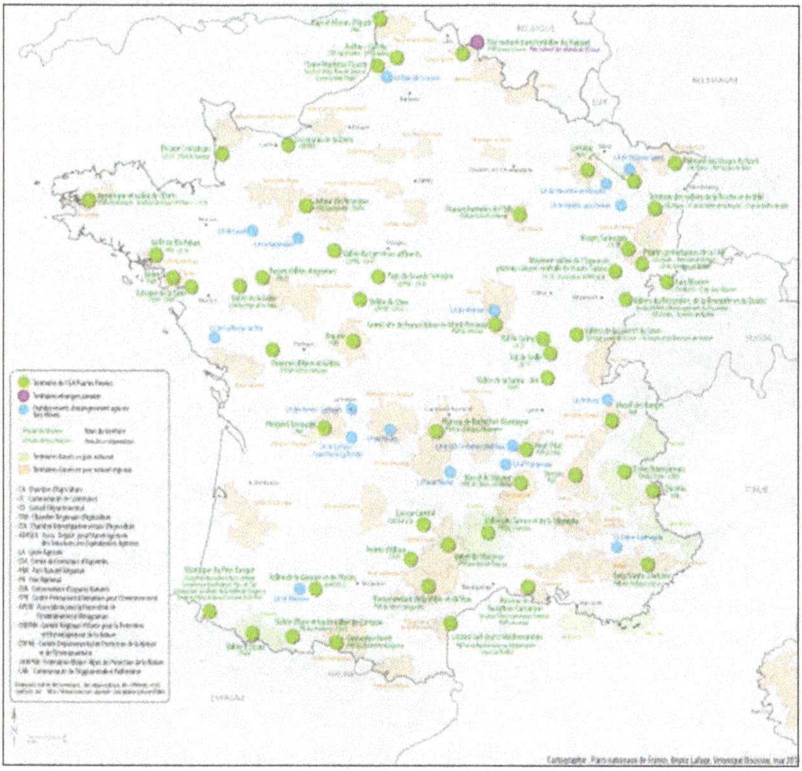

Figure 4 Localization in 2016 of the 57 French 'flowered meadows competition'. Source: Comité National d'Organisation du concours 'Prairies Fleuries'.

4–500 farmers each year. Local prizes are awarded, and the winners in each region compete for a national prize, awarded annually during the national agriculture fair ('salon de l'agriculture') jointly by the ministers of agriculture and the environment.

A multidisciplinary jury (consisting of farmers, naturalists (botanists, entomologists), beekeepers and scientists (agronomists and ecologists)) assesses each grassland. The jury uses a scoring system developed nationally by a college of experts (Magda et al., 2015). This system evaluates not only the level of flora and fauna biodiversity of the grassland and its agronomic value (the level of production and forage quality), but also as to what extent forage products correspond to the needs of the farmer (e.g. quality of forage, timing production, impact on animal product quality). The jury begins with an interview with the farmer about his/her practices and expectations. The jury then inspects the grassland along a diagonal cutting across the field. Ecological

Figure 5 List of species to assess the agri-ecological equilibrium in grasslands. Source: Journal des Prairies Fleuries 2017, APCA-FNPNR-INRA.

and agronomic criteria are measured at three points along this diagonal. At these points the grassland must contain at least four species from a regional list of species developed by experts to ensure the agri-ecological value of the grassland (Fig.5). At the end of the visit, which lasts about an hour, the jury tells the farmer the outcome of its assessment.

The flowering meadows competition has been run in France for more than 10 years. Many farmers participating in the competition were already convinced of the interest of species-rich grasslands, but the competition has made it possible to convince a wider audience such as other farmers, farm advisors and agricultural students who will be future farmers and managers of natural areas. This competition has also played a role in training individual members of each jury: agronomists have learned to know more about animal and plant species, and naturalists to understand how to evaluate the agronomic value. Public broadcasts (national and regional TV, radio and print media) have shown society that it is possible to reconcile a profitable agriculture with biodiversity preservation.

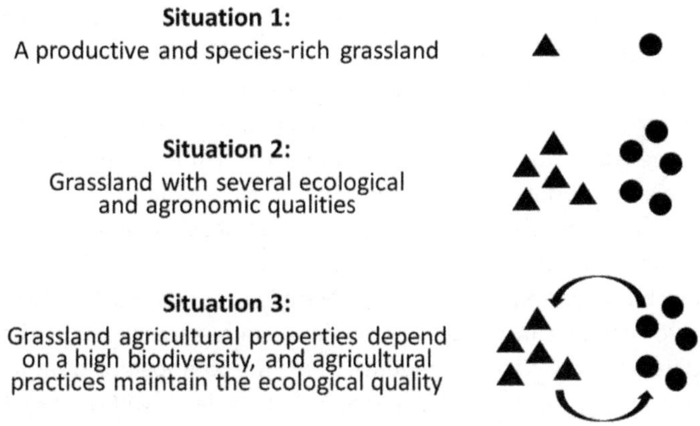

Figure 6 Three possible definitions of agri-ecological equilibrium of grasslands. Each triangle represents an agronomic criterion, and each circle an ecological criterion. Source: Sylvain Plantureux, Univ Lorraine, France.

10 Summary and future trends

Figure 6 summarizes different definitions of the concept of 'agri-ecological equilibrium':

- Situation 1: a productive and species-rich grassland,
- Situation 2: a grassland with several ecological and agronomic qualities, and
- Situation 3: a grassland where agricultural properties depend on high biodiversity, and agricultural practices maintain ecological quality.

Situation 1 is the most widely used definition. It is usually based on a criterion for defining ecological value (usually plant species richness) and another for defining agronomic value (usually the annual return). The problem with this definition is that it is always a compromise that satisfies neither the farmer nor the naturalist. Finding highly productive grasslands (i.e. annual biomass production > 8–10 T DM $ha^{-1}.year^{-1}$ in temperate conditions) with very high species richness is impossible in practice.

Situation 2 depends on a wider range of criteria. The agronomic value may be evaluated both by annual biomass production and by the timing of production, or the quality of forage from a nutritional, organoleptic or medicinal perspective. Ecological value can be defined by plant species richness or the quality of the bloom for pollinating insects, and the shelter role of sward for birds and mammals. It is becoming more common to find compromises combining these different qualities.

Situation 3 offers the opportunity for an agri-ecological balance, which allows reconciliation between biodiversity and production. In this situation, the agronomic value of grassland is totally dependent on its biodiversity. For example, maintaining forage energy value is related to the diversity of grasses with different development cycles. At the same time, agricultural practices are essential for the conservation of biodiversity. For example, mowing keeps open swards, or the type of grazing promotes biodiversity-friendly heterogeneity.

Reconciling biodiversity and production of grasslands is an issue for society as a whole, which requires the production of safe animal products in an environmentally friendly way. The aspect which has been studied the most is the relationship between the level of taxonomic or functional biodiversity and biomass production and production qualities such as resilience and stability. Most studies are controlled experiments, where the composition of vegetation is managed artificially. These results cannot be extrapolated without care to semi-natural species-rich grasslands. This relationship needs further investigation, and the relationship between biodiversity and biomass production should be studied in a larger number of semi-natural communities.

Recent research has focused on the quality of production rather than quantity. The effect of biodiversity on forage qualities (such as energy, protein and mineral content) is becoming clearer, even if further work is needed. Research is now focusing on the secondary metabolite content of meadow species and its effect on the organoleptic quality of animal products, animal and human health. Finally, in order to convince farmers, their advisers, governments and policy makers of the need to preserve the biodiversity of grasslands, it is necessary to conduct an appropriate economic and social assessment. This assessment needs to address both the micro-economic scale (farm-scale) and the macro-economic scale (regions and countries).

A major challenge for research in agronomy and ecology is to take into account a greater diversity of grassland properties. The aim is to exceed the usual criteria of production, energy or protein value of the grass, or the basic diversity criteria such as species richness. This will help to better understand when grasslands with high biodiversity may be of interest to farmers.

Researchers in agronomy and ecology are studying this issue. In Europe, the work of these researchers is communicated in particular within the European Grassland Federation.

11 References

Abernathy, J. E., Graham, D. R. J., Sherrard, M. E. and Smith, D. D. 2016. Productivity and resistance to weed invasion in four prairie biomass feedstocks with different diversity. *GCB Bioenergy* 8(6), 1082–92. doi:10.1111/gcbb.12317.

Ashworth, A. J., Toler, H. D., Allen, F. L. and Augé, R. M. 2018. Global meta-analysis reveals agro-grassland productivity varies based on species diversity over time. *PLoS ONE* 13(7), e0200274. doi:10.1371/journal.pone.0200274.

Assaf, T. A., Beychlag, W. and Isselstein, J. 2011. The relationship between plant diversity and productivity in natural and in managed grasslands. *Applied Ecology and Environmental Research* 9(2), 157–66. doi:10.15666/aeer/0902_157166.

Balmford, A., Green, R. E., and Scharlemann, J. P. W. 2005. Sparing land for nature: exploring the potential impact of changes in agricultural yield on the area needed for crop production. *Global Change Biology* 11(10), 1594–605. doi:10.1111/j.1365-2486.2005.001035.x.

Balmford, A. and Rodrigues, S. L., Walpole, M., ten Brink, P., Kettunen, M., Braat, L. and de Groot, R. S. 2008. *The Economics of Biodiversity and Ecosystems: Scoping the Science*. Research report. European Commission, Cambridge, UK, 305 p.

Baumont, R. 1996. Palatability and feeding behaviour in ruminants. A review. *Annales de Zootechnie* 45(5), 385–400. doi:10.1051/animres:19960501.

Bengtsson, J., Bullock, J. M., Egoh, B., Everson, C., Everson, T., O'Connor, T., O'Farrell, P. J., Smith, H. G. and Lindborg, R. 2019. Grasslands-more important for ecosystem services than you might think. *Ecosphere* 10(2), e02582. doi:10.1002/ecs2.2582.

Benizri, E. and Amiaud, B. 2005. Relationship between plants and soil microbial communities in fertilized grasslands. *Soil Biology and Biochemistry* 37(11), 2055–64. doi:10.1016/j.soilbio.2005.03.008.

Binder, S., Isbell, F., Polasky, S., Catford, J. A. and Tilman, D. 2018. Grassland biodiversity can pay. *Proceedings of the National Academy of Sciences of the United States of America* 115(15), 3876–81. doi:10.1073/pnas.1712874115.

Boeuf-Tremblay, V., Plantureux, S. and Guckert, A. 1995. Influence of mechanical impedance on root exudation of maize seedlings at two development stages. *Plant and Soil* 172(2), 279–87. doi:10.1007/BF00011330.

Bovolenta, S., Romanzin, A., Corazzin, M., Spanghero, M., Aprea, E., Gasperi, F. and Piasentier, E. 2014. Volatile compounds and sensory properties of Montasio cheese made from the milk of Simmental cows grazing on alpine pastures. *Journal of Dairy Science* 97(12), 7373–85. doi:10.3168/jds.2014-8396.

Brett Mattingly, W., Hewlate, R. and Reynolds, H. L. 2007. Species evenness and invasion resistance of experimental grassland communities. *Oikos* 116(7), 1164–70. doi:10.1111/j.0030-1299.2007.15406.x.

Bruinenberg, M. H., Valk, H., Korevaar, H. and Struik, P. C. 2002. Factors affecting digestibility of temperate forages from seminatural grasslands: a review. *Grass and Forage Science* 57(3), 292–301. doi:10.1046/j.1365-2494.2002.00327.x.

Buchin, S., Martin, B., Dupont, D., Bornard, A. and Achilleos, C. 1999. Influence of the composition of Alpine highland pasture on the chemical, rheological and sensory properties of cheese. *Journal of Dairy Research* 66(4), 579–88. doi:10.1017/S0022029999003842.

Caldeira, M. C., Hector, A., Loreau, M. and Pereira, J. S. 2005. Species richness, temporal variability and resistance of biomass production in a Mediterranean grassland. *Oikos* 110(1), 115–23. doi:10.1111/j.0030-1299.2005.13873.x.

Cannon, P. G., Gilroy, J. J., Tobias, J. A., Anderson, A., Haugaasen, T. and Edwards, D. P. 2019. Land-sparing agriculture sustains higher levels of avian functional diversity than land sharing. *Global Change Biology* 25(5), 1576–90. doi:10.1111/gcb.14601.

Cardinale, B. J., Gross, K., Fritschie, K., Flombaum, P., Fox, J. W., Rixen, C., van Ruijven, J., Reich, P. B., Scherer-Lorenzen, M. and Wilsey, B. J. 2013. Biodiversity simultaneously enhances the production and stability of community biomass, but the effects are independent. *Ecology* 94(8), 1697–707. doi:10.1890/12-1334.1.

Chevassus-Au-Louis, B., Salles, J. M., Pujol, J. L., Bielsa, S., Martin, G. and Richard, D. 2009. *Approche économique de la biodiversité et des services liés aux écosystèmes: contribution à la décision publique*. [Rapport de recherche] Ministère de l'Alimentation, de l'Agriculture et de la Pêche, France, 400 p.

Collomb, M., Bütikofer, U., Sieber, R., Jeangros, B. and Bosset, J.-O. 2002. Correlation between fatty acids in cows' milk fat produced in the Lowlands, Mountains and Highlands of Switzerland and botanical composition of the fodder. *International Dairy Journal* 12(8), 661–6. doi:10.1016/S0958-6946(02)00062-6.

Coppa, M., Verdier-Metz, I., Ferlay, A., Pradel, P., Didienne, R., Farruggia, A., Montel, M. C. and Martin, B. 2011. Effect of different grazing systems on upland pastures compared with hay diet on cheese sensory properties evaluated at different ripening times. *International Dairy Journal* 21(10), 815–22. doi:10.1016/j.idairyj.2011.04.006.

Coppa, M., Ferlay, A., Chassaing, C., Agabriel, C., Glasser, F., Chilliard, Y., Borreani, G., Barcarolo, R., Baars, T., Kusche, D., Harstad, O. M., Verbič, J., Golecký, J. and Martin, B. 2013. Prediction of bulk milk fatty acid composition based on farming practices collected through on-farm surveys. *Journal of Dairy Science* 96(7), 4197–211. doi:10.3168/jds.2012-6379.

Costanza, R., d'Arge, R., de Groot, R., Farber, S., Grasso, M., Hannon, B., Limburg, K., Naeem, S., O'Neill, R. V., Paruelo, J., Raskin, R. G., Sutton, P. and van den Belt, M. 1998. The value of the world's ecosystem services and natural capital. *Ecological Economics* 25(1), 3–15. doi:10.1016/S0921-8009(98)00020-2.

Coulon, J.-B., Delacroix-Buchet, A., Martin, B. and Pirisi, A. 2004. Relationships between ruminant management and sensory characteristics of cheeses: a review. *Lait* 84(3), 221–41. doi:10.1051/lait:2004008.

Couvreur, S., Hurtaud, C., Lopez, C., Delaby, L. and Peyraud, J. L. 2006. The linear relationship Between the proportion of fresh grass in the cow diet, milk fatty acid composition, and butter properties. *Journal of Dairy Science* 89(6), 1956–69. doi:10.3168/jds.S0022-0302(06)72263-9.

Craven, D., Isbell, F., Manning, P., Connolly, J., Bruelheide, H., Ebeling, A., Roscher, C., van Ruijven, J., Weigelt, A., Wilsey, B., Beierkuhnlein, C., de Luca, E., Griffin, J. N., Hautier, Y., Hector, A., Jentsch, A., Kreyling, J., Lanta, V., Loreau, M., Meyer, S. T., Mori, A. S., Naeem, S., Palmborg, C., Polley, H. W., Reich, P. B., Schmid, B., Siebenkäs, A., Seabloom, E., Thakur, M. P., Tilman, D., Vogel, A. and Eisenhauer, N. 2016. Plant diversity effects on grassland productivity are robust to both nutrient enrichment and drought. *Philosophical Transactions of the Royal Society Series B: Biological Sciences* 371, 20150277. doi:10.1098/rstb.2015.0277.

Daily, G. C., Alexander, S., Ehrlich, P. R., Goulder, L. and Lubchenco, J. 1997. Ecosystem services: benefits supplied to human societies by natural ecosystems. *Ecology* 2, 2–16.

De Keersmaecker, W., van Rooijen, N., Lhermitte, S., Tits, L., Schaminée, J., Coppin, P., Honnay, O. and Somers, B. 2016. Species-rich semi-natural grasslands have a higher resistance but a lower resilience than intensively managed agricultural grasslands in response to climate anomalies. *Journal of Applied Ecology* 53(2), 430–9. doi:10.1111/1365-2664.12595.

de Vries, F. T., Manning, P., Tallowin, J. R. B., Mortimer, S. R., Pilgrim, E. S., Harrison, K. A., Hobbs, P. J., Quirk, H., Shipley, B., Cornelissen, J. H. C., Kattge, J. and Bardgett, R. D.

2012. Abiotic drivers and plant traits explain landscape-scale patterns in soil microbial communities. *Ecology Letters* 15(11), 1230–9. doi:10.1111/j.1461-0248.2012.01844.x.

Diaz, S., Lavorel, S., de Bello, F., Quetier, F., Grigulis, K. and Robson, T. M. 2007. Incorporating plant functional diversity effects in ecosystem service assessments. *Proceedings of the National Academy of Sciences of the United States of America* 104(52), 20684–9. doi:10.1073/pnas.0704716104.

Díaz, S., Demissew, S., Carabias, J., Joly, C., Lonsdale, M., Ash, N., Larigauderie, A., Adhikari, J. R., Arico, S., Báldi, A., Bartuska, A., Baste, I. A., Bilgin, A., Brondizio, E., Chan, K. M., Figueroa, V. E., Duraiappah, A., Fischer, M., Hill, R., Koetz, T., Leadley, P., Lyver, P., Mace, G. M., Martin-Lopez, B., Okumura, M., Pacheco, D., Pascual, U., Pérez, E. S., Reyers, B., Roth, E., Saito, O., Scholes, R. J., Sharma, N., Tallis, H., Thaman, R., Watson, R., Yahara, T., Hamid, Z. A., Akosim, C., Al-Hafedh, Y., Allahverdiyev, R., Amankwah, E., Asah, S. T., Asfaw, Z., Bartus, G., Brooks, L. A., Caillaux, J., Dalle, G., Darnaedi, D., Driver, A., Erpul, G., Escobar-Eyzaguirre, P., Failler, P., Fouda, A. M. M., Fu, B., Gundimeda, H., Hashimoto, S., Homer, F., Lavorel, S., Lichtenstein, G., Mala, W. A., Mandivenyi, W., Matczak, P., Mbizvo, C., Mehrdadi, M., Metzger, J. P., Mikissa, J. B., Moller, H., Mooney, H. A., Mumby, P., Nagendra, H., Nesshover, C., Oteng-Yeboah, A. A., Pataki, G., Roué, M., Rubis, J., Schultz, M., Smith, P., Sumaila, R., Takeuchi, K., Thomas, S., Verma, M., Yeo-Chang, Y. and Zlatanova, D. 2015. The IPBES Conceptual Framework – connecting nature and people. *Current Opinion in Environmental Sustainability* 14, 1–16. doi:10.1016/j.cosust.2014.11.002.

Duncan, A. J., Ginane, C., Elston, D. A., Kunaver, A. and Gordon, I. J. 2006. How do herbivores trade-off the positive and negative consequences of diet selection decisions? *Animal Behaviour* 71(1), 93–9. doi:10.1016/j.anbehav.2005.03.035.

Duru, M., Cruz P, P., Raouda, A. H. K., Ducourtieux, C. and Theau, J. P. 2008. Relevance of plant functional types based on leaf dry matter content for assessing digestibility of native grass species and species-rich grassland communities in Spring. *Agronomy Journal* 100(6), 1622–30. doi:10.2134/agronj2008.0003.

Duru, M., Adam, M., Cruz, P., Martin, G., Ansquer, P., Ducourtieux, C., Jouany, C., Theau, J. P. and Viegas, J. 2009. Modelling above-ground herbage mass for a wide range of grassland community types. *Ecological Modelling* 220(2), 209–25. doi:10.1016/j.ecolmodel.2008.09.01510.1016/j.ecolmodel.2008.09.015.

Duru, M. 2016. Consommer des produits dont les animaux ont été alimentés à l'herbe est-il suffisant pour équilibrer notre alimentation en acides gras poly-insaturés ? *Fourrages* 228, 301–12.

Eisenhauer, N., Hines, J., Isbell, F., van der Plas, F., Hobbie, S. E., Kazanski, C. E., Lehmann, A., Liu, M., Lochner, A., Rillig, M. C., Vogel, A., Worm, K. and Reich, P. B. 2018. Plant diversity maintains multiple soil functions in future environments. *eLife* 7. doi:10.7554/eLife.41228.

Elmqvist, T., Folke, C., Nyström, M., Peterson, G., Bengtsson, J., Walker, B. and Norberg, J. 2003. Response diversity, ecosystem change, and resilience. *Frontiers in Ecology and the Environment* 1(9), 488–94. doi:10.1890/1540-9295(2003)001[0488:RDECAR]2.0.CO;2.

Fantinato, E., Del Vecchio, S. and Buffa, G. 2019. The co-occurrence of different grassland communities increases the stability of pollination networks. *Flora* 255, 11–7. doi:10.1016/j.flora.2019.03.017.

FAO. 2019. Available at: http://www.fao.org/agriculture/crops/core-themes/theme/spi/grasslands-rangelands-and-forage-crops/en/.

Farruggia, A., Martin, B., Baumont, R., Prache, S., Doreau, M., Hoste, H., Durand, D. and Champanelle, S.-G. 2008. Quels intérêts de la diversité floristique des prairies permanentes pour les ruminants et les produits animaux ? *INRA Productions Animales* 21(2), 181–200.

Finger, R. and Buchmann, N. 2015. An ecological economic assessment of risk-reducing effects of species diversity in managed grasslands. *Ecological Economics* 110, 89–97. doi:10.1016/j.ecolecon.2014.12.019.

Fischer, M. and Wipf, S. 2002. Effect of low-intensity grazing on the species-rich vegetation of traditionally mown subalpine meadows. *Biological Conservation* 104(1), 1–11. doi:10.1016/S0006-3207(01)00149-5.

Frankow-Lindberg, B. E. 2012. Grassland plant species diversity decreases invasion by increasing resource use. *Oecologia* 169(3), 793–802. doi:10.1007/s00442-011-2230-7.

Fraser, M. D., Davies, D. A., Wright, I. A., Vale, J. E., Nute, G. R., Hallett, K. G. and Richardson, R. I. 2007. Effect on upland beef production of incorporating winter feeding of red clover silage or summer grazing of Molinia-dominated semi-natural pastures. *Grass and Forage Science* 62(3), 284–300. doi:10.1111/j.1365-2494.2007.00581.x.

French, K. E. 2017. Species composition determines forage quality and medicinal value of high diversity grasslands in lowland England. *Agriculture, Ecosystems and Environment* 241, 193–204. doi:10.1016/j.agee.2017.03.012.

Fridley, J. D. 2002. Resource availability dominates and alters the relationship between species diversity and ecosystem productivity in experimental plant communities. *Oecologia* 132(2), 271–7. doi:10.1007/s00442-002-0965-x.

Ganguli, A. C., Engle, D. M., Mayer, P. M. and Hellgren, E. C. 2008. Plant community diversity and composition provide little resistance to *Juniperus* encroachment. *Botany* 86(12), 1416–26. doi:10.1139/B08-110.

Gaujour, E., Amiaud, B., Mignolet, C. and Plantureux, S. 2012. Factors and processes affecting plant biodiversity in permanent grasslands. A review. *Agronomy for Sustainable Development* 32(1), 133–60. doi:10.1007/s13593-011-0015-3.

Gessner, D. K., Ringseis, R. and Eder, K. 2017. Potential of plant polyphenols to combat oxidative stress and inflammatory processes in farm animals. *Journal of Animal Physiology and Animal Nutrition* 101(4), 605–28. doi:10.1111/jpn.12579.

Gorlier, A., Lonati, M., Renna, M., Lussiana, C., Lombardi, G. and Battaglini, L. M. 2012. Changes in pasture and cow milk compositions during a summer transhumance in the western Italian Alps. *Journal of Applied Botany and Food Quality* 85, 216–23.

Grace, J. B., Anderson, T. M., Seabloom, E. W., Borer, E. T., Adler, P. B., Harpole, W. S., Hautier, Y., Hillebrand, H., Lind, E. M., Pärtel, M., Bakker, J. D., Buckley, Y. M., Crawley, M. J., Damschen, E. I., Davies, K. F., Fay, P. A., Firn, J., Gruner, D. S., Hector, A., Knops, J. M. H., MacDougall, A. S., Melbourne, B. A., Morgan, J. W., Orrock, J. L., Prober, S. M. and Smith, M. D. 2016. Integrative modelling reveals mechanisms linking productivity and plant species richness. *Nature* 529(7586), 390–3. doi:10.1038/nature16524.

Green, R. E., Cornell, S. J., Scharlemann, J. P. and Balmford, A. 2005. Farming and the fate of wild nature. *Science* 307(5709), 550–5. doi:10.1126/science.1106049.

Grevilliot, F., Krebs, L. and Muller, S. 1998. Comparative importance and interference of hydrological conditions and soil nutrient gradients in ˉoristic biodiversity in ˉood meadows. *Biodiversity and Conservation* 7, 1495–1520.

Grime, J. P. 1973. Control of species density in herbaceous vegetation. *Journal of Environmental Management* 1, 151–67.

Gross, K., Cardinale, B. J., Fox, J. W., Gonzalez, A., Loreau, M., Wayne Polley, H. W., Reich, P. B. and van Ruijven, J. 2014. Species richness and the temporal stability of biomass production: A new analysis of recent biodiversity experiments. *The American Naturalist* 183(1), 1–12. doi:10.1086/673915.

Gustavsson, E., Lennartsson, T. and Emanuelsson, M. 2007. Land use more than 200 years ago explains current grassland plant diversity in a Swedish agricultural landscape. *Biological Conservation* 138(1–2), 47–59. doi:10.1016/j.biocon.2007.04.004.

He, J.-S., Wolfe-Bellin, K. S., Schmid, B. and Bazzaz, F. A. 2005. Density may alter diversity-productivity relationships in experimental plant communities. *Basic and Applied Ecology* 6(6), 505–17. doi:10.1016/j.baae.2005.04.002.

Hector, A., Schmid, B., Beierkuhnlein, C., Caldeira, M. C., Diemer, M., Dimitrakopoulos, P. G., Finn, J. A., Freitas, H., Giller, P. S., Good, J., Harris, R., Hogberg, P., Huss-Danell, K., Joshi, J., Jumpponen, A., Korner, C., Leadley, P. W., Loreau, M., Minns, A., Mulder, C. P., O'Donovan, G., Otway, S. J., Pereira, J. S., Prinz, A., Read, D. J., Scherer-Lorenzen M., Schulze, E.-D., Siamantziouras, A.-S. D., Spehn, E. M., Terry, A. C., Troumbis, A. Y., Woodward, F. I., Yachi, S., and Lawton, J. H. 1999. Plant diversity and productivity experiments in European grasslands. *Science* 286(5442), 1123–7. doi:10.1126/science.286.5442.1123.

Hector, A., Dobson, K., Minns, A., Bazeley-White, E. and Hartley Lawton, J. 2001. Community diversity and invasion resistance: an experimental test in a grassland ecosystem and a review of comparable studies. *Ecological Research* 16(5), 819–31. doi:10.1046/j.1440-1703.2001.00443.x.

Hillebrand, H., Bennett, D. M. and Cadotte, M. W. 2008. Consequences of dominance: a review of evenness effects on local and regional ecosystem processes. *Ecology* 89(6), 1510–20. doi:10.1890/07-1053.1.

Huguenin-Elie, O., Nemecek, T., Plantureux, S., Jeanneret, P. and Lüscher, A. 2012. Environmental impacts of grassland management at the plot and the farm scale. In: *EGF*. Presented at the 24th General Meeting of the European-Grassland-Federation, Lublin, Poland.

Huston, M. A. 1997. Hidden treatments in ecological experiments: re-evaluating the ecosystem function of biodiversity. *Oecologia* 110(4), 449–60. doi:10.1007/s004420050180.

Isbell, F., Calcagno, V., Hector, A., Connolly, J., Harpole, W. S., Reich, P. B., Scherer-Lorenzen, M., Schmid, B., Tilman, D., van Ruijven, J., Weigelt, A., Wilsey, B. J., Zavaleta, E. S. and Loreau, M. 2011. High plant diversity is needed to maintain ecosystem services. *Nature* 477(7363), 199–202. doi:10.1038/nature10282.

Isbell, F., Tilman, D., Polasky, S., Binder, S. and Hawthorne, P. 2013. Low biodiversity state persists two decades after cessation of nutrient enrichment. *Ecology Letters* 16(4), 454–60. doi:10.1111/ele.12066.

Isbell, F., Craven, D., Connolly, J., Loreau, M., Schmid, B., Beierkuhnlein, C., Bezemer, T. M., Bonin, C., Bruelheide, H., de Luca, E., Ebeling, A., Griffin, J. N., Guo, Q., Hautier, Y., Hector, A., Jentsch, A., Kreyling, J., Lanta, V., Manning, P., Meyer, S. T., Mori, A. S., Naeem, S., Niklaus, P. A., Polley, H. W., Reich, P. B., Roscher, C., Seabloom, E. W., Smith, M. D., Thakur, M. P., Tilman, D., Tracy, B. F., van der Putten, W. H., van Ruijven, J., Weigelt, A., Weisser, W. W., Wilsey, B. and Eisenhauer, N. 2015. Biodiversity increases the resistance of ecosystem productivity to climate extremes. *Nature* 526(7574), 574–7. doi:10.1038/nature15374.

Isselstein, J. 2005. Enhancing grassland biodiversity consequences grassland management utilisation. *Grassland: A Global Resource*, 305-21.

Isselstein, J. 2018. Protecting biodiversity in grasslands. In: Marshall, A. and Collins, R. (Eds), *Improving Grassland and Pasture Management in Temperate Agriculture*. Burleigh Dodds Series in Agricultural Science. Burleigh Dodds Science Publishing, Cambridge, pp. 381-96. doi:10.19103/AS.2017.0024.18.

Janssens, F., Peeters, A., Tallowin, J. R. B., Bakker, J. P., Bekker, R. M. and Fillat, F. 1998. Relationship between soil chemical factors and grassland diversity. *Plant and Soil* 202, 69-78.

Jayanegara, A., Togtokhbayar, N., Makkar, H. P. S. and Becker, K. 2009. Tannins determined by various methods as predictors of methane production reduction potential of plants by an in vitro rumen fermentation system. *Animal Feed Science and Technology* 150(3-4), 230-7. doi:10.1016/j.anifeedsci.2008.10.011.

Jouven, M., Carrere, P. and Baumont, R. 2006. Model predicting dynamics of biomass, structure and digestibility of herbage in managed permanent pastures. 1. Model description. *Grass and Forage Science* 61(2), 112-24. doi:10.1111/j.1365-2494.2006.00515.x.

Joyce, C. 2001. The sensitivity of a species-rich flood-meadow plant community to fertilizer nitrogen: the Lužnice river floodplain, Czech Republic. *Plant Ecology* 155, 47-60.

Kirkham, F. W., Tallowin, J. R. B., Sanderson, R. A., Bhogal, A., Chambers, B. J. and Stevens, D. P. 2008. The impact of organic and inorganic fertilizers and lime on the species-richness and plant functional characteristics of hay meadow communities. *Biological Conservation* 141(5), 1411-27. doi:10.1016/j.biocon.2008.03.010.

Klimek, S., Richter gen. Kemmermann, A., Hofmann, M. and Isselstein, J. 2007 Plant species richness and composition in managed grasslands: the relative importance of field management and environmental factors. *Biological Conservation* 134(4), 559-70. doi:10.1016/j.biocon.2006.09.007.

Kruess, A. and Tscharntke, T. 2002a. Grazing intensity and the diversity of grasshoppers, butterflies, and trap-nesting bees and wasps. *Conservation Biology* 16(6), 1570-80. doi:10.1046/j.1523-1739.2002.01334.x.

Kruess, A. and Tscharntke, T. 2002b. Contrasting responses of plant and insect diversity to variation in grazing intensity. *Biological Conservation* 106(3), 293-302. doi:10.1016/S0006-3207(01)00255-5.

Küchenmeister, F., Küchenmeister, K., Wrage, N., Kayser, M. and Isselstein, J. 2012. Yield and yield stability in mixtures of productive grassland species: does species number or functional group composition matter?: diversity and functional groups. *Grassland Science* 58(2), 94-100. doi:10.1111/j.1744-697X.2012.00242.x.

Lepš, J. 2004. Variability in population and community biomass in a grassland community affected by environmental productivity and diversity. *Oikos* 107(1), 64-71. doi:10.1111/j.0030-1299.2004.13023.x.

Li, Y., Dong, S., Liu, S., Su, X., Wang, X., Zhang, Y., Zhao, Z., Gao, X., Li, S. and Tang, L. 2019. Relationships between plant diversity and biomass production of alpine grasslands are dependent on the spatial scale and the dimension of biodiversity. *Ecological Engineering* 127, 375-82. doi:10.1016/j.ecoleng.2018.12.015.

Loucougaray, G., Dobremez, L., Gos, P., Pauthenet, Y., Nettier, B. and Lavorel, S. 2015. Assessing the effects of grassland management on forage production and environmental quality to identify paths to ecological intensification in mountain grasslands. *Environmental Management* 56(5), 1039–52. doi:10.1007/s00267-015-0550-9.

Lüscher, A., Mueller-Harvey, I., Soussana, J. F., Rees, R. M. and Peyraud, J. L. 2014. Potential of legume-based grassland-livestock systems in Europe: a review. *Grass and Forage Science: the Journal of the British Grassland Society* 69(2), 206–28. doi:10.1111/gfs.12124.

MacArthur, R. 1955. Fluctuations of animal populations and a measure of community stability. *Ecology* 36(3), 533. doi:10.2307/1929601.

Macheboeuf, D., Coudert, L., Bergeault, R., Lalière, G. and Niderkorn, V. 2014. Screening of plants from diversified natural grasslands for their potential to combine high digestibility, and low methane and ammonia production. *Animal: An International Journal of Animal Bioscience* 8(11), 1797–806. doi:10.1017/S1751731114001785.

Magda, D., de Sainte Marie, C., Plantureux, S., Agreil, C., Amiaud, B., Mestelan, P. and Mihout, S. 2015. Integrating agricultural and ecological goals into the management of species-rich grasslands: learning from the flowering meadows competition in France. *Environmental Management* 56(5), 1053–64. doi:10.1007/s00267-015-0553-6.

Makkar, H. P. S. 2003. Effects and fate of tannins in ruminant animals, adaptation to tannins, and strategies to overcome detrimental effects of feeding tannin-rich feeds. *Small Ruminant Research* 49(3), 241–56. doi:10.1016/S0921-4488(03)00142-1.

Marini, L., Scotton, M., Klimek, S., Isselstein, J. and Pecile, A. 2007. Effects of local factors on plant species richness and composition of Alpine meadows. *Agriculture, Ecosystems and Environment* 119(3–4), 281–8. doi:10.1016/j.agee.2006.07.015.

Marriott, C. A., Fothergill, M., Jeangros, B., Scotton, M. and Louault, F. 2004. Long-term impacts of extensification of grassland management on biodiversity and productivity in upland areas. A review. *Agronomie* 24(8), 447–62. doi:10.1051/agro:2004041.

Martin, B., Verdier-Metz, I., Buchin, S., Hurtaud, C. and Coulon, J. -B. 2005. How do the nature of forages and pasture diversity influence the sensory quality of dairy livestock products? *Animal Science* 81(2), 205–12. doi:10.1079/ASC50800205.

Marzetti, S., Disegna, M., Villani, G. and Speranza, M. 2011. Conservation and recreational values from semi-natural grasslands for visitors to two Italian parks. *Journal of Environmental Planning and Management* 54(2), 169–91. doi:10.1080/09640568.2010.505792.

Michaud, A., Plantureux, S., Pottier, E. and Baumont, R. 2015. Links between functional composition, biomass production and forage quality in permanent grasslands over a broad gradient of conditions. *The Journal of Agricultural Science* 153(5), 891–906. doi:10.1017/S0021859614000653.

Mokany, K., Ash, J. and Roxburgh, S. 2008. Functional identity is more important than diversity in influencing ecosystem processes in a temperate native grassland: determinants of ecosystem processes. *Journal of Ecology* 96(5), 884–93. doi:10.1111/j.1365-2745.2008.01395.x.

Niderkorn, V. and Baumont, R. 2009. Associative effects between forages on feed intake and digestion in ruminants. *Animal: An International Journal of Animal Bioscience* 3(7), 951–60. doi:10.1017/S1751731109004261.

Nijs, I. and Roy, J. 2000. How important are species richness, species evenness and interspecific differences to productivity? A mathematical model. *Oikos* 88(1), 57–66. doi:10.1034/j.1600-0706.2000.880107.x.

Olff, H. and Ritchie, M. E. 1998. Effects of herbivores on grassland plant diversity. *Trends in Ecology and Evolution* 13(7), 261-5. doi:10.1016/S0169-5347(98)01364-0.

Oppermann, R. 2003. Nature balance scheme for farms–evaluation of the ecological situation. *Agriculture, Ecosystems and Environment* 98(1-3), 463-75. doi:10.1016/S0167-8809(03)00105-1.

Orford, K. A., Murray, P. J., Vaughan, I. P. and Memmott, J. 2016. Modest enhancements to conventional grassland diversity improve the provision of pollination services. *Journal of Applied Ecology* 53(3), 906-15. doi:10.1111/1365-2664.12608.

Pallett, D. W., Pescott, O. L. and Schäfer, S. M. 2016. Changes in plant species richness and productivity in response to decreased nitrogen inputs in grassland in southern England. *Ecological Indicators* 68, 73-81. doi:10.1016/j.ecolind.2015.12.024.

Peeters, A., Beaufoy, G., Canals, R. M., De Vliegher, A., Huyghe, C., Isselstein, J., Jones, G., Kessler, W., Kirilov, A., Mosquera-Losada, M. R., Nillsdotter-Linde, N., Parente, G., Peyraud, J. L., Pickert, J., Plantureux, S., Porqueddu, C., Rataj, D., Stypinski, D. Tonn, B., Van den Pol - Van Dasselaar, A., Vintu, V. and Wilkins, R. J., 2014. Grassland term definitions and classifications adapted to the diversity of european grassland-based systems. *Grassland Science in Europe* 19, 743-50.

Piotrowska, K., Connolly, J., Finn, J., Black, A. and Bolger, T. 2013. Evenness and plant species identity affect earthworm diversity and community structure in grassland soils. *Soil Biology and Biochemistry* 57, 713-9. doi:10.1016/j.soilbio.2012.06.016.

Plantureux, S. 1996a. Biodiversité, type de sol et intensité de l'exploitation de prairies permanentes du Plateau lorrain. *Acta Botanica Gallica* 143(4-5), 339-48. doi:10.1080/12538078.1996.10515730.

Plantureux, S. 1996b. Flora-sys: système informatique de gestion et d'aide à l'interprétation des relevés floristiques. *Acta Botanica Gallica* 143(4-5), 403-10. doi:10.1080/12538078.1996.10515736.

Plantureux, S., Peeters, A. and McCracken, D. 2005. Biodiversity in intensive grasslands: effect of management, improvement and challenges. *Agronomy Research* 3(2), 153-64.

Plantureux, S., de Sainte Marie, C., Agreil, C., Amiaud, B., Dombremez, L., Fargier, J., Fleury, P., Fritz, H., Langlois, J. L., Magda, D., Mestelan, P., Meuret, M., Mouget, T., Nettier, B., Sérès, T. and Vansteelant, J. Y. 2011. Analyse de la mise en oeuvre des mesures agri-environnementales à obligation de résultat sur les surfaces herbagères. *Fourrages* 208, 271-81.

Pokorny, M. L., Sheley, R. L., Zabinski, C. A., Engel, R. E., Svejcar, T. J. and Borkowski, J. J. 2005. Plant Functional Group Diversity as a mechanism for invasion resistance. *Restoration Ecology* 13(3), 448-59. doi:10.1111/j.1526-100X.2005.00056.x.

Polley, H. W., Isbell, F. I. and Wilsey, B. J. 2013. Plant functional traits improve diversity-based predictions of temporal stability of grassland productivity. *Oikos* 122(9), 1275-82. doi:10.1111/j.1600-0706.2013.00338.x.

Pontes, L. S., Carrère, P., Andueza, D., Louault, F. and Soussana, J. F. 2007. Seasonal productivity and nutritive value of temperate grasses found in semi-natural pastures in Europe: responses to cutting frequency and N supply. *Grass and Forage Science* 62(4), 485-96. doi:10.1111/j.1365-2494.2007.00604.x.

Potts, S. G., Vulliamy, B., Dafni, A., Ne'eman, G. and Willmer, P. 2003. Linking bees and flowers: how do floral communities structure pollinator communities? *Ecology* 84(10), 2628-42. doi:10.1890/02-0136.

Poutaraud, A., Michelot-Antalik, A. and Plantureux, S. 2017. Grasslands: a Source of secondary metabolites for livestock health. *Journal of Agricultural and Food Chemistry* 65(31), 6535–53. doi:10.1021/acs.jafc.7b00425.

Prieto, I., Violle, C., Barre, P., Durand, J. L., Ghesquiere, M. and Litrico, I. 2015. Complementary effects of species and genetic diversity on productivity and stability of sown grasslands. *Nature Plants* 1(4), 15033. doi:10.1038/nplants.2015.33.

Pykala, J. 2005. Plant species responses to cattle grazing in mesic semi-natural grassland. *Agriculture, Ecosystems and Environment* 108(2), 109–17. doi:10.1016/j.agee.2005.01.012.

Quétier, F., Rivoal, F., Marty, P., de Chazal, J., Thuiller, W. and Lavorel, S. 2010. Social representations of an alpine grassland landscape and socio-political discourses on rural development. *Regional Environmental Change* 10(2), 119–30. doi:10.1007/s10113-009-0099-3.

Ravetto Enri, S., Renna, M., Probo, M., Lussiana, C., Battaglini, L. M., Lonati, M. and Lombardi, G. 2017. Relationships between botanical and chemical composition of forages: a multivariate approach to grasslands in the Western Italian Alps: Composition of forages in the Italian Alps. *Journal of the Science of Food and Agriculture* 97(4), 1252–9. doi:10.1002/jsfa.7858.

Reynaud, A., Fraisse, D., Cornu, A., Farruggia, A., Pujos-Guillot, E., Besle, J. M., Martin, B., Lamaison, J. L., Paquet, D., Doreau, M. and Graulet, B. 2010. Variation in content and composition of phenolic compounds in permanent pastures according to botanical variation. *Journal of Agricultural and Food Chemistry* 58(9), 5485–94. doi:10.1021/jf1000293.

Richardson, D. M. and Pyšek, P. 2007. Elton, C. S. 1958. The ecology of invasions by animals and plants. London: methuen. *Progress in Physical Geography: Earth and Environment* 31, 659–66. doi:10.1177/0309133307087089.

Richmond, A. S., Wylie, A. R. G., Laidlaw, A. S. and Lively, F. O. 2015. Methane emissions from beef cattle grazing on semi-natural upland and improved lowland grasslands. *Animal: An International Journal of Animal Bioscience* 9(1), 130–7. doi:10.1017/S1751731114002067.

Riesch, F., Stroh, H. G., Tonn, B. and Isselstein, J. 2018. Soil pH and phosphorus drive species composition and richness in semi-natural heathlands and grasslands unaffected by twentieth-century agricultural intensification. *Plant Ecology and Diversity* 11(2), 239–53. doi:10.1080/17550874.2018.1471627.

Rosa García, R., Fraser, M. D., Celaya, R., Ferreira, L. M. M., García, U. and Osoro, K. 2013. Grazing land management and biodiversity in the Atlantic European heathlands: a review. *Agroforestry Systems* 87(1), 19–43. doi:10.1007/s10457-012-9519-3.

Sabatier, R., Doyen, L. and Tichit, M. 2014. Heterogeneity and the trade-off between ecological and productive functions of agro-landscapes: a model of cattle-bird interactions in a grassland agroecosystem. *Agricultural Systems* 126, 38–49. doi:10.1016/j.agsy.2013.02.008.

Schellberg, J., Möseler, B. M., Kübauch, W. and Rademacher, I. F. 1999. Long-term effects of fertilizer on soil nutrient concentration, yield, forage quality and floristic composition of a hay meadow in the Eifel mountains, Germany. *Grass and Forage Science* 54(3), 195–207. doi:10.1046/j.1365-2494.1999.00166.x.

Schläpfer, F., Tucker, M. and Seidl, I. 2002. Returns from hay cultivation in fertilized low diversity and non-fertilized high diversity grassland. *Environmental and Resource Economics* 21, 89–100.

Schmid, B. 2002. The species richness-productivity controversy. *Trends in Ecology and Evolution* 17(3), 113–4. doi:10.1016/S0169-5347(01)02422-3.

Schulze, E. D., Bouriaud, O., Weber, U., Roscher, C., Hessenmoeller, D., Kroiher, F. and Schall, P. 2018. Management breaks the natural productivity-biodiversity relationship in forests and grassland: an opinion. *Forest Ecosystems* 5(3). doi:10.1186/s40663-017-0122-y.

Seither, M. and Wrage, M. N. 2014. Biomass production of LolioCynosuretum grassland is not increased by plantspecies richness. *Journal of Plant Nutrition and Soil Science* 177, 613–23. doi:10.1002/jpln.201100349.

Silvertown, J., Poulton, P., Johnston, E., Edwards, G., Heard, M. and Biss, P. M. 2006. The Park Grass Experiment 1856 -2006: its contribution to ecology. *Journal of Ecology* 94(4), 801–14. doi:10.1111/j.1365-2745.2006.01145.x.

Smith, R. S., Shiel, R. S., Millward, D., Corkhill, P. and Sanderson, R. A. 2002. Soil seed banks and the effects of meadow management on vegetation change in a 10-year meadow field trial. *Journal of Applied Ecology* 39(2), 279–93. doi:10.1046/j.1365-2664.2002.00715.x.

Spehn, E. M., Joshi, J., Schmid, B., Alphei, J. and Körner, C. 2000. Plant diversity effects on soil heterotrophic activity in experimental grassland ecosystems. *Plant and Soil* 224, 217–30.

Strecker, T., Barnard, R. L., Niklaus, P. A., Scherer-Lorenzen, M., Weigelt, A., Scheu, S. and Eisenhauer, N. 2015. Effects of plant diversity, Functional Group Composition, and Fertilization on Soil Microbial Properties in Experimental Grassland. *PLoS ONE* 10(5), e0125678. doi:10.1371/journal.pone.0125678.

Tälle, M., Deák, B., Poschlod, P., Valkó, O., Westerberg, L. and Milberg, P. 2016. Grazing vs. mowing: A meta-analysis of biodiversity benefits for grassland management. *Agriculture, Ecosystems and Environment* 222, 200–12. doi:10.1016/j.agee.2016.02.008.

Teillard, F., Doyen, L., Dross, C., Jiguet, F. and Tichit, M. 2017. Optimal allocations of agricultural intensity reveal win-no loss solutions for food production and biodiversity. *Regional Environmental Change* 17(5), 1397–408. doi:10.1007/s10113-016-0947-x.

Tichit, M., Durant, D. and Kerneıs, E. 2005. The role of grazing in creating suitable sward structures for breeding waders in agricultural landscapes. *Livestock Production Science* 96(1), 119–28. doi:10.1016/j.livprodsci.2005.05.010.

Tilman, D., Wedin, D. and Knops, J. 1996. Productivity and sustainability influenced by biodiversity in grassland ecosystems. *Nature* 379(6567), 718–20. doi:10.1038/379718a0.

Tilman, D., Reich, P. B., Knops, J., Wedin, D., Mielke, T. and Lehman, C. 2001. Diversity and productivity in a long-term grassland experiment. *Science* 294(5543), 843–5. doi:10.1126/science.1060391.

van Ruijven, J. and Berendse, F. 2007. Contrasting effects of diversity on the temporal stability of plant populations. *Oikos* 116, 1323–30. doi: 10.1111/j.2007.0030-1299.16005.x.

Veen, P., Jefferson, R., De Smidt, J. and Van den Straaten, J. 2009. *Grasslands in Europe of High Nature Value*. KNNV Publishing, Zeist.

Venail, P., Gross, K., Oakley, T. H., Narwani, A., Allan, E., Flombaum, P., Isbell, F., Joshi, J., Reich, P. B., Tilman, D., van Ruijven, J. and Cardinale, B. J. 2015. Species richness, but not phylogenetic diversity, influences community biomass production and temporal stability in a re-examination of 16 grassland biodiversity studies. *Functional Ecology* 29, 615–26.

Vogel, A., Scherer-Lorenzen, M. and Weigelt, A. 2012. Grassland resistance and resilience after drought depends on management intensity and species richness. *PLoS ONE* 7(5), e36992. doi:10.1371/journal.pone.0036992.

Xu, X., Polley, H. W., Hofmockel, K. and Wilsey, B. J. 2017. Species composition but not diversity explains recovery from the 2011 drought in Texas grasslands. *Ecosphere* 8(3), e01704. doi:10.1002/ecs2.1704.

Yang, Y., Tilman, D., Furey, G. and Lehman, C. 2019. Soil carbon sequestration accelerated by restoration of grassland biodiversity. *Nature Communications* 10(1), 718. doi:10.1038/s41467-019-08636-w.

Yoshihara, Y., Furusawa, S. and Sato, S. 2016. Recent pasture management determines biodiversity and productivity, and past management determines forage quality. *Écoscience* 23(3–4), 89–96. doi:10.1080/11956860.2016.1266832.

Zechmeister, H. G., Schmitzberger, I., Steurer, B., Peterseil, J. and Wrbka, T. 2003. The influence of land-use practices and economics on plant species richness in meadows. *Biological Conservation* 114(2), 165–77. doi:10.1016/S0006-3207(03)00020-X.

Chapter 8

The importance of agroforestry systems in supporting biodiversity conservation and agricultural production: a European perspective

M. R. Mosquera-Losada, J. J. Santiago-Freijanes, A. Rigueiro-Rodríguez, F. J. Rodríguez-Rigueiro, D. Arias Martínez, A. Pantera and N. Ferreiro-Domínguez, University of Santiago de Compostela, Spain

1 Introduction

Biodiversity has become a key international environmental and production issue in recent decades, as shown by the Convention on Biological Diversity (CBD) agreed at the Rio de Janeiro Summit in 1992 (UN 1992). More recently, action on biodiversity has been reinforced with a global strategic plan to combat biodiversity losses (Aichi targets (CBD 2010)) as well as other initiatives such as the EU's 2020 biodiversity strategy (EC 2011b).

Biodiversity is part of the natural capital that needs to be preserved for the following generations. Millions of years of evolution are represented in many different types of organisms that may be useful for both food and health. As an example, most traditional medicines are based on the extraction of existing plant compounds. This means that most pharmaceutical compounds are not

http://dx.doi.org/10.19103/AS.2020.0071.14

'created' by the pharmaceutical industry. Instead they are 'copied' from nature which means we need the 'originals' to be preserved.

Biodiversity is crucial to sustainably increase crop production per unit of land by optimal combination of plants that are best able to use soil and other natural resources in a complementary way and thus contribute to the provision of ecosystem services. The co-evolution of different species adapted to a specific habitat has been ignored by intensive agricultural practices and therefore the complementary capacity they have to use resources optimally and provide ecosystem services has also been ignored.

The reduction in biodiversity can be seen, for example, in Europe where only 17% of assessed habitat and 17% of assessed species have a favourable conservation status (Maes (2017). It has been estimated that this loss of biodiversity is causing large economic losses, for example, the loss of pollination is estimated as causing losses of €15 billion per year in the EU) (EC 2011b). Based on these facts and figures, the EU developed a biodiversity strategy based on six main targets (EC 2011a) that include the:

(1) development of a network of natural habitats;
(2) ecosystem services restoration based on the development of a green infrastructure;
(3) ensuring sustainability of agriculture and forestry activities;
(4) safeguarding and protecting EU fish stocks;
(5) controlling invasive species; and
(6) stepping up the EU's contribution to concerted global action to avert biodiversity loss.

The EU strategy is in line with two major commitments made by EU leaders in March 2010:

- halting the loss of biodiversity in the EU by 2020; and
- protecting, valuing and restoring EU biodiversity and ecosystem services by 2050.

As an integral part of the Europe 2020 Strategy, the biodiversity strategy contributes towards the EU's resource efficiency objectives by ensuring that Europe's natural capital is managed sustainably, as well as contributing to climate change mitigation and adaptation goals by improving the resilience of ecosystems and the services they provide.

Agroforestry can directly contribute to targets 1, 2, 3 and 6 of the EU biodiversity strategy. Agroforestry is a sustainable land use system where woody perennials (trees or shrubs) are combined with agricultural production (Mosquera-Losada et al. 2018a, b). Agroforestry can provide solutions to meet

European and global biodiversity strategies for reducing biodiversity losses and increasing biodiversity, while sustainably increasing the efficiency of agricultural systems through the use of the natural capital (Rigueiro-Rodríguez et al. 2010). This chapter describes the main reasons for agroforestry being able to enhance biodiversity and fulfil European and global biodiversity targets.

2 The contribution of agroforestry to global biodiversity goals

The main contribution of agroforestry to biodiversity is land heterogeneity due to the presence of woody perennials in agricultural systems or the understorey in forestry systems (Rigueiro-Rodríguez et al. 2009). Agroforestry is a type of land use that can be linked to all type of agricultural/forestry lands. The main types of agroforestry are (Mosquera-Losada et al. 2018a):

- silvopasture;
- silvoarable cultivation;
- riparian buffer strips;
- forest farming; and
- homegardens.

Silvopasture and silvoarable practices are implemented when farmers combine woody perennials with grazing animals and arable lands, respectively (Mosquera-Losada et al. 2017, 2018a). Woody perennials combined with agricultural production and adjacent to water courses can be categorized as riparian buffer strips. Forest farming is associated with understory agricultural production. When woody perennials usually associated with fruit trees are combined with agricultural production (i.e. vegetables) in urban/semiurban areas, we talk about 'homegardens' (Table 1).

Agroforestry improves biodiversity and therefore the potential provision of ecosystem services. The presence of isolated trees in an arable land clearly creates a spatial and temporal micro-environment with different water and nutrient pools that favours some varieties/species over the others, resulting in differences in species composition between the areas influenced by woody perennials and the open sites. Different understory varieties/species are differently adapted to different woody perennials-influenced areas. The heterogeneity caused by the presence of woody perennials affects not only plants but also other taxa, resulting in a cascade effect promoting the provision of ecosystem services.

As an example of heterogeneity, the level of adaptation of cereal crops to shade has been found to vary upon variety (Mosquera-Losada et al. 2019). The use of different woody perennials species in hedgerows, when associated

Table 1 Agroforestry practices associated with policy land use: agriculture, forestry and urban and peri-urban areas

Land use and agroforestry practice		Common name	Brief description
AGRICULTURE	Silvopasture	Wood pasture and parkland	Typically areas of widely spaced trees that are also used for forage and animal production
		Meadow orchards	This practice includes fruit orchards, shrubs which are grazed or sown with pastures, but also olive groves and vineyards
	Silvoarable	Hedgerows and windbreak systems	Here the woody components are planted to provide shelter, shade or parcel demarcation to a crop and/or livestock production system
		Alley-cropping systems	Widely spaced woody perennials inter-cropped with annual or perennial crops. It comprises alley cropping, scattered trees and orchards and line belts within the plots. These practices are sometimes found only during the first few years of the plantation
	Riparian buffer strips	Riparian buffer strips	Areas of tree and shrubs allowed to establish croplands/pastures and water sources such as streams, lakes, wetlands and ponds to protect water quality can be identified as silvoarable or silvopasture
FOREST	Silvopasture	Forest grazing	Although the land cover is described as forest, the understory is grazed
	Forest farming	Forest farming	Forested areas used for production or harvest of naturally standing specialty crops for medicinal, ornamental or culinary uses
URBAN AND PERI-URBAN	Homegardens	Homegardens	Combining trees/shrubs with vegetable production usually associated with peri-urban or urban areas

with wetlands or croplands, has been described by Nieto et al. (2014) as a good long-term strategy to promote pollination. When the presence of different species of trees or shrubs is assessed at farm scale, woody perennials have been found to enlarge and complement the overall plant flowering period, facilitating pollination and insect abundance that could be profitably exploited (honey production). In addition, the presence of trees and shrubs reduces water pollution in aquatic environments where good water quality is of paramount importance for the survival of aquatic species as well as birds and other animals. At landscape level, the presence of woody perennials in different farms enhances the impact that trees or shrubs can have on biodiversity inside and outside the farm by modifying the environment through wind speed reduction, as well as affecting water cycling and therefore increasing micro-habitat heterogeneity that contributes to biodiversity promotion. The presence of woody perennials also positively affects soil microorganisms, insect and vascular plants, acting in a holistic way through the whole ecosystem.

When agroforestry is implemented in permanent grasslands, differences in plant species composition between the area below the tree and the open areas occur, due to the higher level of nutrients below trees, as a result of leaf litter accumulation, favouring the presence of monocots over that of dicots, including legumes (Pardini and Rigueiro-Rodríguez 2010). Permanent grasslands including woody perennials support livestock production, providing a source of feed by enlarging the feeding season from grasses which continue to grow below the tree when grasses cease to grow in open sites (Fig. 1), or by directly providing feed through their leaves/stems or their fruits. Europe hosts half of the autochthonous livestock breeds, half of which are at risk of extinction (Rois et al. 2006), and these breeds find their ideal habitat in permanent grasslands with woody perennials, highlighting their role in animal biodiversity conservation.

The same advantages apply where fruit trees are part of the system and herbaceous vegetation is used by livestock. Through keeping grassy

Figure 1 Tree shade effect on below tree vegetation at the end of the growing season in the dehesa (Spanish agroforestry system where low tree (*Quercus ilex* and *Q. suber*) densities are combined with agricultural and/or pastoral activities), promoting the extension of the growing season to feed animals. Photo: Gerardo Moreno.

areas under control (reducing weeds and potential insect pest habitats), sheep grazing has been described as a good practice to reduce the need of pesticides for some fruit trees such as chestnuts, maintaining biodiversity across different scales. When silvopastoralism is used in permanent grasslands with woody perennials, permanent crops and forestry, the presence of animals enhances biodiversity through the creation of microsites caused by trampling, faeces deposition and plant species selection made by animals (Buttler et al. 2009). All these features increase niche heterogeneity and therefore biodiversity. The use of grazing animals in forestland where fire risk is high is also one of the cheapest solutions to reduce fire risk and consequently the risk of biodiversity losses (Etienne 1996; Castro 2009; Casals et al. 2009; Papanastasis et al. 2009).

3 Agroforestry and the protection of species and habitats

Due to the large proportion of degraded land in Europe, the European Commission has issued directives aiming at protecting habitats that still preserve biodiversity and provide more public goods than degraded and unprotected lands (EC 2011a and b). The Habitat and Birds Directives linked to the Nature 2000 network, together with the high nature value (HNV) systems, are examples of key EU biodiversity protection initiatives. The Nature 2000 Network represents 18% of the EU's land area, being the largest protected area network in the world (EC 2009), while high nature value (HNV) systems are estimated to cover as much as 30% of EU-28 agricultural land (Keenleyside et al. 2014). The first target of the EU biodiversity strategy aims at protecting species and habitats through the EU NATURA 2000 network and aims at better conservation or a secure status for 100% more habitats and 50% more species by the year 2020.

There are four key EU actions for the protection of species and habitats:

(i) the completion of the Natura 2000 Network (already achieved) (EC 2009);
(ii) ensuring its adequate management and funding;
(iii) monitoring and reporting biodiversity through ICT tools at European level; and
(iv) involving society through the increasing biodiversity awareness.

Due to the important role that woody perennials and agroforestry play in biodiversity promotion around Europe, this monitoring should specifically look at the presence of agroforestry practices in the European landscape. A comparison between the map of the higher nature value (HNV) farmland area in Europe (EEA 2012) and the map of the areas allocated to silvopastoralism reveals that most silvopastoral areas are located in the Mediterranean area, and

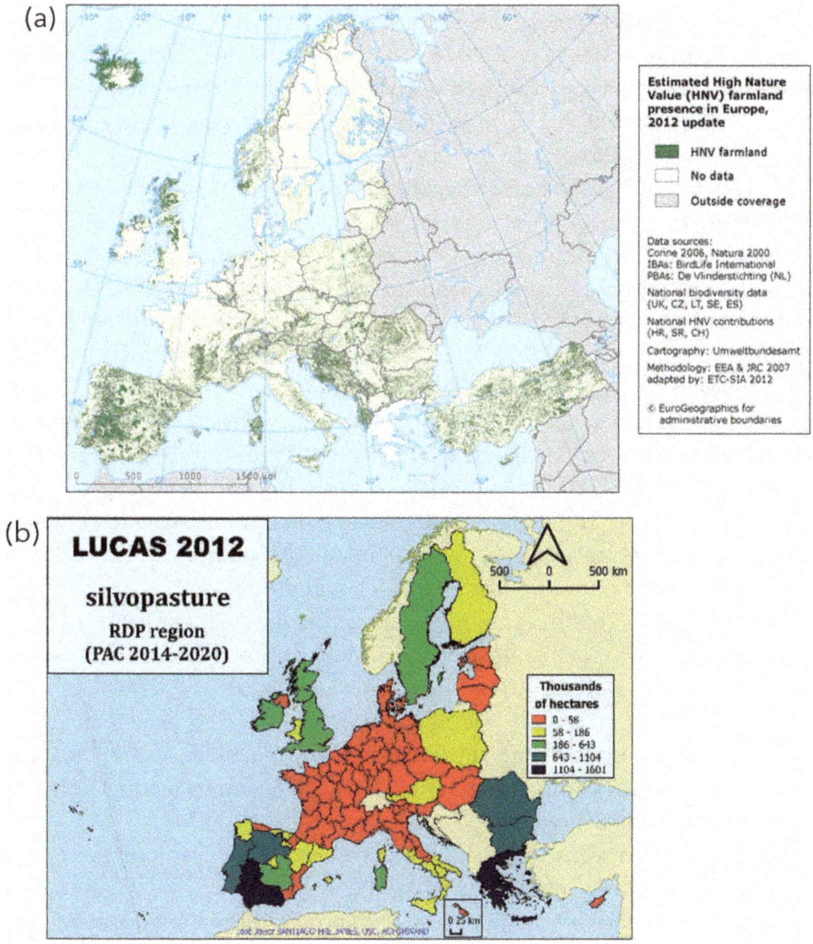

Figure 2 (a) Estimated high nature value (HNV) farmland area in Europe (EEA 2012) and (b) the silvopasture area in Europe (Mosquera-Losada et al. 2018a).

that most of the HNV farmland is managed following agroforestry principles (Fig. 2).

4 Agroforestry and the maintenance and restoration of ecosystems

Target 2 of the EU biodiversity strategy (EC 2011b) aims at maintaining and restoring ecosystems and their services by including green infrastructures in spatial planning and restoring at least 15% of degraded ecosystems by

2020. As a sustainable land-use system, agroforestry could be used to restore degraded ecosystems (Kay et al. 2018). There are good examples of ecosystem services restoration in lands formerly used for mining and in mountain areas that have been carried out through the implementation of agroforestry practices (Quinkenstein et al. 2011).

Target 2 of the EU biodiversity strategy also aims at contributing to the EU's sustainable growth objectives as well as helping to mitigate and adapt to climate change. Both objectives can be achieved with agroforestry, which has been recognized by the IPCC and the EU (Decision 529/2013/EU) as one of the most important strategies for climate change mitigation and adaptation (Mosquera-Losada et al. 2017). The presence of woody perennials in arable systems makes it possible to produce more biomass and therefore carbon per unit of land that can be stored for a long time both above and below ground. Below ground storage is especially important, as tree/shrub roots are able to introduce biomass and therefore carbon into the soil at a deeper depth, reducing CO_2 losses to the atmosphere as a result of land management (e.g. ploughing). The increase of biomass has been estimated to be at least 40% more (with a range between 20% and 80%) in agroforestry than in open sites (Dupraz and Liagre 2008), promoting a range of ecosystem services.

The EU biodiversity strategy includes three actions in order to tackle its second target:

(i) mapping and assessing the state and economic value of ecosystems and their services in the entire EU territory, linked to the principle that 'what gets measured gets saved';
(ii) restoration of ecosystems fostering ecosystem services, and promoting the use of green infrastructures; and
(iii) assessing the efficacy of EU funding to foster biodiversity, and developing schemes to pay for ecosystem services.

Measurements of agroforestry land have been carried out by den Herder et al. (2017) and Mosquera-Losada et al. (2018a) looking respectively at only trees and trees or/and shrubs. Kay et al. (2018) has provided a methodology to measure the state of ecosystem services in Europe, associating agroforestry with those areas where there is a higher need to enhance ecosystem services. It identifies the positive influence of agroforestry systems on the supply of regulating services and their role in enhancing landscape structure. Mosquera-Losada et al. (2018b) have evaluated the effects of the main funding agriculture scheme in Europe, the common agriculture policy (CAP), on agroforestry, highlighting that agroforestry is promoted by the EU but not always recognized as such. As an example, homegarden agroforestry practices should be considered within the concept of smart-villages trying to better connect rural and urban areas. As a result they are not usually directly funded by the CAP, as urban areas are out of the scope of the CAP.

5 Achieving more sustainable agriculture and forestry

Agroforestry is linked to integration of agriculture and forestry at landscape level (Mosquera-Losada 2018a; Santiago-Freijanes et al. 2018a). By introducing woody perennials in agricultural lands or agricultural activities in forest lands and connecting both types of land use (green corridors connecting forest and agricultural areas), biodiversity is enhanced. This can be done through the use of understory forage in forest land when forage availability is scarce in open grasslands. The connection between forest and open grasslands through the involvement of livestock has a long tradition and can involve short (transtermitance) or long distances (transhumance) (Bunce et al. 2009). Moreover, the existence of scattered trees in arable lands can be an excellent corridor connecting forestland biodiversity (large mammals and birds). Sustainable agriculture and forestry should be linked to the use of the relevant resources, either the understory in forests or the trees or shrubs in agricultural lands. The recently established European bioeconomy strategy aims at connecting these two types of land use as a source of goods and services.

Management of shrub hedges is time-consuming but helps to reduce the effect of winds on croplands while delivering products (usually biomass) that can, for example, be used as fuel for heating. Some animals can also consume the shrubs during periods of shortage of grassland, contributing to increase the resilience of livestock farming systems. Shrubs can also be used as a source of nutrients for the soil. In Galicia, for example, legume shrubs are traditionally sown to increase soil fertility and, after processing, as fertilizer. The most used legume species was *Ulex europapaeus* with a productivity that can be over 100 Mg DM ha^{-1} after 8 years. Plants were harvested and used as animal bedding which, when enriched with faeces and urine, reduced the C/N relationship, and therefore improved the release of nutrients when applied to arable lands. Improvement of nutrients also increases biodiversity through the application of organic matter into those soils that have a low organic-matter content.

Target 3 of the EU biodiversity strategy includes 5 actions (from action 8 to action 12 of the strategy) to which agroforestry can contribute. Action 8 is related to the enhancement of CAP direct payments to reward environmental public goods such as crop rotation and permanent pastures. The strategy also allows the improvement of cross-compliance standards for GAEC (good agricultural and environmental conditions), by also including the provisions of the Water Framework directive. International and European policies dealing with agroforestry have been described by Mosquera-Losada et al. (2018a) and Santiago-Freijanes et al. (2018b), while excellent recommendations to foster agroforestry across the world has been reviewed by Buttoud (2013). CAP payments to agroforestry have been included in the last (2007–2013) and current (2014–2020) CAP with different degree of success, where agroforestry is promoted through different policy measures. The next CAP (2021-2027) aims at providing funds based on evidence or result-based payments linked to the provision of ecosystem services. Agroforestry is associated with GAEC as agroforestry practices foster hedgerows and trees (Santiago-Freijanes et al. 2018c) that also improve water quality.

Action 9 of the EU biodiversity strategy is related to better orientation of rural development policies towards achieving biodiversity needs and developing tools to help farmers and foresters work together towards biodiversity conservation. The creation of knowledge reservoirs AFINET (http://www.eurafagroforestry.eu/afinet) and handbooks (www.agroforestrynet.eu) integrates knowledge from research projects and farmer practices. Action 10 is related to the conservation and support of genetic diversity in Europe's agriculture that can be linked to the preservation of crop varieties better adapted to shading conditions (Mosquera-Losada et al. 2019) or autochthonous breeds (Rois et al. 2006) as mentioned before.

Action 11 of the EU biodiversity strategy is related to the encouragement of forest owners to protect and enhance forest biodiversity through adequate activities such as agroforestry (Mosquera-Losada et al. 2006) as well as the use of animal breeds that are adapted to forest areas (Mosquera-Losada et al. 2017). This is also related to action 12, related to the promotion of the integration between biodiversity measures such as fire prevention and the preservation of wilderness areas in forest management plans.

6 Stopping the loss of global biodiversity

The EU is committed to stepping up its contribution in the fight to protect global biodiversity in line with international commitments made in October 2010, when the UN Convention on Biological Diversity adopted a strategic plan to address global biodiversity loss over the coming decade. This objective is based on four actions (from action 17 to action 20 of the EU biodiversity strategy). Action 17 aims at reducing the impacts of EU consumption patterns on biodiversity and to make sure that the EU initiative on resource efficiency and trade negotiations all reflect this objective. Consumer education and the promotion of extensive use of local food was one of the challenges mentioned by the farmers in the AFINET project to foster the transition from conventional to agroforestry systems. The use of local food and resources is key to maintain biodiversity at both the EU and international scale. Action 18 (target more EU funding towards global biodiversity and make this funding more effective), action 19 (systematically screen EU action for development cooperation to reduce any negative impacts on biodiversity) and action 20 (make sure that the benefits of nature's genetic resources are shared fairly and equitably) are also related with agroforestry.

7 Future trends

Future trends in research should focus on the most appropriate woody perennials to be integrated with agricultural production in different environments. Recognition of agroforestry as part of the way to fulfil sustainable development goals should be researched in order to provide standardized indicators that should be suitable to each type of land cover such as forestry, permanent grasslands, permanent crops and arable lands. Silvopasture increases the biodiversity of forest, permanent grasslands and permanent crops. Stocking rate should be the primary factor to evaluate the effect of livestock on both biodiversity and woody perennials. Evaluation of arable crops biodiversity should be conducted to find the most suitable arable crops varieties that are able to grow under the different types of shading considering light intensity and tree canopy cover. Special attention should be given to the evaluation of the advantages of agroforestry in HNV areas where biodiversity should be protected. Research on land restoration degraded by different factors such as mining or fires should be specifically targeted due to the large amount of existing degraded areas. Research should focus on the use of pioneer species able to cope with particular types of degraded land. Moreover, the comparison of different woody perennial alternatives in provision of different ecosystem services should be evaluated. CAP agroforestry accountability is key to understand the contribution of this type of land-use system in halting biodiversity losses. More research on agroforestry accountability at CAP, but also at global level, is needed.

8 Where to look for further information

Below is a list of papers for more information on agroforestry and biodiversity:

- Fernández-Núñez, E., Rigueiro-Rodríguez, A. and Mosquera-Losada, M. R. 2014. Silvopastoral systems established with *Pinus radiata* D. Don and *Betula pubescens* Ehrh.: tree growth, understorey biomass and vascular plant biodiversity. *Forestry* 87, 512–24.
- Ferreiro-Domínguez, N., Rigueiro-Rodríguez, A. and Mosquera-Losada, M. R. 2011. Response to sewage sludge fertilisation in a *Quercus rubra* L. silvopastoral system: soil, plant biodiversity and tree and pasture production. *Agriculture, Ecosystems and Environment* 141, 49–57.

- Mosquera-Losada, M. R., Rodríguez-Barreira, S., López-Díaz, M. L., Fernández-Núñez, E. and Rigueiro-Rodríguez, A. 2009. Biodiversity and silvopastoral system use change in very acid soils. *Agriculture, Ecosystems and Environment* 131, 315–24.
- Rigueiro-Rodríguez, A., Ferreiro-Domínguez, N. and Mosquera-Losada, M. R. 2010. The effects of fertilization with anaerobic, composted and pelletized sewage sludge on soil, tree growth, pasture production and biodiversity in a silvopastoral system under ash (*Fraxinus excelsior* L). *Grass and Forage* 65, 248–59.
- Rigueiro-Rodríguez, A., Mouhbi, R., Santiago-Freijanes, J. J., González-Hernández, M. P. and Mosquera-Losada, M. R. 2012. Horse grazing systems: understory biomass and plant biodiversity of a *Pinus radiata* stand. *Scientia Agricola* 69(1).

9 Acknowledgements

This chapter was partially carried out under the H2020 AFINET project: Agroforestry Innovation Networks; Grant agreement 727872. This work was undertaken by the University of Santiago de Compostela (USC) to support the objectives of the Global Research Alliance on agricultural greenhouse gases (www.globalresearchalliance.org). The information contained within should not be taken to represent the views of the Alliance as a whole or its partners.

10 References

Bunce, R. G. H., Pérez-Soba, M. and Smith, M. 2009. *Assessment of the Extent of Agroforestry Systems in Europe and Their Role Whiting Transhumance Systems*, pp. 321–42.

Buttler, A., Kohler, F. and Gillet, F. 2009. The Swiss mountain wooded pastures: patterns and processes. In: Rigueiro-Rodriguez, A., McAdam, J. and Mosquera-Losada, M. R. (Eds.), *Agroforestry in Europe*. Springer, Dordrecht, pp. 377-96.

Buttoud, G. 2013. *Advancing Agroforestry on the Policy Agenda*. FAO. http://www.fao.org/3/a-i3182e.pdf (accessed 05 May 2019).

Casals, T., Baiges, T., Bota, G., Chocarro, C., Bello, F., Fanlo, R., Sebastiá, M. T. and Taull, M. 2009. Silvopastoral systems in the North-eastern Iberian Peninsula: a multifunctional perspective. In: Rigueiro-Rodriguez, A., McAdam, J. and Mosquera-Losada, M. R. (Eds.), *Agroforestry in Europe*. Springer, Dordrecht, pp. 377–96.

Castro, M. 2009. Silvopastoal systems in Portugal: current status and future prospects. In: Rigueiro-Rodriguez, A., McAdam, J. and Mosquera-Losada, M. R. (Eds.), *Agroforestry in Europe*. Springer, Dordrecht, pp. 161–82.

CBD 2010. Strategic plan for biodiversity 2011–2020, including Aichi targets. Available at: https://www.cbd.int/sp/default.shtml (accessed 05 May 2019).

Den Herder, M., Moreno, G., Mosquera-Losada, R. M., Palma, J. H. N., Sidiropoulou, A., Santiago-Freijanes, J. J., Crous-Duran, J., Paulo, J. A., Tomé, M., Pantera, A.,

Papanastasis, V. P., Mantzanas, K., Pachana, P., Papadopoulos, A., Plieninger, T. and Burgess, P. J. 2017. Current extent and stratification of agroforestry in the European Union. *Agriculture, Ecosystems and Environment* 241, 121–32. doi:10.1016/j.agee.2017.03.005.

Dupraz, C. and Liagre, F. 2008. *Agroforesterue: des arbres et des cultures. France Agricole.*

EC 2009. *The EU's Protected Areas – Natura 2000*. Available at: http://ec.europa.eu/environment/basics/natural-capital/natura2000/index_en.htm (accessed 05 May 2019).

EC 2011a. Communication from the commission to the European Parliament, the Council, the economic and social committee and the committee of the regions. Our life insurance, our natural capital: an EU biodiversity strategy to 2020. Available at: https://eur-lex.europa.eu/legal-content/EN/TXT/PDF/?uri=CELEX:52011DC0244&from=EN (accessed 05 May 2019).

EC 2011b. The EU biodiversity Strategy to 2020. Available at: http://ec.europa.eu/environment/nature/info/pubs/docs/brochures/2020%20Biod%20brochure%20final%20lowres.pdf (accessed 05 May 2019).

EEA 2012. Estimated HNV farmland presence in Europe. Available at: https://www.eea.europa.eu/data-and-maps/figures/estimated-high-nature-hnv-presence (accessed 05 May 2019).

Etienne, M. 1996. *Western European Silvopasatoral Systems. INRA Editions*.

Kay, S., Crous-Duran, J., Ferreiro-Domínguez, N., García de Jalón, S., Graves, A., Moreno, G., Mosquera-Losada, M. R., Palma, J. H. N., Roces-Díaz, J. V., Santiago-Freijanes, J. J., Szerencsits, E., Weibel, R. and Herzog, F. 2018. Spatial similarities between European agroforestry systems and ecosystem services at the landscape scale. *Agroforestry Systems* 92(4), 1075–89. (doi:10.1007/s10457-017-0132-3).

Keenleyside, C., Beaufoy, G., Tucker, G. and Jones, G. 2014. High Nature Value farming throughout EU-27 and its financial support under the CAP. Institute for European Environmental Policy. Available at: http://ec.europa.eu/environment/agriculture/pdf/High%20Nature%20Value%20farming.pdf (accessed 05 May 2019).

Maes 2017. A model for the assessment of habitat conservation status in the EU. JRC Scientific and Policy reports. Available at: https://core.ac.uk/download/pdf/38626714.pdf (accessed 05 May 2019).

Mosquera-Losada, M. R., McAdam, J. and Rigueiro-Rodríguez, A. 2006. *Silvopastoralism and Sustainable Land Management*. CAB International, Wallingford.

Mosquera-Losada, M. R., Santiago Freijanes, J. J., Pisanelli, A., Rois, M., Smith, J., den Herder, M., Moreno, G., Malignier, N., Mirazo, J. R., Lamersdorf, N., Ferreiro Domínguez, N., Balaguer, F., Pantera, A., Rigueiro-Rodríguez, A., Gonzalez-Hernández, P., Fernández-Lorenzo, J. L., Romero-Franco, R., Chalmin, A., Garcia de Jalon, S., Garnett, K., Graves, A. and Burgess, P. J. 2016. Extent and success of current policy measures to promote agroforestry across Europe. Deliverable 8.23 for EU FP7 Research Project: AGFORWARD 613520, 95 pp. Available at: https://www.agforward.eu/index.php/es/extent-and-success-of-current-policy-measures-to-promote-agroforestry-across-europe.html?file=files/agforward/documents/Deliverable%208.23%20Extent%20and%20Success%20of%20Current%20Policy%20Measures%208%20Dec%202016.pdf (accessed 05 May 2019).

Mosquera-Losada, M. R., Borek, R., Balaguer, F. and Mezzarila, G. and Ramos-Font, M. E. 2017. *Agroforestry as a Mitigation and Adaptation Tool*. Available at: https://ec.euro

pa.eu/eip/agriculture/sites/agri-eip/files/fg22_mp9_cc_adaptation_mitigation_2017_en.pdf (accessed 05 May 2019).

Mosquera-Losada, M. R., Santiago-Freijanes, J. J., Rois-Díaz, M., Moreno, G., den Herder, M., Aldrey-Vázquez, J. A., Ferreiro-Domínguez, N., Pantera, A., Pisanelli, A. and Rigueiro-Rodríguez, A. 2018a. Agroforestry in Europe: a land management policy tool to combat climate change. *Land Use Policy* 78, 603–13. (doi:10.1016/j.landusepol.2018.06.052).

Mosquera-Losada, M. R., Santiago-Freijanes, J. J., Pisanelli, A., Rois, M., Smith, J., den Herder, M., Moreno, G., Ferreiro-Domínguez, N., Malignier, N., Lamersdorf, N., Balaguer, F., Pantera, A., Rigueiro-Rodríguez, A., Aldrey, J. A., Gonzalez-Hernández, M. P., Fernández-Lorenzo, J. L., Romero-Franco, R. and Burgess, P. J. 2018b. Agroforestry in the European common agricultural policy. *Agroforestry Systems* 92(4), 1117–27. (doi:10.1007/s10457-018-0251-5).

Mosquera-Losada, M. R., López-Díaz, M. L., Ferreiro-Domínguez, N., Rodríguez-Rigueiro, F. J., Arias Martínez, D., Santiago-Freijanes, J. J., Coello, J., Papadopoulou, P. and Rigueiro-Rodríguez, A. 2019. Silvoarable practices as a mechanism to enhance climate adaptation. North American Agroforestry conference. Agroforestry for sustainable production and resilient landscapes. AFTA 2019 Conference. https://afta2019.org/program/ (accessed 05 May 2019).

Nieto, A., Roberts, S. P. M., Kemp, J., Rasmont, P., Kuhlmann, M., Criado, M. G., Biesmeijer, J. C., Bogusch, P., Dathe, H. H., De la Rúa, P., De Meulemeester, T., Dehon, M., Dewulf, A., Ortiz-Sánchez, F. J., Lhomme, P., Pauly, A., Potts, S. G., Praz, C., Quaranta, M., Radchenko, V. G., Scheuchl, E., Smit, J., Straka, J., Terzo, M., Tomozii, B., Window, J. and Michez, D. 2014. European Red List of Bees, European Commission. https://ec.europa.eu/environment/nature/conservation/species/redlist/downloads/European_bees.pdf

Papanastasis, V. P., Mantzanas, K., Dini-Papanastasi, O. and Ispikoudis, I. 2009. Traditional agroforestry systems and their evolution in Greece. In: Rigueiro-Rodríguez, A., McAdam, J. and Mosquera-Losada, M. R. (Eds.), *Agroforestry in Europe*. Springer, Dordrecht, pp. 89–109. doi:10.1007/978-1-4020-8272-6_5.

Pardini, Mori S. and Rigueiro-Rodríguez, M.-L. 2010. *Efecto del arbolado en la producción de pasto y trigo ("Triticum aestivum") ecológicos en la maremma toscana* (Italia central). *Revista de la Sociedad Española para el Estudio de los Pastos* 40(2), 211–33.

Quinkenstein, A., Böhm, C., Silva Matos, E., Freese, D. and Hüttl, R. 2011. Assessing the carbon sequestration in short rotation coppices of Robinia pseudoacacia L. on marginal sites in Northeast Germany. In: Kumar, B. M. and Nair, P. K. R. (Eds.), *Carbon Sequestration Potential of Agroforestry Systems: Opportunities and Challenges*. Springer, pp. 201–16.

Rigueiro-Rodriguez, A., McAdam, J. and Mosquera-Losada, M. R. 2009. *Agroforestry in Europe*. Springer, Dordrecht.

Rigueiro-Rodríguez, A., Rois-Díaz, M. and Mosquera-Losada, M. R. 2010. Integrating silvopastoralism and biodiversity conservation. In: Lichfouse, E. (Ed.), *Biodiversity, Biofuels, Agroforestry and Conservsation Agriculture*. Springer, pp. 359–73. doi:10.1007/978-90-481-9513-8_12.

Rois-Díaz, M., Mosquera-Losada, M. R. and Rigueiro-Rodríguez, A. 2006. Biodiversity Indicators on Silvopastoralism across Eruope. European Forest Institute Technical report 21, 66 pp. Available at: file:///C:/Users/Usuario/Downloads/AS3452428396011521459323715063_content_1%20(1).pdf (accessed 05 May 2019).

Santiago-Freijanes, J. J., Mosquera-Losada, M. R., Rois-Díaz, M., Ferreiro-Domínguez, N., Pantera, A., Aldrey, J. A. and Rigueiro-Rodríguez, A. 2018a. Global and European policies to foster agricultural sustainability: agroforestry. *Agroforestry Systems*, 1-16. doi:10.1007/s10457-018-0215-9.

Santiago-Freijanes, J. J., Pisanelli, A., Rois-Díaz, M., Aldrey-Vázquez, J. A., Rigueiro-Rodrígueza, A., Pantera, A., Vity, A., Lojka, B., Ferreiro-Domínguez, N. and Mosquera-Losada, M. R. 2018b. Agroforestry development in Europe: policy issues. *Land Use Policy* 76, 144–56. doi:10.1016/j.landusepol.2018.03.014.

Santiago-Freijanes, J. J., Rigueiro-Rodríguez, A., Aldrey, J. A., Moreno, G., den Herder, M., Burgess, P. and Mosquera-Losada, M. R. 2018c. Understanding Agroforestry practices in Europe through landscape features policy promotion. *Agroforestry Systems* 92(4), 1105–115. doi:10.1007/s10457-018-0212-z.

UN 1992. The Rio declaration on environment and development. Available at: http://www.unesco.org/education/pdf/RIO_E.PDF (accessed 05 May 2019).

Index

Lightning Source UK Ltd.
Milton Keynes UK
UKHW021848100920
369673UK00001B/20

9 781786 763488